THE PHYSICS AND MODELING OF MOSFETS
Surface-Potential Model HiSIM

International Series on Advances in Solid State Electronics and Technology (ASSET)

Founding Editor: Chih-Tang Sah

ASSET

International Series on Advances in Solid State Electronics and Technology

Founding Editor: Chih-Tang Sah

THE PHYSICS AND MODELING OF MOSFETS
Surface-Potential Model HiSIM

Mitiko Miura-Mattausch
Hans Jürgen Mattausch
Tatsuya Ezaki

Hiroshima University, Japan

World Scientific

NEW JERSEY · LONDON · SINGAPORE · BEIJING · SHANGHAI · HONG KONG · TAIPEI · CHENNAI

Published by

World Scientific Publishing Co. Pte. Ltd.
5 Toh Tuck Link, Singapore 596224
USA office: 27 Warren Street, Suite 401-402, Hackensack, NJ 07601
UK office: 57 Shelton Street, Covent Garden, London WC2H 9HE

British Library Cataloguing-in-Publication Data
A catalogue record for this book is available from the British Library.

First published 2008 (Hardcover)
Reprinted 2016 (in paperback edition)
ISBN 978-981-3203-31-0

International Series on Advances in Solid State Electronics and Technology
THE PHYSICS AND MODELING OF MOSFETS
Surface-Potential Model HiSIM

ISBN-13 978-981-256-864-9
ISBN-10 981-256-864-6

Desk edtior: Tjan Kwang Wei

Printed in Singapore

Foreword

The idea of a monograph series on compact transistor modeling came about when I was looking into the literature to write the keynote address, on the history of MOS transistor compact modeling, invited by the Founder of the Workshop on Compact Modeling, Professor Xing Zhou of ˙Nanyang Technology University, and his program committee, to be presented at its 4th Workshop on May 10, 2005. The purpose is to provide an archival reference series, described by the originators or the veterans, of the compact models of the MOS and bipolar-junction transistors and passive components. A second purpose is to serve as textbooks for graduate students and reference books for practicing engineers, to rapidly disseminate to them the detailed design methodologies and underlying physics, in order to meet the ever faster advances in the computer-aided-design of integrated circuits containing thousands to billions of silicon semiconductor MOS and bipolar-junction transistors, passive diode components (capacitors, inductors, resistors), and interconnects. It is also the objective of this monograph series to provide timely updates via website and internet exchanges between the readers and authors, for public dissemination, and for new editions when sufficient materials are accumulated. I am especially thankful to the invited authors of the four startup and later monograph volumes who concurred with me and agreed to take up the chore to write their books.

It is most timely to document the second generation compact MOS transistor model, evolved from my 1964 threshold-voltage model to my 1966 surface-potential model that is being employed by the circuit design engineers to simulate the transistor and by the foundries to fabricate integrated circuits in this decade and beyond. This monograph is a device-physics-based in-depth description of one of the three industry-consensus compact MOS transistor models. It is known as HiSIM, developed by Professor Mitiko Miura-Mattausch, her two collaborators, Professor Hans Jurgen Mattausch and Assistant Professor Tatsuya Ezaki, and her graduate students and industrial associates at the Hiroshima University in Japan. We are extremely pleased to publish this timely monograph to serve as ·a reference for practicing engineers and a textbook for graduate students. Its depth and thoroughness provide the clarity and rigor to grasp the approximations that are necessary to compact the MOS transistor model for simulation of integrated circuits containing thousands to billions of MOS transistors.

I would like to thank all the WSPC editors and this monograph volume's copy editor Mr. Tjan Kwang Wei (Tjian Guangwei) at Singapore and the acquisition editor Dr. Yubing Zhai at New Jersey, for their and her timely efforts, and Professor Kok-Khoo Phua, Founder and Chairman of WSPC whose support and publishing philosophy have made all these possible. I am also indebted to Professor Binbin Jie of Peking University for his helpful efforts as my collaborator and also personal assistant, without which my reentry into device modeling would not have happened after 40 years of absence since writing the threshold-voltage and then two surface-potential-based MOS transistor modeling articles in the IEEE Electron Devices in 1964 and 1966. I also acknowledge my late wife Linda Su-Nan Chang Sah for founding the CTSAH Associates in the 1970's to provide me the financial independence to undertake engineering and science investigations in later years, and her younger brother Fred Tsang and our children, Dinah and Robert Sah for their continued technical supports and mental encouragements.

Chih-Tang Sah (Tom Sah)
Gainesville. Florida, USA
February 27, 2008

Preface

In recent years the field of compact MOSFET modeling for circuit simulation has entered into a remarkable transition phase. The traditional approach, represented by the series of the BSIM model standards and based on the threshold voltage of the MOSFET combined with a piece-wise modeling of different operating regions, has been widely displaced from the agenda of active research groups in compact modeling. All currently pursued new approaches are based instead on the drift-diffusion approximation and are centered on either the inversion charge or the surface potential of the MOSFET channel. Some of the resulting new compact MOSFET models are already on the verge of productive usage in industrial applications with the surface-potential-based models taking the lead position. Consequently, surface-potential-based models are predicted to become the future main stream in the productive application of the IC industry.

Aggressive scaling down of device sizes is causing enhanced complexities in device characteristics, which require very sophisticated device physics to understand as well as to model. The main complexity is caused by the 2-dimensional effects, which were negligible in the past. The quantum mechanical effect is also becoming increasingly important. To model all these newly important effects, the most straightforward approach is to follow their underlying physics. For consistency of this well understandable approach, the basis of the model has to follow the device physics as well. All device features are induced by the potential distribution along the channel, which is represented by the surface potential as the physical basis of a compact model.

The described changes represent indeed a revolutionary movement for the compact modeling community, which has been rather stable and conservative with respect to the applied modeling approaches. This change of compact modeling concepts opens at the same time the possibility of a merger with the technology CAD (TCAD) field, which usually targets the problems of technology and device optimization. Thus the newly emerging compact models are expected to increasingly become real bridges between fabrication technology and resulting integrated circuits, because compact model parameters are indeed closely corresponding to the physical parameters of the fabrication technology.

This textbook describes the device physics of observed MOSFET phenomena and the modeling approach of the surface-potential-based MOSFET model HiSIM (Hiroshima-university STARC IG-FET Model). HiSIM is the result of research work pioneering the application of the surface-potential modeling to MOSFETs fabricated with advanced technologies which was carried out as a cooperation between Hiroshima University and the Semiconductor Technology Academic Research Center (STARC).

Contributors for the cooperation between Hiroshima University and STARC are

D. Navarro, N. Sadachika, M. Miyake, Y. Takeda, G. Suzuki, S. Hosokawa, T. Mizoguchi, S. Jinbou, N. Nakayama, S. Chiba, K. Konno, Y. Mizukane, T. Warabino, K. Machida, T. Yoshida, H. Ueno, K. Hisamitsu, H. Kawano, D. Kitamaru, T. Honda, S. Matsumoto, S. Mitani, D. Miyawaki, H. Nagakura, S. Nara, M. Nishizawa, T. Okagaki, S. Ohshiro, Y. Shiraga, K. Suematsu, M. Suetake, M. Tanaka, Y. Tatsumi, T. Yamaoka at Hiroshima University

and

T. Ohguro, T. Iizuka, M. Taguchi, S Kumashiro, T. Yamaguchi, K. Yamashita, S. Odanaka, N. Shigyo, R. Inagaki, Y. Furui, S. Miyamoto, S. Hazama at STARC

The basic concepts of the modeling approach used for developing HiSIM and the resulting modeling equations up to the version HiSIM2.4.0, are explained in detail. For this purpose the physics behind HiSIM is illustrated with the help of 2D-device simulations and is used to extract the resulting essential MOSFET properties in simple analytical form.

There is of course still room for further improvements and remaining insufficiencies exist, which require continuing modeling efforts. We analyse such problems and give an overview of resulting future development requirements.

Hopefully, this textbook will serve the developers as well as the users of compact models in industry and academia as a reference source for the practical application of the surface-potential approach to the development of MOSFET models for circuit simulation. In this way we together with all contributors wish to help and facilitate the development of even more reliable and powerful compact models for present and future advanced device structures.

The authors want to express their gratitude and thanks to EDA vendors for their suggestions and support. A special acknowledgement goes to Professor Chin-Tang Sah, who is the Founding Editor of the ASSET international series of Monographs on Advances in Solid State Electronics and Technology, for his diligent support of the preparation of this monograph by Providing a thorough editing of the whole manuscript with valuable in-depth technical advice. We also are in debt to this monograph's copy editor Mr. Tjan Kwang Wei and Jessie Tan for creating the final printable monograph version under a tight time schedule.

February 2006
Higashi-Hiroshima

Contents

Definition of Symbols Used for Variables and Constants

Table 0.1 List of symbols. Subscripts S, D, B and G represent source, drain, bulk and gate terminals of a MOSFET, respectively.

Symbol	Description	Unit
B	Magnetic induction	Wb/m^2
C_{dep}	Depletion capacitance	F
C_{diode}	Diode capacitance	F
C_{ef}	Edge-fringing capacitances	F
C_{fring}	Fringing capacitance	F
C_{gate}	Gate capacitance	F
$C_{\mathrm{gbo\_loc}}$	Gate-to-bulk overlap capacitance	F
C_{int}	Intrinsic capacitance	F
C_{m}	Transcapacitance	F
C_{ov}	Overlap capacitance	F
C_{over}	Overlap capacitance	F
C_{ox}	Gate-oxide capacitance	F
$C_{\mathrm{ox,STI}}$	Gate-oxide capacitance for the STI case	F
C_{Qy}	Lateral-field-induced capacitance	F
C_{gxo}	Source or drain overlap capacitance	F
	x = s: source, d: drain	
$C_{\mathrm{gs}}, C_{\mathrm{sg}}$	Gate-source capacitance	F
$C_{\mathrm{ds}}, C_{\mathrm{sd}}$	Drain-source capacitance	F
$C_{\mathrm{bs}}, C_{\mathrm{sb}}$	Bulk-source capacitance	F
$C_{\mathrm{bd}}, C_{\mathrm{db}}$	Bulk-drain capacitance	F
$C_{\mathrm{bg}}, C_{\mathrm{gb}}$	Gate-bulk capacitance	F
$C_{\mathrm{ds}}, C_{\mathrm{sd}}$	Drain-source capacitance	F
C_{ss}	Source capacitance	F
C_{gg}	Gate capacitance	F
C_{dd}	Drain capacitance	F
C_{bb}	Bulk capacitance	F

Table 0.2   List of symbols. Subscripts S, D, B and G represent source, drain, bulk and gate terminals of a MOSFET, respectively.

Symbol	Description	Unit
D_n	Electron diffusion coefficient	cm^2/s
D_p	Hole diffusion coefficient	cm^2/s
$\boldsymbol{E}$	Electric field	V/m
E_c	Bottom of conduction band	eV
E_{eff}	Effective field	V/m
$E_{eff,ph}$	Effective field for phonon scattering	V/m
$E_{eff,sr}$	Effective field for surface roughness scattering	V/m
E_g	Band gap	eV
E_{gb}		
E_v	Top of valence band	eV
E_{ox}	Vertical electric field in the gate oxide	V/m
E_{pg}	Electric field in poly silicon at the upper gate oxide interface	V/m
E_s	Vertical electric field of the surface of substrate	V/m
E_y	Electric field along the channel direction	V/m
E_x	Electric field perpendicular to the channel direction	V/m
E_D		
E_F	Fermi energy level	eV
E_{Si}	Vertical electric field perpendicular to the channel surface	V/m
E_0		V/m
E_{max}		V/m
g_{ds}	Channel conductance	S
g_{ds0}	Channel conductance at $V_{ds} = 0V$	S
g_m	Transconductance	S
$\hbar$	Reduced Planck constant (1.05458×10^{-34})	Js
I_{bs}	Bulk-source current	A
I_{bd}	Bulk-drain current	A
I_{diode}	Diode current	A
I_{ds}	Drain current	A
$I_{ds,STI}$	STI leakage current	A
$I_{ds,QS}$	Drain current under the QS approximation	A
$I_{ds,2D}$	Drain current obtained from a 2-D device simulator	A
I_{gb}	Current between gate and bulk	A
I_{gd}	Current between gate and drain	A
I_{gs}	Current between gate and source	A
I_x	Terminal current (x= S, D, B, or G)	A
I_{x0}	Terminal current (x= S, D, B, or G)	A
I_T	Transport current	A
I_{sub}	Substrate current	A
I_{gate}	Gate current	A
$I_{ds,sat}$	Saturation current	A
$I_{ds,STI}$	STI leakage current	A
I_{GIDL}	GIDL current	A
j_n	Electron current density	A/cm^2
j_p	Hole current density	A/cm^2
k	Boltzmann constant (1.38066×10^{-23})	J/K

Table 0.3 List of symbols. Subscripts S, D, B and G represent source, drain, bulk and gate terminals of a MOSFET, respectively.

L_{ch}		cm or μm
L_{eff}	Effective channel length	cm or μm
L_{gate}	Gate length	cm or μm
L_{over}	Overlap length	cm or μm
L'_{over}	Effective overlap length	cm or μm
L_{poly}	Length of gate poly silicon region	cm or μm
L_n	Electron diffusion length	cm or μm
L_p	Hole diffusion length	cm or μm
$L_{D,STI}$		cm or μm
L_P	Pocket-penetration length	cm or μm
m	Electron mass (0.91095×10^{-30})	kg
m^*	Effective mass	kg
n	Density of free electrons	cm^{-3}
n_{av}	Average sheet carrier concentration	cm^{-2}
n_c	Sheet carrier concentration	cm^{-2}
$n_{c,th}$		cm^{-2}
n_i	Intrinsic density	cm^{-3}
n_{i0}	Intrinsic density at $T = 300$K	cm^{-3}
n_{n0}	Electron density in an n-type semiconductor under thermal equilibrium	cm^{-3}
n_{p0}	Electron density in a p-type semiconductor under thermal equilibrium	cm^{-3}
n_p		cm^{-3}
$n_{p0,pg}$	Intrinsic density of electrons in poly silicon	cm^{-3}
$p_{p0,pg}$	Intrinsic density of holes in poly silicon	cm^{-3}
n_{th}	Threshold sheet concentration	cm^{-2}
N_{dep}		
N_{pg}	Impurity concentration in the gate poly silicon	cm^{-3}
N_{sub}	Substrate impurity concentration	cm^{-3}
N_{subb}		cm^{-3}
N_{subc}		cm^{-3}
N_{subp}		cm^{-3}
N_{subeff}	Averaged impurity concentration	cm^{-3}
N_D	Donor impurity concentration	cm^{-3}
N_A	Acceptor impurity concentration	cm^{-3}
N_C	Effective density of states in conduction band	cm^{-3}
N_V	Effective density of states in valence band	cm^{-3}
$N_{sub,STI}$		cm^{-3}
N_{STI}		cm^{-3}
N_{vtm}		
p	Density of free holes	cm^{-3}
p_{n0}	Hole density in an n-type semiconductor under thermal equilibrium	cm^{-3}
p_{p0}	Hole density in a p-type semiconductor under thermal equilibrium	cm^{-3}
q	Elementary charge (1.60218×10^{-19})	C
q_S	Charge associated with source of a MOSFET	C
q_D	Charge associated with drain of a MOSFET	C

Table 0.4   List of symbols. Subscripts S, D, B and G represent source, drain, bulk and gate terminals of a MOSFET, respectively.

q_c	Carrier density under the NQS approximation	C/cm^{-2}
Q_c	Carrier density under the QS approximation	C/cm^{-2}
Q_b	Depletion charge	C
Q_{bmod}		
Q_{dep}	Depletion charge of diode	C
Q_g	Charge induced in the gate	C
Q_i	Inversion charge	C
$Q_{gate,D}$		C
$Q_{gate,S}$		C
Q_{gos}	Gate charge above the overlapped source-side contact	C
Q_{god}	Gate charge above the overlapped drain-side contact	C
Q_x	Charge associated with terminal x (x= S, D, B, or G)	C
Q_{STI}		C
$Q_{N,STI}$		C
$Q_{i,STI}$		C
Q_{SP}	Space charge	C
R_s	Source resistance	Ω
R_d	Drain resistance	Ω
R_g	Gate resistance	Ω
S_{ij}	Elements of S-parameter	Ω
S_{I_d}	Drain current noise density	A^2/Hz
S_{I_g}	Induced gate current noise density	A^2/Hz
S_v	Voltage spectral intensity	V^2/Hz
t	Time	s
T	Absolute temperature	K
T_{fox}	Oxide thickness at the trench edge	nm
T_{ox}	Physical gate-oxide thickness	nm
$T_{ox,eff}$	Effective oxide thickness	nm
V_{bi}	Built-in potential	eV
V_{bs}	Bulk-source voltage	V
V_{bd}	Bulk-drain voltage	V
V_{dsat}	Saturation drain voltage	V
V_{fb}	Flat-band voltage	V
$V'_{gs,STI}$		
V_{max}	Maximum velocity for high field mobility	cm/s
V_{th}	Threshold voltage	V
$V_{th,long}$	Threshold voltage of a long channel transistor	V
V_x	Terminal voltage (x= S, D, B, or G)	V
V_{xs}	Voltage between terminal x and source (x = b, d, or s)	V
V_p	Voltage amplitude of a sinusoidal signal	V
V_{DC}	DC voltage	V

Table 0.5   List of symbols. Subscripts S, D, B and G represent source, drain, bulk and gate terminals of a MOSFET, respectively.

W_d	Depletion-layer thickness	cm or μm
W_{eff}	Effective channel width	cm or μm
W_{gate}	Gate width	cm or μm
W_{STI}	Width of STI edge	cm or μm
$W_{d,STI}$	Depletion-layer thickness near STI	cm or μm
W_n	Depletion width extended in n-Si	cm or μm
W_p	Depletion width extended in p-Si	cm or μm
X_j	Junction depth	cm or μm
X_{Qy}	Position of the maximum lateral field relative to the channel/drain junction	cm or μm
Y_{ij}	Elements of Y-parameter	S
ΔV_{th}	Threshold voltage shift	V
$\Delta V_{th,P}$	Threshold voltage shift arising from a pocket implantation	V
$\Delta V_{th,R}$	Threshold voltage shift arising from a retrograded implantation	V
$\Delta V_{th,PSC}$	Reverse-short-channel effect (pocket implant)	V
$\Delta V_{th,RSC}$	Threshold voltage shift induced by a reverse short channel effect	V
$\Delta V_{th,SC}$	Threshold voltage shift induced by a short channel effect	V
$\Delta V_{th,SCSTI}$	Threshold voltage shift arising from an STI region	V
$\Delta V_{th,W}$	Threshold voltage shift arising from an STI region	V
ΔT_{ox}	Additional oxide thickness caused by QM effect	nm
ϵ_0	Permittivity in vacuum (8.85418×10^{-14})	F/cm
ϵ_{ox}	Permittivity of SiO$_2$ ($3.9\epsilon_0$)	F/cm
ϵ_{pg}	Permittivity of poly silicon	F/cm
ϵ_{Si}	Permittivity of silicon ($11.9\epsilon_0$)	F/cm
β	Thermal voltage ($kT/q = 0.0259$V)	V
μ	Mobility	cm^2/Vs
μ_s	Mobility at source side	cm^2/Vs
μ_d	Mobility at drain side	cm^2/Vs
μ_{av}	Averaged mobility of μ_s and μ_d	cm^2/Vs
μ_n	Electron mobility	cm^2/Vs
μ_p	Hole mobility	cm^2/Vs
μ_0	Low-field mobility	cm^2/Vs
μ_{CB}	Mobility arising from Coulomb scattering	cm^2/Vs
μ_{PH}	Mobility arising from phonon scattering	cm^2/Vs
μ_{SR}	Mobility arising from surface roughness scattering	cm^2/Vs
ϕ	Electric potential	V
ϕ_f	Quasi-Fermi potential	V
ϕ_{sc}		V
ϕ_{th}	Surface potential at threshold	V
ϕ_{thc}	Surface potential at threshold	V
ϕ_S	Surface potential	V
ϕ_{S0}	Surface potential at the source side	V
ϕ_{SL}	Surface potential at the drain side	V

Table 0.6   List of symbols. Subscripts S, D, B and G represent source, drain, bulk and gate terminals of a MOSFET, respectively.

Φ_B	Potential difference between the Fermi level and the intrinsic level	V
Φ_{BC}		V
$\Phi_{B,STI}$		V
$\phi_{S,STI}$		V
ϕ_{Spg}		V
ω	Angular frequency	rad/s
τ	Delay of a channel formation	s
τ_{cond}	Carrier conduction delay	s
τ_{diff}	Carrier diffusion delay	s
τ_{sup}	Delay caused by carrier supply mechanism	s
τ_{tran}	Carrier transit delay in a MOSFET channel	s

Chapter 1

Semiconductor Device Physics

1.1 Band Structure Concept

1.1.1 *Energy Bands and Quasi Particles*

The electronic properties of semiconductor devices are basically understood by quantum mechanics, since an electron is a quantum mechanical particle. One of the most important concepts for understanding the electric characteristics of semiconductor devices is the energy band structure which is composed of the electronic states from many atoms (i.e. 10^{23}cm^{-3}). The mechanisms, when the band structures are formed, can be explained only by quantum mechanics. Once we know the energy bands, the electrons in a semiconductor can be treated as "quasi" particles with properties similar to those of the classical particles, and we don't have to get back explicitly to quantum mechanics any more. Thanks to this feature we can then easily capture device characteristics within the framework of classical mechanics and electro dynamics. In this section, the energy band structures of semiconductor materials are briefly explained based on the quantum mechanical background.

We first discuss the electronic states of isolated atoms. Table 1.1 shows the electron configurations of atoms from hydrogen to argon. Quantum mechanics tells us that the possible energy states of electrons in atoms are discretized, or quantized, because otherwise electrons would continuously lose parts of their energy by emitting electromagnetic waves, and would be sucked into the atomic nucleus. Quantized electronic energy levels under the central electrical force field of an atomic nucleus are classified as s, p, d, $\cdots$ orbitals according to the spatial probability distribution of the electron as denoted by the principle quantization number. This is listed in Table 1.1. Tables 162.1 and 162.2 in [5] give more complete data. See

Table 1.1   Electron configurations of atoms.

Atomic Number	Name	Symbol	Atomic Shells and Orbitals					
			K	L		M		
			1s	2s	2p	3s	3p	3d
1	Hydrogen	H	1					
2	Helium	He	2					
3	Lithium	Li	2	1				
4	Beryllium	Be	2	2				
5	Boron	B	2	2	1			
6	Carbon	C	2	2	2			
7	Nitrogen	N	2	2	3			
8	Oxygen	O	2	2	4			
9	Fluorine	F	2	2	5			
10	Neon	Ne	2	2	6			
11	Sodium	Na	2	2	6	1		
12	Magnesium	Mg	2	2	6	2		
13	Aluminum	Al	2	2	6	2	1	
14	Silicon	Si	2	2	6	2	2	
15	Phosphorus	P	2	2	6	2	3	
16	Sulfur	S	2	2	6	2	4	
17	Chlorine	Cl	2	2	6	2	5	
18	Argon	Ar	2	2	6	2	6	

Fig. 156.2 of [5] for the geometrical picture of these atomic orbitals or electron probability density distribution around an isolated positive point charge. Each geometrically distinct orbital can be occupied by at most 2 electrons with opposite spin. Thus, the single s orbital can be occupied by 2 electrons, the triple p orbitals can be occupied by 6 electrons, and the 5 d orbitals can be occupied by 10 electrons. Atomic orbitals can be grouped as atomic shells, such as the K, L and M shells, a classification used textbook of chemistry [1,2]. For example, the L shell consists of one $2s$ and three $2p$ orbitals and thus can contain up to 8 electrons. The number "2" of the L shell's s and p orbitals is called the principal quantum numbers. Further explanation of the atomic quantum numbers goes beyond the scope of this book, interested readers should refer to [3–5]. The electrons in the outermost atomic shell are known as the valence electrons. The electrons in the inner shells are known as the core electrons. The valence electrons and the spatial distribution of the core electrons around the positive nuclear point charges determine the properties of an atom as well as a crystal formed by the atoms [5]. For example, the neon atom has the L shell completely filled by electrons, giving the closed-shell configuration which gives the stable or not reactive chemical properties. In Table 1.1, there are

three important elements listed, which contributed significantly to the great success of the semiconductor or electronic industry, namely silicon, boron and phosphorus. Why are these three elements the important players in electronics? To understand the reasons, the principles explained here and derived from quantum mechanics are needed.

Table 1.1 shows that the neutral silicon atom has 14 positively charged protons in its nucleus and 14 negatively charged electrons surrounding the nucleus. The nucleus is so small compared with the sizes of the electron orbits that it can be considered a point charge of $+14$. Ten of these electrons occupy the inner shells, K and L which are close to the nucleus. The $+14$ nucleus and the -10 electrons together is known as the silicon ion core, with a net charge of $+4$, Si^{+14}. The ion core has a finite size rather than a point and its charge is spatially distributed from the spatial distribution of the 10 core electrons. This spatial distribution of the core charge is different for different number of nuclei and core electrons. The core charge distribution is the fundamental origin that gives the different properties (biological, chemical, electrical, mechanical, optical, thermal) of different atoms, molecules, and materials (gas, liquid, solids - crystalline and noncrystalline). See chapters 1 to 2 of reference [5] for further elementary descriptions. The remaining four electrons of Si are in the M shell, 2 occupy the $3s$ orbitals and the other 2, the $3p$ orbitals. They are known as the valence electrons in solid state physics and chemistry and also semiconductor device theory. Their orbitals are spread out in the space configuration determined by the attractive Coulomb electric force between the distributed Si^{+4} ion core charge and the four negatively charged electrons. This is a 5-body problem, reduced from a 15-body problem ($+14$-charge nucleus plus 14 negatively charged electrons). Figure 1.1 schematically shows the energy levels and orbits (in circles, not real shapes) of the 14 electrons in a $1/r$-like potential well of the Si^+ nucleus of a silicon atom, not to scale. The electron configuration is written as $(1s)^2(2s)^2(2p)^6(3s)^2(3p)^2$. For approximate sizes, see Figs. 141.1, 171.1 and 173.1 of reference [5].

The electron configuration explained above is valid only for a spatially isolated many-electron atom. Semiconductor crystal has about 10^{23} atoms per cubic centimeter and the atoms are closely spaced. Therefore, we have to consider the effects of the nearby atoms or their electrons, ion cores, on the electronic states of any electron in a crystal, for example, one of the four valence or outer shell electrons of the silicon atom, moving in a silicon crystal consisting of all the other $10^{23} - 1$ electrons and the 10^{23} silicon ion cores. This gives us the one-electron energy band diagram which is

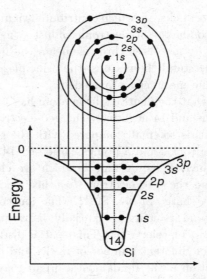

Fig. 1.1   Atomic structure of silicon.

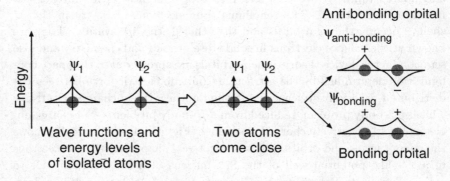

Fig. 1.2   Electron energy levels of an isolated atom, hybrid orbitals and energy bands of covalent semiconductors.

so successfully used to mathematically derive the properties of crystalline solids, such as Si, Ge, GaAs and other semiconductors, and insulators.

We will describe the many-body effects by considering two atoms each with one orbital (the $1s$ orbital). The change of the electron states by bringing the two atoms together is illustrated in a one-dimensional picture or cross sectional view in three parts of Fig. 1.2. In this figrue, the position probability distribution of the $1s$ electron is sketched. It is known as the electron wavefunction, which is the solution of the Schrödinger equation for an electron in a central or $1/r$ Coulomb potential. They are denoted by ψ_1 and ψ_2 on atoms 1 and 2. The amplitude of the square of the wavefunction is the probability of finding the electron in the 3-dimensional space point (x, y, z) and at time t, $P_1(x, y, z, t) = |\psi_1|^2$. When the two isolated atoms come together (center figure) the two atomic wavefunctions overlap. The two wavefunctions of this 2-electron system or 2-electron molecule can be represented mathematically by two other wavefunctions or molecular wavefunctions, which are known as the bonding orbitals and antibonding orbitals (right figure), and they are the linear combinations of the two atomic orbitals. They are known as molecular orbitals and written in normalized form, with the reciproal square root of 2 multiplier, by the following two equatons in order to give unity when the probability is integrated over all space. This is also known as hybrid orbitals in quantum chemistry. They are "hybrid" to bring out the symmetry properties of the actual configurations or spatial variations of the electrons' spatial density distribution on the molecule. See references [1, 2, 6].

$$\psi_{\text{anti-bonding}} = +\frac{1}{\sqrt{2}}\psi_1 - \frac{1}{\sqrt{2}}\psi_2 \qquad (1.1)$$

$$\psi_{\text{bonding}} = +\frac{1}{\sqrt{2}}\psi_1 + \frac{1}{\sqrt{2}}\psi_2. \qquad (1.2)$$

An important point is that when the isolated atoms are brougth together, the isolate-atomic orbitals are changed from their isolated shapes (not shown in the central and rightside parts of Fig. 1.2.

This simple picture of a diatomic molecule consisting of two one-electron atoms can be extended to explain the electrical properties of silicon semiconductor crystal consisting of many silicon ion cores each with four valence electrons, $(3s)^2$ and $(3p)^2$. The key point is that of symmetry (determined again by the spatial distribution of the Si^{+4} core charge), namely, the silicon ion cores forms tetrahedron units in the silicon crystal, or the electron distribution in the silicon crystal has the tetrahedron symmetry. Therefore, to bring out the tetrahedral symmetry of the spatial distribution of the four valence electrons in silicon crystal, instead of the two hybrid linear combnations given byEq. 1.1 and 1.2, there are four hybrid atomic orbitals combinations from the four atomic orbitals, $3s$, $3p_x$, $3p_y$, and $3p_z$. For easier visual observation, the isotropic $1s$ and y-directed $2p_y$ atomic orbitals are shown in Fig. 1.3 (a) and (b), and the sp^3 hybrid orbital is shown in Fig. 1.3 (c). For the $n = 3$ orbitals, see Figs. 156.2 (a), (b) and (c) of [5] and the wavefunctions listed in Table 156 in [5].

The stable silicon crystal configuration is reached (such as cooling from a melt during crystal growth onto a crytalline silicon seed or a birth site) when the total energy of the electrons and the silicon ion cores is a minimum. This stable configuration for silicon crystal has the tetrahedral symmetry. The electron density is localized in the direction from the central silicon ion core towards the 4 corners of the tetrahedron as show in Fig. 1.3 (c). Because of the proximity of the adjacent silicon ion core and its four valence electrons, the atomic orbitals of the four valence electrons on the two adjacent silicon atoms are paired into lower energy bonding and higher energy antibonding configurations. They prefer to stay in the lower energy bonding molecular orbital configuation. The band of electron energy levels with the lower energy bonding molecular orbitals is known as the valence band in the energy band description of silicon semiconductor. The band of electron energy levels with the higher energy antibonding molecular orbitals is known as the conduction band, which is only partially filled by the excited valence band electrons or bonding electrons that have broken away from the bonds of the bonding orbitals of the lower energy valence band by thermal vibration of the silicon ionic cores. In the semiconductor device physics and technology, the missing valence bond electrons are known as holes. The electrons excited into the conduction band are known as conduction electrons or conduction-band electrons, or simply just electrons. These are really quasi particles not the electrons (and positrons) in vacuum with the free mass and without encountering frequent and dense scatterers. These are quasi-particles because the properties of these quasi electron and hole particles are modified by the adjacent electrons and holes, and silicon ion cores in the silicon solid from the fundamental Coulombic electric force between each two (binary) or more of them, generally known as collision, consisting of scattering that determines their drift and diffusion speed or mobility and diffusivity, and generation-recombination-tripping that determines their lifetime.

The foregoing description of the formation of the valence (bonding) and conduction (anti-bonding) energy bands is illustrated in four parts of Fig. 1.4 from the Japanese reference [6]. Similar description can also be found in English references in journal articles and text books such as reference [1] under the general heading of linear combination of atomic orbitals (LCAO), linear combination of molecular orbitals (LCMO), linear combination bonding orbitals (LCBO), $\cdots$ and cluster calculation of the one-electron energy band of solids from applying the LCBO to a cluster of atoms in the crytal symmetry configuration with the atoms on the surface of the finite-size cluster terminated by hydrogen atom to avoid the surface discontinuity that form electron states that are localized on the surface, known as surface states.

When the complete crystal structure is built up, the set of bonding orbitals forms a group of energetic states called the valence band, and the set of antibonding orbitals forms another group of energetic states called the conduction band. An occupied orbital in the conduction band acts as a particle, referred to as a quasi-particle in the condensed matter physics terminology, with similar characteristics of true electrons in vacuum except for its mass which will be discussed in the next section (Sec. 1.1.2). We simply call it electron in the semiconductor physics. In a similar way, an empty orbital in the valence band is referred to as a quasi-particle named hole. E_c, E_v and E_g in Fig. 1.4 are the bottom of the conduction band, the top of the valence band and the energy spacing (energy gap) between E_c and E_v. Note that at the zero temperature limit all the electronic states in the valence band are perfectly loaded with the electrons coming from all atoms, and that the conduction band states are left unoccupied. The electrons in the fully filled band are not able to move through the band, because there is no energetically close empty state for electrons to move to. Therefore, a pure semiconductor at low temperature acts as an insulator.

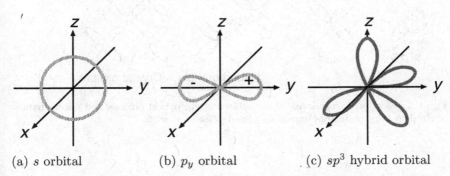

(a) s orbital (b) p_y orbital (c) sp^3 hybrid orbital

Fig. 1.3 Schematic of wavefunctions of (a) s, (b) p_y and (c) sp^3 hybrid orbitals.

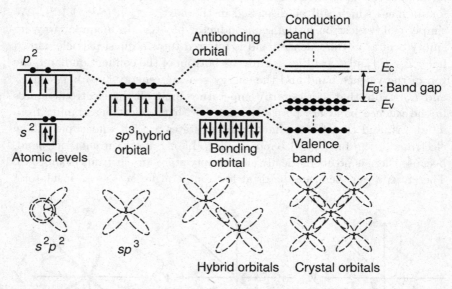

Fig. 1.4   Electron energy levels of an isolated atom, hybrid orbitals and energy bands of covalent semiconductors from a simplified point of view [6].

1.1.2 *Effective Mass Approximation*

The bond picture of a semiconductor described in the previous section is very illustrative to give a mental picture of the electrons and atomic cores or ionic cores in a solid. But it does not offer a simple mathematical formulation that can be readily used to model quantitatively and accurately the electrical properties of semiconductors. (It does but the mathematics is much more tedious and complex.) To get such a compact description useful to develop compact models of transistors and semiconductor devices for computer aided design of integrated circuits, the opposite approach is taken, namely, starting from the electron in free space or vacuum. The motion of such a single electron is described by the continuous energy, E, versus wave number, k, or electron momentum $\boldsymbol{p} = \hbar\boldsymbol{k}$ relation, given by the parabolic formula, $E_k = \hbar^2 k^2 / 2m$ (in one space dimension) where m is the rest mass of an electron in free space or vaccuum. We then ask the question of what happens to the E-$\boldsymbol{k}$ characteristics of such a single electron when the electron is in a solid in the vicinity of many other electrons and many charged nucleii or charged atomic cores or ion cores. Figure 1.5 shows the energy E versus the wave number $\boldsymbol{k}$ of such a single electron (hence

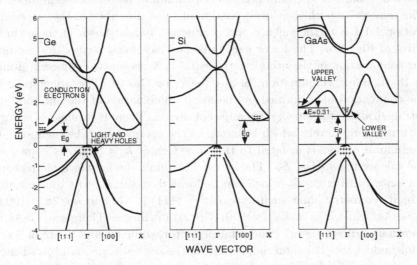

Fig. 1.5   Energy dispersion curves of the band structures for Ge, Si and GaAs in L direction ⟨111⟩ and X direction ⟨100⟩ of k space [7].

called the one-electron energy band diagram or one-quasi-electron energy band diagram) along the two principle directions $L(111)$ and $X(100)$ of propagation of the 1-electron wavefunction in the three-dimensional space of energy versus wavenumber, in Ge, Si, and GaAs. The E-k diagram is presented in a finite range of k from $ka = -\pi$ to $+\pi$ in the one dimension case, because the E-k diagram is periodic in k-space with period of $2\pi/a$. This reduced k-space is known as the lowest Brillion Zone or Jones Zone (for crystals with more than one atom per unit cell), or just the Zone, so $k = (000)$ is known as the Zone Center, labeled by the Greek Letter Γ, while $ka = (2\pi, 0, 0)$ is known as the Zone boundary at the (100) or the X-direction of the k-space, while $ka/2\pi = (111)$ is the Zone boundary at the (111) or L-direction of the k-space. The early solid state and semiconductor device physicists frequently called the momentum of such single electron in a solid, $p = \hbar k$, as the "crystal momentum". This is a rather confusion term, in contrast to the correct term "the momentum of a single electron in a crystal" or just "electron momentum", since it is not the momentum of the entire crystal of 10^{23} electrons and atoms. These E-k diagrams are computed from the three dimensional free electron E-k equation modified by the presence of the other electrons and the ion cores in the crystal with the parameters adjusted to fit the experimental measured energy gaps at different values of k, from optical absoprtion and reflectance experiments. The connection with the simple bond model just described in the preceding section 1.1.1 is in the valence and conduction bands shown in these three parts of Fig. 1.5. The lower part is the valence band formed by the linear combination of the atomic tetrahedral orbitals on two adjacent atoms in the bonding combination, Si-Si, Ge-Ge, or Ga-As. The upper part is the conduction band formed by the linear combination of the antibonding tetrahedral orbitals on the two adjacent atoms. There is a range of energy with no energy levels, which is known as the energy gap. The highest E-k maximum, always translated to the Zone Center $k = (000)$, is known as the valence band edge, E_v. The lowest E-k minimum is known as the conductance band edge, E_c which can be located at any k point, but usually a high symmetry point such as $ka/2\pi = (111)$ in Ge and $ka/2\pi = (000)$ in GaAs, while it is at $ka/2\pi = (0.85, 0, 0)$ in silicon. Therefore, GaAs is known as a direct energy band semiconductor, while Ge and Si are indirect semiconductors. These are important on the optical transition rate, hence the emission of light by a conduction band electron dropped into a valence band hole to give off the energy as light or as a photon. For Si and Ge, the large change of the electron momentum in the electron-hole recombination process to give off light or photon must be carried away by phonons or lattice vibrations which is very inefficient because the energy of the phonon energy (Raman phonon at 63meV for Si, and 42meV for Ge. See Fig.313.3 of [5].) is so much smaller than that of the recombination energy or the energy gap (1165meV for Si and 744meV for Ge) that 19 (Si) and 18 (Ge) phonons must be emitted. This energy gap model also explains the lack of any electrons and holes at low temperatures since fewer valence electrons can be thermally excited to the conductance band, to give a electron-hole pair.

The mathematical expedient analytical device modeling using the E-k or energy band relationship comes from the extrema of the E-k diagrams near the maximum or valence band edge E_v, and the minimum or the conduction band edge E_c, where most holes and electrons at thermal energies ($kT \approx 25$meV). At these extrema, the E-k relationship is quadratic, $E = E(k^2)$, so the E-k surface can be expanded in quadratic k terms. This gives the effective mass concept, when the E-k relationship of the electrons and holes in solid is compared with that of electrons in vaccuum. The general expansion to the quadratic term is given by the following two expressions when the energy is near the band edge and the electron states are not degerate. For the valence band, usually formed by the three degenerate p-type atomic orbitals, additional nonlinear quadratic terms given by the square root of k^4 and $k_x^2 k_y^2$ terms, provides better representation of the holes.

$$E(\mathbf{k}) = E_c + \frac{\hbar^2}{2} \sum_{\mu\nu} k_\mu (\mathbf{M}^{-1})_{\mu\nu} k_\nu \qquad \text{(electrons)}, \qquad (1.3)$$

$$E(\mathbf{k}) = E_v - \frac{\hbar^2}{2} \sum_{\mu\nu} k_\mu (\mathbf{M}^{-1})_{\mu\nu} k_\nu \qquad \text{(holes)}, \qquad (1.4)$$

where $\mathbf{M}^{-1}$ is the inverse effective mass tensor [8]. In Eqs. (1.3) and (1.4), the origin of the k space is taken to be the band minimum or maximum. Since the $\mathbf{M}^{-1}$ is a real and symmetric tensor, we can find an orthogonal coordinate system where $\mathbf{M}^{-1}$ has the diagonal form. Thus the simple spheroid are sufficient approximations near the band edge energy,

$$E(\mathbf{k}) = E_c + \hbar^2 \left(\frac{k_1^2}{2m_1^*} + \frac{k_2^2}{2m_2^*} + \frac{k_3^2}{2m_3^*} \right) \qquad \text{(electrons)}, \qquad (1.5)$$

$$E(\mathbf{k}) = E_v - \hbar^2 \left(\frac{k_1^2}{2m_1^*} + \frac{k_2^2}{2m_2^*} + \frac{k_3^2}{2m_3^*} \right) \qquad \text{(holes)}. \qquad (1.6)$$

For the general case of different m_1^*, m_2^* and m_3^*, the constant energy surfaces are ellipsoidal in shape, and k_1, k_2 and k_3 are the wave number components along the principal axes of the ellipsoids. Basically the band structure near a band extremum like minimum or maximum can be specified by the principal axes, the three effective masses along the principal axes and the position of the band extremum.

For example, silicon has six conduction band minima at points in the $\langle 100 \rangle$ directions of the k space as shown in Fig. 1.6 (a). Another important example is germanium which, as illustrated in Fig. 1.6 (b), has eight conduction band minima at positions along the $\langle 111 \rangle$ directions in k space.

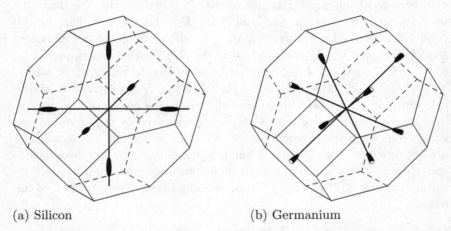

(a) Silicon   (b) Germanium

Fig. 1.6   Constant energy surfaces near the conduction band minima of (a) silicon and (b) germanium. Silicon has six and germanium has eight energetically equivalent conduction band minima [7].

1.2   Carrier Density and Fermi Level in Semiconductors

1.2.1   *Impurities in Semiconductors*

Next we consider the role of impurity atoms introduced into the semiconductor crystal. Figure 1.7 depicts simple schematics of both the crystal structures and the energy bands for pure silicon, boron doped silicon and phosphorus doped silicon. Lines connecting each atom conceptually show the bonding orbitals formed by the sp^3 hybrid orbitals as explained in Fig. 1.4. Solid circles on the bonds represent the electrons in the bonding orbitals. When a boron atom is alternatively placed at a silicon-atom site of the crystal, one empty bonding orbital is generated in the valence band as shown in Fig. 1.7 (b), because the boron atom has only 3 electrons in its outermost atomic shell as shown in Tab. 1.1. Since in the semiconductor crystal, boron atoms tend to form a closed structure with 8 electrons like the silicon atoms, a boron atom can receive 1 electron from its neighboring silicon atoms and thus can become an ion with $-q$ charge. The neighboring silicon atom which looses one of its electrons will replenish the lost electron by taking another electron from one of its neighbors. The essence of the described mechanism is that the cast-off electron which is called a "hole" can move throughout the crystal just as a positively charged particle. In the boron doped semiconductors, "positive" holes can carry electric currents, therefore this type of semiconductor is called "p-type" semiconductor.

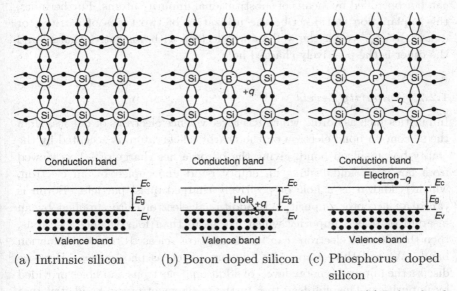

(a) Intrinsic silicon   (b) Boron doped silicon   (c) Phosphorus doped silicon

Fig. 1.7   Electron configurations and band structures of (a) intrinsic, (b) B-doped and (c) P-doped silicon.

Consider the replacement of a silicon atom by a phosphorus atom which is the silicon neighbor to the right in the periodic table (see Tab. 1.1). Such a phosphorus substitutional atom has 5 electrons in the outer shell. Since the valence band can accept only 4 electrons per atom, 4 electrons out of 5 can occupy the valence states, and the remaining 1 electron should be in the conduction band because only these unoccupied states are available as shown in Fig. 1.7 (c). The electron in the conduction band is called conduction electron and can contribute to the conductivity of the silicon crystal. The conduction electron makes the crystal "n-type" which means "negative" charges carry the currents in the semiconductor crystals. Figures 1.7 (b) and (c) are not to scale. See Fig. 222.1 and 222.2 of reference [5] for the to-scale drawings of a trapped electron and hole respectively at the positively charged phosphorus donor and negatively charged boron acceptor in silicon. (See also problem P223.4 on why the schematic diagrams are incorrect and will lead to incorrect device models if applied at face value.)

The above explanation is very primitive, but it can capture the most important property of a semiconductor, namely that the number of carriers in a semiconductor crystal can be mainly determined by the number of impurity atoms. Consequently the electrical characteristics of semiconductors can be controlled by means of substitutional impurity atoms. Furthermore, this explanation can describe the utilization of two types of carriers, or quasi-particles: one is the well known electron with a negative charge, and the other is the positively charged hole.

1.2.2 *Impurity Levels*

Impurities such as boron, which has three valence electrons or one less than the silicon, an bond electron on the nearby silicon atom is accepted by the empty boron-silicon bond, giving the boron a net charge of -1 as viewed from far away, and leaving an empty bond unoccupied by an electron, which is known as a hole or positively charged quasi particle. Boron is called an acceptor impurity. If additional electrons are supplied by an inserted or doped impurity which has more than four valence electrons, then these excess electrons can be donated (or released) to the conduction band. We call these impurities "donors", such as phosphorus. Here we discuss the impurity-energy levels of additional electrons and holes provided by impurities. This will lead to a further refinement of the band structure schematic for doped semiconductors depicted in Fig. 1.7 (b) and 1.7 (c).

First consider a n-type semiconductor. Imagine that we take a crystal of pure silicon, and replace one silicon atom by a phosphorus atom as shown in Fig. 1.8. In a first approximation, we can ignore the difference of the inner shell configurations of silicon and phosphorus atoms. Then we can represent the phosphorus atom by the sum of a silicon atom and the pair of a localized positive charge and a mobile negative charge called electron. Furthermore, the whole silicon crystal can be replaced by a continuous medium which has the dielectric constant of silicon. Finally, the fixed positive charge and the mobile electron are embedded into this continuous medium with the permittivity of silicon as depicted in Fig. 1.8. Since this simple model view is very close to a hydrogen atom in free space, the energy levels of the electron are easily calculated based on the atomic model of hydrogen [5,9].

For a hydrogen atom, the radius of the first Bohr orbit a_0 is

$$a_0 = \frac{4\pi\epsilon_0\hbar^2}{m_e q^2} = 0.529\text{Å}, \tag{1.7}$$

and the ground state binding energy E_0 is expressed as

$$E_0 = \frac{m_e q^4}{(4\pi\epsilon_0)^2 2\hbar^2} = 13.6\text{eV}, \tag{1.8}$$

where m_e and q are the electron mass and the elementary charge, respectively. In the medium of the semiconductor, electrons or holes near the energy band extrema act as free particles (quasi-particles) with effective masses m^* as discussed in Sec. 1.1.2. The permittivity becomes $\epsilon = \kappa\epsilon_0$ where κ is the dielectric constant of the medium. By changing the electron mass and the dielectric constant from their vacuum-space values to the values in semiconductors, the Bohr radius r and the ground state energy E, called impurity level, of an electron quasi-particle, which we will for simplicity refer to as electron in the following, in a semiconductor are determined as follows:

$$r = \frac{4\pi\epsilon\hbar^2}{m^* q^2} = \kappa\frac{m_e}{m^*} \cdot \frac{4\pi\epsilon_0\hbar^2}{m_e q^2} = \kappa\frac{m_e}{m^*} \cdot a_0 = \kappa\frac{m_e}{m^*} \times 0.529\text{Å}, \tag{1.9}$$

$$E = \frac{m^* q^4}{(4\pi\epsilon)^2 2\hbar^2} = \frac{1}{\kappa^2}\frac{m^*}{m_e} \cdot \frac{m_e q^4}{(4\pi\epsilon_0)^2 2\hbar^2} = \frac{1}{\kappa^2}\frac{m^*}{m_e} \cdot E_0 = \frac{1}{\kappa^2}\frac{m^*}{m_e} \times 13.6\text{eV}. \tag{1.10}$$

For silicon, the effective mass of an electron and the dielectric constant are $m^* = 0.45m_e$ and $\kappa = 11.9$, respectively. Substituting these values to Eqs. (1.9) and (1.10), the radius is $r \approx 13.8\text{Å}$ and impurity level is $E \approx 45\text{meV}$ below the conduction band minimum. For p-type silicon, the impurity level for acceptors, i.e. the binding energy of the corresponding hole quasi-particles, can be calculated in the same way as for donor atoms, and is about the same depth above the valence band edge as the phosphorus donor below the conduction band edge. For details see section 222 and 223 of reference [5]. Table 1.2 lists the hole and electron binding energies or energy levels in the Group III and V impurities in Si and Ge.

Figure 1.9 illustrates the impurity levels for both donors and acceptors in the energy-band versus distance or E-x diagram, with the impurity

Table 1.2   Impurity levels measured in meV of group III acceptors and group V donors in silicon and germanium. Table entry
is $E_A - E_v$ for acceptors, and is $E_c - E_D$ for donors. E_A and E_D
are the energy levels of acceptors and donors, respectively. E_D
and E_A are the donor level and the acceptor level, respectively.

	Group III acceptors				Group V donors			
	B	Al	Ga	In	P	As	Sb	Bi
Si	45	57	65	160	44	49	39	69
Ge	10.4	10.2	10.8	11.2	12.0	12.7	9.6	12.0

levels and impurity electrons or holes added. These energy levels listed
in Table 1.2 are small in comparison to the respective energy gaps. For
example, at room temperature the energy gaps of silicon, germanium and
gallium arsenide are about 1.1eV, 0.66eV and 1.4eV, respectively. Therefore, the energy levels of donors and acceptors are very close to the edges
of the conduction band and the valence band, respectively. Since the thermal energy kT at room temperature is about 26meV, electrons bounded at
donor sites, for example, are easily excited to the conduction bands, and
then contribute conductivity to the semiconductor crystals as negatively
charged quasi-particles.

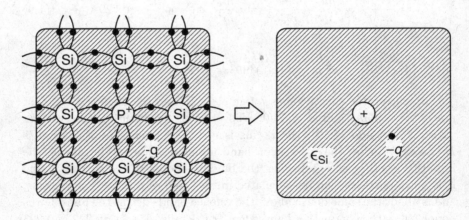

Fig. 1.8   Schematic view of a phosphorus doped silicon crystal as a homogeneous ma-
terial with the dielectric constant of silicon and a fixed positive charge plus a mobile
negative charge representing a phosphorus atom.

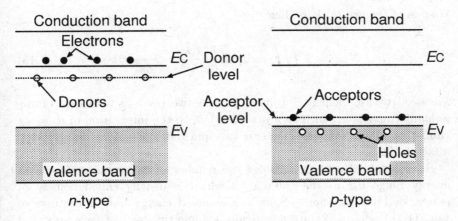

Fig. 1.9 Illustration of the impurity energy levels for both donors and acceptors in schematic band structures. Electrons in the conduction band and holes in the valence band introduced by the donors and acceptors, respectively, are also shown.

1.2.3 *Number of Carriers under Thermal Equilibrium*

The number of electrons and holes at temperature T is the most important measure required for understanding the electrical characteristics of semiconductors. In this section, we derive the carrier density as a function of temperature.

As discussed in section 1.1.2, the carrier energy as a function of momentum at the conduction and valence bands' edges can be approximated in a quadratic form. Let us further simplify this quadratic form of Eqs. (1.5) and (1.6) by assuming a parabolic band with an isotropic effective mass in $\boldsymbol{k}$-space described by

$$E(\boldsymbol{k}) = E_{\mathrm{c}} + \frac{\hbar^2 k^2}{2m_{\mathrm{e}}} \qquad \text{(electrons)}, \qquad (1.11)$$

$$E(\boldsymbol{k}) = E_{\mathrm{v}} - \frac{\hbar^2 k^2}{2m_{\mathrm{p}}} \qquad \text{(holes)}, \qquad (1.12)$$

where m_{e} and m_{p} are the effective masses of electron and hole, respectively. k is the wave vector magnitude in the three dimensional $\boldsymbol{k}$-space. Taking into account of the 2 electron spin orientations at one energy level, there are $2\,dx\,dy\,dz\,d(\hbar k_x)\,d(\hbar k_y)\,d(\hbar k_z)$ quantum states in the volume element shown in Fig. 1.10. The density of the electronic state $N(\boldsymbol{k})$ in the elemental volume of the crystal V is then

$$N(\boldsymbol{k}) = 2 \iiint_V dx\, dy\, dz\, \frac{\hbar^3 d^3\boldsymbol{k}}{(2\pi\hbar)^3} = 2\frac{V}{8\pi^3} d^3\boldsymbol{k}. \tag{1.13}$$

We need to use elemental volume V since we need to take care of the nonuniformity later, such as a p/n junction, so the integration of $dx\, dy\, dx$ cannot be a large volume. There are two spin states per energy level, giving a factor of 2.

It is helpful for us to consider the number of electronic states in an energy range dE in the volume V, which is usually called density of states, or DOS for short. Since the assumed energy band structures of Eqs. (1.11) and (1.12) are spherically symmetric, we can take spherical coordinates to calculate the DOS. The volume element of the momentum space in spherical coordinates is $d^3\boldsymbol{k} = k^2\, dk\, \sin\theta\, d\theta\, d\phi$, so that $N(\boldsymbol{k})$ becomes

$$N(\boldsymbol{k}) = \frac{V}{4\pi^3} k^2\, dk\, \sin\theta\, d\theta\, d\phi. \tag{1.14}$$

Integration of Eq. (1.14) over θ and ϕ gives the number of states at energy E, since the carrier energy depends only on the magnitude k. Then we obtain

$$N(E)\, dE = \frac{V}{4\pi^3} \int_0^{2\pi} d\phi \int_0^\pi d\theta \sin\theta\, k^2\, dk = \frac{V}{4\pi^3} \cdot 2\pi \cdot 2\, k^2\, dk = \frac{V}{\pi^2} k^2\, dk \tag{1.15}$$

$$g(E) = \frac{k^2}{\pi^2} \left(\frac{dE}{dk}\right)^{-1}, \tag{1.16}$$

where $g(E) \equiv N(E)/V$ is the DOS at energy E. The derivative of energy with respect to k is computed by using Eqs. (1.11) and (1.12). For electrons, we obtain

$$k^2 \left(\frac{dE}{dk}\right)^{-1} = k^2 \left(\frac{\hbar^2 k}{m_e}\right)^{-1} = \frac{m_e k}{\hbar^2} = \frac{1}{2} \left(\frac{2m_e}{\hbar^2}\right)^{3/2} \sqrt{E - E_c} \tag{1.17}$$

The final expression of the DOS or $g(E)$ in our terminology for the energy dependence can thus be written as follows.

$$g(E) = \begin{cases} \dfrac{1}{2\pi^2}\left(\dfrac{2m_{\mathrm{e}}}{\hbar^2}\right)^{3/2}\sqrt{E - E_{\mathrm{c}}} & \text{(for electrons)} \\[3mm] \dfrac{1}{2\pi^2}\left(\dfrac{2m_{\mathrm{p}}}{\hbar^2}\right)^{3/2}\sqrt{E_{\mathrm{v}} - E} & \text{(for holes).} \end{cases} \qquad (1.18)$$

For the ellipsoidal energy band of the conduction band, m_{e} can be replaced by $m_{\mathrm{d}} = (m_1^* m_2^* m_3^*)^{1/3}$ where m_1^*, m_2^* and m_3^* are the effective masses along the three principal axes of the ellipsoid. m_{d} is knows as the density-of-state effective mass. For silicon conduction band, there are six valleys in the (100) directions, so a multiplier of 6 is added to Eq. (1.18) and $6^{-3/2}$ is added to m_{d}. For the degenerate silicon valence band, assuming only the two upper valence bands are occupied by holes and neglecting the split-off valence band due to spin-orbit interaction, instead of 6, a multiplier 2 is added to Eq. (1.18). For further details, see reference [5] by Sah.

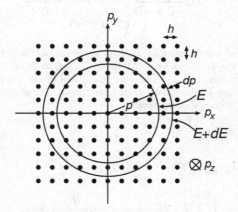

Fig. 1.10 Positions of electronic states (solid circles) aligned in a lattice pattern in the momentum space. The lattice constant is $h = 2\pi\hbar$. Each state can contain up to 2 electrons with opposite spin. Two spheres in the momentum space with radii of p and $p + dp$ show the constant energy surfaces with energy of E and $E + dE$, respectively. The number of solid circles in the spherical shell corresponds to the density of states at energy E. Since the volume of the spherical shell is $4\pi p^2\, dp$, the number of electronic states including spins is given by $2 \cdot 4\pi p^2\, dp/(2\pi\hbar)^3 = p^2\, dp/\pi^2\hbar^3 = (k^2/\pi^2)\, dk$.

Since electrons and holes obey the Fermi-Dirac statistics $f_{\mathrm{FD}}(E)$

(Fig. 1.11)

$$f_{\mathrm{FD}}(E) = \frac{1}{\exp[(E - E_{\mathrm{f}})/kT] + 1} \tag{1.19}$$

where E_{f} is the chemical potential, or the Fermi energy, which is determined by the carrier concentration, and vice versa. Of course, the temperature T is also an important quantity which determines the carrier concentrations. The total concentrations of electrons n and holes p in the band at

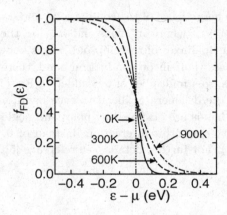

Fig. 1.11 Fermi-Dirac distribution for various temperatures as a function of the energy measured from the Fermi energy. At the point where $E = E_{\mathrm{f}}$, the Fermi-Dirac distribution takes the value of 0.5, more specifically half of the electronic states at the Fermi energy are occupied by carriers. In the zero temperature limit, the Fermi-Dirac distribution function becomes a step function.

temperature T are given by

$$n(T) = \int_{E_{\mathrm{c}}}^{\infty} dE\, g(E) f_{\mathrm{FD}}(E)$$

$$= \frac{1}{2\pi^2} \left(\frac{2m_{\mathrm{e}}}{\hbar^2} \right)^{3/2} \int_{E_{\mathrm{c}}}^{\infty} dE \frac{\sqrt{E - E_{\mathrm{c}}}}{\exp[(E - E_{\mathrm{f}})/kT] + 1}, \tag{1.20}$$

$$p(T) = \int_{-\infty}^{E_{\mathrm{v}}} dE\, g(E)[1 - f_{\mathrm{FD}}(E)]$$

$$= \frac{1}{2\pi^2} \left(\frac{2m_{\mathrm{p}}}{\hbar^2} \right)^{3/2} \int_{-\infty}^{E_{\mathrm{v}}} dE \sqrt{E_{\mathrm{v}} - E} \left(1 - \frac{1}{\exp[(E - E_{\mathrm{f}})/kT] + 1} \right)$$

$$= \frac{1}{2\pi^2} \left(\frac{2m_{\mathrm{p}}}{\hbar^2} \right)^{3/2} \int_{-\infty}^{E_{\mathrm{v}}} dE \frac{\sqrt{E_{\mathrm{v}} - E}}{\exp[(E_{\mathrm{f}} - E)/kT] + 1}. \tag{1.21}$$

The integrals containing the Fermi-Dirac distribution function can be carried out analytically, if the Fermi energy satisfies the following conditions:

$$E_c - E_f \gg kT, \tag{1.22}$$

$$E_f - E_v \gg kT. \tag{1.23}$$

For the conduction band, the electron energy E satisfies $E - E_f \gg kT$ under these conditions, so that the Fermi-Dirac function closely approaches to a simple exponential Maxwell-Boltzmann statistics function as

$$f_{\mathrm{FD}}(E) = \frac{1}{\exp[(E - E_f)/kT] + 1} \approx \exp\left(-\frac{E - E_f}{kT}\right). \tag{1.24}$$

Figure 1.12 shows how the Fermi-Dirac function comes close to the Maxwell-Boltzmann function at $E - E_f \gg kT$. As similar arguments are true for

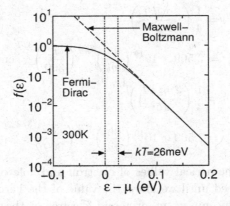

Fig. 1.12 Comparison between the Fermi-Dirac and Maxwell-Boltzmann distribution at $T = 300K$. The lateral axis is the energy measured from the Fermi energy E_f. The Fermi-Dirac statistics (solid line) is well approximated by the Maxwell-Boltzmann distribution (broken line) at energies sufficiently greater than the Fermi energy $E - E_f \gg kT$ ($kT \approx 26\text{meV}$ for $T = 300K$).

the valence band, Eqs. (1.20) and (1.21) reduce to

$$n(T) = \frac{1}{2\pi^2}\left(\frac{2m_e}{\hbar^2}\right)^{3/2}\int_{E_c}^{\infty} dE\sqrt{E - E_c}\exp\left(-\frac{E - E_f}{kT}\right), \tag{1.25}$$

$$p(T) = \frac{1}{2\pi^2}\left(\frac{2m_p}{\hbar^2}\right)^{3/2}\int_{-\infty}^{E_v} dE\sqrt{E_v - E}\exp\left(-\frac{E_f - E}{kT}\right). \tag{1.26}$$

Now the integrations can be easily carried out by using the following relation [10]:

$$\int_0^\infty dx \, \sqrt{x} \exp(x) = \frac{\sqrt{\pi}}{2}. \tag{1.27}$$

Hereby Eqs. (1.25) and (1.26) become

$$n(T) = N_c(T) \exp\left(-\frac{E_c - E_f}{kT}\right), \tag{1.28}$$

$$p(T) = N_v(T) \exp\left(-\frac{E_f - E_v}{kT}\right), \tag{1.29}$$

where $N_c(T)$ and $N_v(T)$ are the effective densities of states as a function of temperature in the conduction and valence bands, respectively, and are expressed as

$$N_c(T) = \frac{1}{4}\left(\frac{2m_e kT}{\pi\hbar^2}\right)^{3/2} \tag{1.30}$$

$$= 2.509 \times 10^{19}\left(\frac{m_e}{m}\right)^{3/2}\left(\frac{T}{300}\right)^{3/2} \mathrm{cm}^{-3}, \tag{1.31}$$

$$N_v(T) = \frac{1}{4}\left(\frac{2m_p kT}{\pi\hbar^2}\right)^{3/2} \tag{1.32}$$

$$= 2.509 \times 10^{19}\left(\frac{m_p}{m}\right)^{3/2}\left(\frac{T}{300}\right)^{3/2} \mathrm{cm}^{-3}. \tag{1.33}$$

Although the numerical values of electron and hole concentration still cannot be calculated until we know the value of the Fermi energy E_f, we can now obtain the important principle features of the carrier densities. When we focus on the product of the electron and the hole concentration, the Fermi energy in Eqs. (1.28) and (1.29) is canceled out and the product is thus given by

$$n(T)p(T) = N_c(T)N_v(T)\exp\left(-\frac{E_c - E_v}{kT}\right) = N_c(T)N_v(T)\exp\left(-\frac{E_g}{kT}\right). \tag{1.34}$$

This relationship, which is often called the mass action law, enables us to determine the density of one carrier type at equilibrium, if we know the carrier density of the other type. Equation (1.34) is very useful when calculating carrier densities of both intrinsic, pure and impure or doped semiconductors.

1.2.3.1 *Carrier Density in Pure Semiconductors*

In the case of a pure infinitely large semiconductor at thermal equilibrium without external perturbations (such as spot of light or a nearby electrode), the electron and hole densities are equal, and any perturbation transient would decay to zero in a short time, known as the dielectric relaxation time (See Chapter 2 of reference [5] by Sah for in-depth discussion on the concept of equilibrium.). Let us denote the equal electron and hole concentration of the pure semiconductor by $n_i(T)$ at temperatue T as

$$n(T) = p(T) \equiv n_i(T). \tag{1.35}$$

Applying Eq. (1.35) to the mass action law, expressed by Eq. (1.34), the intrinsic carrier concentration is determined as

$$n_i(T) = [N_c(T)N_v(T)]^{1/2} \exp\left(-\frac{E_g}{2kT}\right) \tag{1.36}$$

$$= \frac{1}{4}\left(\frac{2kT}{\pi\hbar^2}\right)^{3/2}(m_e m_p)^{3/4}\exp\left(-\frac{E_g}{2kT}\right). \tag{1.37}$$

By using the intrinsic carrier concentration n_i, the mass action law (Eq. (1.37)) can be rewritten in a very simple form as

$$n(T)p(T) = n_i^2(T). \tag{1.38}$$

Since the n_i is equal to the values of n (Eq. (1.28)) and p (Eq. (1.29)), the Fermi level in the intrinsic case E_i can be determined by equating n_i with n or p. Therefore E_i is derived as

$$E_i = E_v + \frac{1}{2}E_g + \frac{1}{2}kT\log\left(\frac{N_v}{N_c}\right), \tag{1.39}$$

or

$$E_i = E_v + \frac{1}{2}E_g + \frac{3}{4}kT\log\left(\frac{m_p}{m_e}\right). \tag{1.40}$$

From Eq. (1.40) it can be seen that when the temperature T approaches 0K, the Fermi level E_i lies in the middle of the energy gap. Further, let us recall the conditions imposed on the Fermi level as described by Eqs. (1.22) and (1.23). Since $\log(m_p/m_e)$ is of the order of unity, and furthermore the energy gap E_g is typically at least one order of magnitude larger than kT, E_i is far from both E_c and E_v on the energy scale of kT. Therefore, the intrinsic Fermi level E_i will satisfy the conditions described by Eqs. (1.22) and (1.23) which are required to simplify the integrals which contain the Fermi-Dirac statistics function (Eqs. (1.20) and (1.21)).

1.2.3.2   *Carrier Density in Impure or Doped Semiconductors*

In an impure semiconductor the carrier densities in both the conduction and valence bands are mostly determined by the impurity concentration as explained in the section 1.2.1. The carrier densities can be calculated by using the mass action law (Eq. (1.34)) as in the case of an intrinsic semiconductor, because the mass action law is valid regardless of the existence of impurities.

Charge neutrality is satisfied in homogeneous semiconductor at thermal equilibrium [5]. Therefore,

$$p + N_D^+ = n + N_A^-, \qquad (1.41)$$

is valid, where N_D^+ and N_A^- are the ionized donor and acceptor concentrations, respectively. Equation (1.41) is a general formula applicable also to the compensated semiconductor where both the donor and acceptor impurities are located in the same region of the semiconductor as long as there is no spatial variation of the impurity concentrations and there is no temperature and illumination gradient, so that electrical neutrality is maintained through the entire semiconductor. For further discussion of equilibrium, see Chapter 2 of [5] given by Sah.

For an n-type compensated semiconductor where the donor concentration N_D is larger than the acceptor concentration N_A, as shown in Fig. 1.13, the amount of holes is so small that the hole density p can be approximated to be zero. So we can mainly concentrate on number of electrons, or majority carriers. Furthermore, almost all the acceptors are ionized by additional electrons from the donors as illustrated in Fig. 1.13, which is to say $N_A^- = N_A$. Hence, we obtain

$$N_D^+ = n + N_A. \qquad (1.42)$$

Here the number of ionized donors is given by

$$N_D^+ = N_D \left\{ 1 + \left[1 + \frac{1}{g_D} \exp\left(\frac{E_D - E_f}{kT} \right) \right]^{-1} \right\}. \qquad (1.43)$$

$E_c - E_D$ is the binding energy of the electron trapped at the donor impurity. g_D is the number of different spatial configurations of the bound orbitals which the electron can be trapped at the donor impurity [5]. For the simple hydrogen level, $g_D = 2$ for the $1s$ ground and $2s$ excited states, $g_D = 2 \times 3 = 6$ for the three $2p$ excited states. For the six-valley Si conduction band, these are multiplied by 6. Thus, for phosphorus donor in Si, and electron bound to the $1s$ bound state with $E_c - E_D = 45\text{meV}$, $g_D = 2 \times 6 = 12$. For holes trapped at the boron acceptor in Si, there are two valence bands which are degenerate at the valence band maximum, $k = 0$, thus, with the spherical approximation of the E-k relation of the two Si valence bands near the band edge, Eq. (1.12) and Fig. 1.5, $g_A = 2 \times 2 = 4$. These are sufficient approximations since the excited bound states at these shallow-level impurity centers are usually empty or not occupied by a trapped electron or hole respectively at and above room temperatures or even liquid nitrogen temperature. If the excited states are considered, then the g_D and g_A and E_D and E_A in these expressions are effective values or state sum values summed over all the ground and excited bound states.

By using Eq. (1.28), Eq. (1.43) is rewritten as

$$N_D^+ = N_D \left[1 + \frac{g_D n}{N_c} \exp\left(\frac{\Delta E_D}{kT}\right)\right]^{-1} = N_D \left[\frac{n_{1D}}{n_{1D} + n}\right] \quad (1.44)$$

where $n_{1D} = n(E_f = E_D)$ and $\Delta E_D = E_c - E_D$ is the binding energy of the electron trapped at the donor impurity ion.

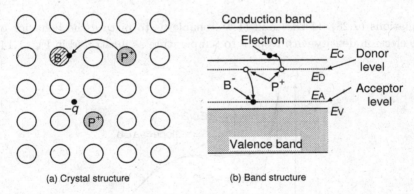

(a) Crystal structure   (b) Band structure

Fig. 1.13   Illustration of (a) crystal and (b) band structure for an n-type compensated semiconductor containing boron and phosphorus atoms located in the same region of the semiconductor crystal. Open circles are atoms of a host semiconductor, and small solid circles are representing electrons provided by phosphorus atoms. In the n-type semiconductor, almost all the acceptors are ionized by the electrons from the donor atoms.

For the case of a nearly uncompensated n-type semiconductor where

$n \gg n_i \gg N_A$, Eq. (1.42) becomes

$$n = N_D^+ \approx \frac{N_D}{\dfrac{g_D n}{N_c} \exp\left(\dfrac{\Delta E_D}{kT}\right)}, \tag{1.45}$$

and the electron density is given approximately by

$$n \approx \sqrt{\frac{N_D N_c}{g_D}} \exp\left(\frac{\Delta E_D}{2kT}\right). \tag{1.46}$$

In the low temperature limit, on the other hand, a so-called freeze-out of carriers takes place resulting in $N_D > N_A \gg n \gg p$. Dropping n from the right hand side of Eq. (1.42) yields

$$N_A = N_D^+ = N_D \left[1 + \frac{g_D n}{N_c} \exp\left(\frac{\Delta E_D}{kT}\right)\right]^{-1}, \tag{1.47}$$

and solving Eq. (1.47) for n results in

$$n = \frac{N_c}{g_D} \frac{N_D - N_A}{N_A} \exp\left(\frac{\Delta E_D}{kT}\right). \tag{1.48}$$

Equations (1.28), (1.183) and (1.48) enable us to picture the behavior of the electron density with respect to temperature as illustrated in Fig. 1.14.

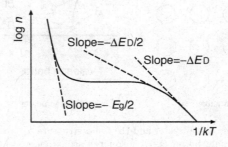

Fig. 1.14 Electron density as a function of reciprocal temperature for the case of an n-type semiconductor ($N_D > N_A$).

For a p-type semiconductor, by simply swapping the subscripts "D" for "A" in Eqs. (1.44), (1.183) and (1.48), we obtain the following equations for the ionized acceptor concentration and the hole density:

$$N_A^+ = N_A \left[1 + \frac{g_A n}{N_v} \exp\left(\frac{\Delta E_A}{kT} \right) \right]^{-1}, \tag{1.49}$$

$$p \approx \sqrt{\frac{N_A N_v}{g_A}} \exp\left(\frac{\Delta E_A}{2kT} \right) \qquad \text{(nearly uncompensated p-type region)}, \tag{1.50}$$

$$p = \frac{N_v}{g_A} \frac{N_A - N_D}{N_D} \exp\left(\frac{\Delta E_A}{kT} \right) \qquad \text{(freeze-out region)}. \tag{1.51}$$

In the discussion above, the minority carriers (holes in the n-type semiconductor, electrons in the p-type semiconductor) have been neglected. The minority carrier concentration can be determined from the majority carrier concentration by the mass action law (see Eq. (1.34)).

1.2.4 *Fermi Level*

This section is dedicated to the discussion of Fermi levels in impure semiconductors. The Fermi level is basically determined by the carrier densities as can be understood from Eqs. (1.28) and (1.29). Therefore we first try to obtain the majority carrier densities. Combining the mass action law (Eq. (1.34)) and the charge neutrality condition (Eq. (1.41)) gives the equations for the electron and hole densities as

$$n + N_A^- = p + N_D^+ = \frac{n_i^2}{n} + N_D^+, \tag{1.52}$$

$$\Rightarrow n^2 - (N_D^+ - N_A^-)n - n_i^2 = 0, \tag{1.53}$$

$$p + N_D^+ = n + N_A^- = \frac{n_i^2}{p} + N_D^+, \tag{1.54}$$

$$\Rightarrow p^2 - (N_A^- - N_D^+)p - n_i^2 = 0. \tag{1.55}$$

At around room temperature or at higher temperatures, impurities are normally fully ionized because of the thermal exitation. Then the ionized impurity densities are almost identical to the total impurity densites, which means $N_D^- = N_D$ and $N_A^- = N_A$. Thus the carrier densities are given by the solutions of the quadratic equations Eqs. (1.53) and (1.55) with these replacements:

$$n = \frac{1}{2}\left[(N_\mathrm{D} - N_\mathrm{A}) + \sqrt{(N_\mathrm{D} - N_\mathrm{A})^2 + 4n_\mathrm{i}^2}\right] \tag{1.56}$$

$$p = \frac{1}{2}\left[(N_\mathrm{A} - N_\mathrm{D}) + \sqrt{(N_\mathrm{A} - N_\mathrm{D})^2 + 4n_\mathrm{i}^2}\right]. \tag{1.57}$$

When the impurity densities are sufficiently low, so that $|N_\mathrm{D} - N_\mathrm{A}| \ll n_\mathrm{i}$ is satisfied, expanding of the square roots in Eqs. (1.56) and (1.57) yields

$$n \approx n_\mathrm{i} + \frac{1}{2}(N_\mathrm{D} - N_\mathrm{A}), \tag{1.58}$$

$$p \approx n_\mathrm{i} + \frac{1}{2}(N_\mathrm{A} - N_\mathrm{D}). \tag{1.59}$$

For the case of $|N_\mathrm{D} - N_\mathrm{A}| \gg n_\mathrm{i}$, which we are interested in, the carrier densities are given by

$$\left. \begin{array}{l} n \approx N_\mathrm{D} - N_\mathrm{A} \\[2mm] p \approx \dfrac{n_\mathrm{i}^2}{N_\mathrm{D} - N_\mathrm{A}} \end{array} \right\} \quad N_\mathrm{D} > N_\mathrm{A}, \tag{1.60}$$

$$\left. \begin{array}{l} n \approx \dfrac{n_\mathrm{i}^2}{N_\mathrm{A} - N_\mathrm{D}} \\[2mm] p \approx N_\mathrm{A} - N_\mathrm{D} \end{array} \right\} \quad N_\mathrm{A} > N_\mathrm{D}. \tag{1.61}$$

A fact of particular importance, derived from Eqs. (1.60) and (1.61), is that the excess carriers ($N_\mathrm{D} - N_\mathrm{A}$ for electrons and $N_\mathrm{A} - N_\mathrm{D}$ for holes) which originate from the doped impurities contribute almost all to the conduction and valence carriers. On the contrary, the densities of the minority carrier type are really quite small as for example $p \approx n_\mathrm{i}^2/(N_\mathrm{D} - N_\mathrm{A})$ for an n-type semiconductor, which is derived with the mass action law.

From Eqs. (1.28) and (1.29), the Fermi energy of an impure semiconductor with respect to the conduction and valence band edges are expressed as

$$E_\mathrm{c} - E_\mathrm{f} = kT \log\left(\frac{N_\mathrm{c}}{n}\right), \tag{1.62}$$

$$E_\mathrm{f} - E_\mathrm{v} = kT \log\left(\frac{N_\mathrm{v}}{p}\right). \tag{1.63}$$

Furthermore, we can also derive the Fermi energies measured from the intrinsic Fermi energy. By using Eq. (1.39), $E_f - E_i$ is transformed as

$$E_f - E_i = E_f - E_v - \frac{1}{2}E_g - \frac{1}{2}kT\log\left(\frac{N_v}{N_c}\right) \tag{1.64}$$

$$= E_f - E_v - \frac{E_c - E_v}{2} - \frac{1}{2}kT\log\left(\frac{N_v}{N_c}\right) \tag{1.65}$$

$$= E_f - \frac{E_c + E_v}{2} - \frac{1}{2}kT\log\left(\frac{N_v}{N_c}\right) \tag{1.66}$$

$$= -\frac{E_c - E_f}{2} + \frac{E_f - E_v}{2} - \frac{1}{2}kT\log\left(\frac{N_v}{N_c}\right). \tag{1.67}$$

Applying Eqs. (1.62) and (1.63) to Eq. (1.67), we obtain

$$E_f - E_i = \frac{1}{2}kT\left[-\log\left(\frac{N_c}{n}\right) + \log\left(\frac{N_v}{p}\right) - \log\left(\frac{N_v}{N_c}\right)\right] \tag{1.68}$$

$$= \frac{1}{2}kT\log\left(\frac{n}{p}\right). \tag{1.69}$$

When we recall the mass action law $np = n_i^2$, the Fermi energy measured from the intrinsic Fermi energy is expressed by using n and n_i as

$$E_f - E_i = \frac{1}{2}kT\log\left(\frac{n^2}{np}\right) = \frac{1}{2}kT\log\left(\frac{n^2}{n_i^2}\right) = kT\log\left(\frac{n}{n_i}\right). \tag{1.70}$$

Eliminating the electron density n from Eq. (1.69) yields

$$E_i - E_f = kT\log\left(\frac{p}{n_i}\right). \tag{1.71}$$

Figure 1.15 shows the Fermi energies E_f calculated by using Eqs. (1.70) and (1.71) for both n- and p-type silicon as a function of impurity densities. The temperature is set to 300K to meet the assumption that the impurities are fully ionized by the thermal excitation, which was used to derive Eqs. (1.60) and (1.61). It should be noticed that the Fermi energy comes close to energy band edges with increasing impurity concentrations.

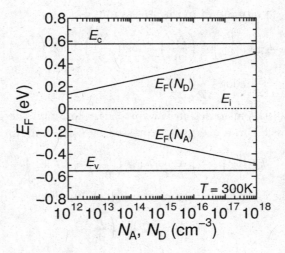

Fig. 1.15   Fermi energies $E_f(N_D)$ and $E_f(N_A)$ of doped silicon as a function of impurity densities as calculated by Eqs. (1.70) and (1.71). The energy is measured from the intrinsic Fermi energy, and the temperature is 300K. Lines labeled by E_c and E_v are the conduction and valence band edges, respectively.

1.3   *P-N* Junction

In this section we focus on *p-n* junctions which are really important not
only for realizing advanced electronic devices, but also for understanding
the electronic characteristics of semiconductor devices. Figure 1.16 shows
a very primitive picture of the *p-n* junction in which the silicon ion cores
are not shown and only the ionized impurities are shown. Circled + and

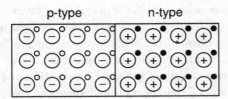

(a) p- and n-type semiconductor just stuck together

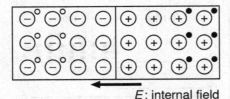

E: internal field
(b) Charge distribution under thermal equilibrium

Fig. 1.16   Schematic of the spatial distributions of charges at a *p-n* junction. (a) Illustration of *p*- and *n*-type semiconductors just stuck together. Figure (a) is just for
explanation and doesn't represent a really possible state. (b) Due to the concentration
gradient, electrons from the *n*-region diffuse across the junction into the *p*-region. In the
same way, holes from the *p*-region diffuse across the junction into the *n*-region. Due to
this diffusion, electron concentration in the *n*-type layer and hole concentration in the
p-type layer next to *p/n* junction boundary are reduced below their equilibrium value,
which give a positive charged layer on the *n*-side and negative charged layer on the *p*-
side. The area where ionized impurities are left causes an internal electric field *E* which
acts as a blocking force against the injection of electrons and holes into *p* and *n* region,
reapectively.

− signs to show their net charge compared with that of the +4 charged
silicon ion core, for example, phosphorous with +5 core and +5 electron
is shown by a circled + with a electron next to it as a solid dot. Hole
is represented by hollow circle next to a circled − for an acceptor such
as boron. When a junction of *p*- and *n*-type semiconductors is formed,
both electrons from the *n*-region and holes from the *p*-region diffuse across

the junction into the other region, due to their concentration gradient. Due to this diffusion, electron concentration in the n-type layer and hole concentration in the p-type layer next to p/n junction boundary are reduced below their equilibrium value which give a positive charged layer on the n-side and negative charged layer on the p-side, which are shown by the bare donor and acceptor ions. This double layer space charge creates an electric field in the opposite direction of electron and hole diffusion which prevents their further diffusion across the p/n boundary. Under thermal equilibrium and no applied voltage across to the terminals to the end of the p and n region, the drift current from electrons in the electric field just balances the diffusion current of electrons from the concentration gradient to give zero net electron current. Similar balance is attained for holes.

Figure 1.17 explains the same process of formation of the p-n junction from the point of view of the energy band diagram. We begin with the

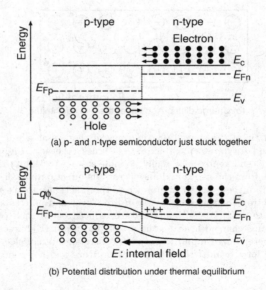

(a) p- and n-type semiconductor just stuck together

(b) Potential distribution under thermal equilibrium

Fig. 1.17   Band diagrams of a p-n junction for (a) just attached p- and n-type semiconductors, and (b) under thermal equilibrium. The broken lines show Fermi energies of p and n region. (a) Both electrons in the n region and holes in the p region diffuse into the other region, and the ionized impurities are left behind. (b) The depletion layer, where fixed ions exist, can create an internal electric field E across the p-n junction. The internal field can dam up diffusion of both electrons and holes. The Fermi energies on both sides must lie on a straight line under thermal equilibrium. ϕ is the position dependent electrostatic potential across the p-n junction.

band diagram of p and n semiconductors just glued together (Fig. 1.17 (a)). The number of electrons in the n-region is larger than that in the p-region resulting in a diffusion current of electrons from n to p region. For the same reason, a diffusive hole current flows in the opposite direction. These two diffusive currents can be dammed up by the internal electric field created due to the depletion region. As shown in Fig. 1.17 (b), the Fermi energies along the complete crossing between n and p region are required to lie on a straight line under thermal equilibrium, which is the condition for which the electrostatic potential and the carrier densities have to be calculated.

1.3.1 *P-N Junction in Thermal Equilibrium*

In this section we aim to determine the potential profile and the carrier densities of the p-n junction in thermal equilibrium. We again assume that the nondegenerated conditions defined in Eqs. (1.22) and (1.23) are valid so that the Maxwell-Boltzmann distribution function can be applied to the carrier statistics. In principle a nonuniform doping concentration can be expected to cause a variation of electrostatic potential $\phi(x)$. Here $\phi(x)$ is defined as

$$-q\phi(x) \equiv E_{\mathrm{c}}(x) - E_{\mathrm{c}}(0) = E_{\mathrm{v}}(x) - E_{\mathrm{v}}(0) \tag{1.72}$$

where $E_{\mathrm{c}}(0)$ and $E_{\mathrm{v}}(0)$ are the conduction and valence band edges at the p-n junction as shown in Fig. 1.17. Here we define $x = -d_{\mathrm{p}}$ and $x = d_{\mathrm{n}}$ as the positions of the depletion layer edges in p and n regions, respectively. So the depletion layer lies within a region $-d_{\mathrm{p}} < x < d_{\mathrm{n}}$, and its width is $W = d_{\mathrm{p}} + d_{\mathrm{n}}$.

We now begin to determine the built-in potential V_{bi} which is the potential drop between the p and n regions expressed as $qV_{\mathrm{bi}} = q[\phi(\infty) - \phi(-\infty)]$. Basically the carrier densities are determined by a modification of Eqs. (1.28) and (1.29), which takes account of the nonuniformity of the bands across the p-n junction. Thus we can rewrite these equations by using $\phi(x)$ as follows:

$$n(x) = N_c(T) \exp\left(-\frac{E_c(x) - E_f}{kT}\right) \tag{1.73}$$

$$= N_c(T) \exp\left(-\frac{-q\phi(x) + E_c(0) - E_f}{kT}\right), \tag{1.74}$$

$$p(x) = N_v \exp\left(-\frac{E_f - E_v(x)}{kT}\right) \tag{1.75}$$

$$= N_v(T) \exp\left(-\frac{E_f + q\phi(x) - E_v(0)}{kT}\right). \tag{1.76}$$

Since the carrier densities far away from the p-n junction are not influenced by the potential variation near the p-n junction, the carrier concentrations are nearly equal to the homogeneous case as in Eqs. (1.28) and (1.29). Therefore, in the limit of $x \to \pm\infty$

$$n_{n0} = n(\infty) = N_c \exp\left(-\frac{-q\phi(\infty) + E_c(0) - E_f}{kT}\right), \tag{1.77}$$

$$p_{p0} = p(-\infty) = N_v \exp\left(-\frac{E_f + q\phi(-\infty) - E_v(0)}{kT}\right) \tag{1.78}$$

where subscripts n and p mean the type of semiconductors, and subscript 0 refers to the thermal equilibrium. Under the thermal equilibrium the Fermi energy in the entire system does not vary with position. Therefore the Fermi energies in Eqs. (1.77) and (1.78) must have the same value, and hence we obtain

$$n_{n0}\, p_{p0} = N_c N_v \exp\left(-\frac{-q[\phi(\infty) - \phi(-\infty)] + E_c(0) - E_v(0)}{kT}\right), \tag{1.79}$$

$$= N_c N_v \exp\left(-\frac{-qV_{bi} + E_g}{kT}\right), \tag{1.80}$$

$$qV_{bi} = E_g + kT \log\left(\frac{n_{n0}\, p_{p0}}{N_c N_v}\right). \tag{1.81}$$

Applying Eq. (1.36), derived from the mass action law (Eq. (1.34)), to eliminate the energy gap E_g in Eq. (1.81) yields

$$qV_{bi} = -kT \log\left(\frac{n_i^2}{N_c N_v}\right) + kT \log\left(\frac{n_{n0}\, p_{p0}}{N_c N_v}\right), \tag{1.82}$$

$$= kT \log\left(\frac{n_{n0}\, p_{p0}}{n_i^2}\right). \tag{1.83}$$

The dependence of the built-in potential on the acceptor density N_A is plotted for various donor densities N_D in Fig. 1.18.

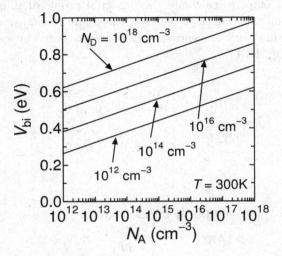

Fig. 1.18   Built-in potentials of the silicon *p-n* junctions as a function of acceptor density.

Let us move on to the calculation of the potential profile in the *p-n* junction. The Poisson equation for the *p-n* junction is given by

$$-\nabla^2\phi = -\frac{d^2\phi(x)}{dx^2} = \frac{\rho(x)}{\epsilon}, \tag{1.84}$$

$$\rho(x) = q[N_D(x) - N_A(x) - n(x) + p(x)] \tag{1.85}$$

where $N_D(x)$ and $N_A(x)$ are the impurity concentrations along the x direction. Combination of Eqs. (1.73) and (1.77) gives the electron density $n(x)$ as:

$$n(x) = n_{n0} \exp\left(-\frac{q[\phi(\infty) - \phi(x)]}{kT}\right). \tag{1.86}$$

In the same way, by using Eqs. (1.75) and (1.78), the hole density $p(x)$ is expressed as

$$p(x) = p_{p0} \exp\left(-\frac{q[\phi(x) - \phi(-\infty)]}{kT}\right). \tag{1.87}$$

It should be noticed that the charge density $\rho(x)$ also contains the potential $\phi(x)$ which has to be determined by the Poisson equation. This means the

Poisson equation (Eq. (1.84)) is a nonlinear differential equation for $\phi(x)$, so that numerical techniques will be required for obtaining an exact solution.

Here we attempt to calculate the potential profile of simple "abrupt" junctions where the transition between the p and n semiconductors is so steep that the impurity concentration can be represented by a step function as shown in Fig. 1.19.

$$N_D(x) = \begin{cases} N_D, & (x > 0) \\ 0, & (x < 0) \end{cases} \tag{1.88}$$

$$N_A(x) = \begin{cases} 0, & (x > 0) \\ N_A, & (x < 0) \end{cases} \tag{1.89}$$

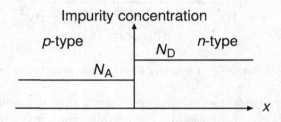

Fig. 1.19   Schematic of impurity concentration as a function of position in an abrupt p-n junction. The impurity densities changes rapidly in a step-like form at the junction.

Even for this simplest case of an abrupt junctions, the Poisson equation cannot be solved analytically. In order to obtain an analytical solution we have to simplify the charge density term $\rho(x)$. As illustrated in Fig. 1.20 (c), the potential change can be assumed to occur only in the depletion layer $(-d_p < x < d_n)$, that is to say the potential far away from the junction has the value under the thermal equilibrium condition. This means, outside of the depletion layer, $n = n_{n0} = N_D$ and $p = p_{p0} = N_A$ are valid, resulting in $\rho = 0$. Inside the depletion layer, the change in potential energy from the thermal equilibrium value $(q[\phi(0) - \phi(-\infty)]$ and $-q[\phi(\infty) - \phi(0)]$ for the valence and conduction bands, respectively) is of the order $E_g \gg kT$ as indicated in Fig. 1.17, so that $n \ll n_{n0} = N_D$ and $p \ll p_{p0} = N_A$. Therefore, the charge density can be well approximated by $\rho(x) = q[N_D(x) - N_A(x)]$ in the depletion layer $-d_p < x < d_n$ except for the area extremely close to $x = -d_p$ and $x = d_n$. With the above approximations, the Poisson equation

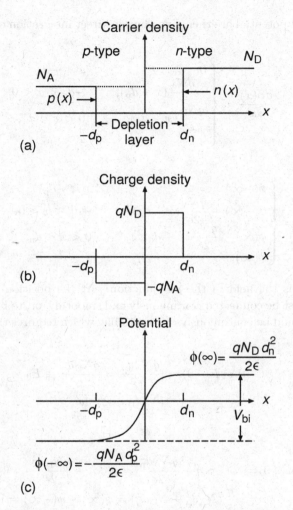

Fig. 1.20 Summary of the approximations employed to obtain analytical solutions of the potential and electric field profiles across the abrupt *p-n* junction. (a) Carrier density, (b) charge density and (c) potential profile are schematically plotted as a function of position across the *p-n* junction.

can be reduced to the form

$$\frac{d^2\phi(x)}{dx^2} = -\frac{\rho(x)}{\epsilon} = \begin{cases} 0, & x < -d_{\mathrm{p}} \\ \dfrac{qN_A}{\epsilon}, & -d_{\mathrm{p}} < x < 0, \\ -\dfrac{qN_D}{\epsilon}, & 0 < x < d_{\mathrm{n}}, \\ 0, & x > d_{\mathrm{n}}. \end{cases} \tag{1.90}$$

The field and potential are then given by the direct integration of Eq. (1.90) as

$$E(x) = -\frac{d\phi(x)}{dx} = \begin{cases} 0, & x < -d_\mathrm{p} \\ -\dfrac{qN_A}{\epsilon}(x + d_\mathrm{p}), & -d_\mathrm{p} < x < 0, \\ \dfrac{qN_D}{\epsilon}(x - d_\mathrm{n}), & 0 < x < d_\mathrm{n}, \\ 0, & x > d_\mathrm{n}, \end{cases} \tag{1.91}$$

$$\phi(x) = \begin{cases} \phi(-\infty), & x < -d_\mathrm{p} \\ \phi(-\infty) + \dfrac{qN_A}{2\epsilon}(x + d_\mathrm{p})^2, & -d_\mathrm{p} < x < 0, \\ \phi(\infty) - \dfrac{qN_D}{2\epsilon}(x - d_\mathrm{n})^2, & 0 < x < d_\mathrm{n}, \\ \phi(\infty), & x > d_\mathrm{n} \end{cases} \tag{1.92}$$

where $E(x)$ is the field in the *p-n* junction. At the position $x = 0$, the potential must be connected continuously and smoothly, or, to be exact, the potential should be continuously differentiable, which requires the following conditions

$$E(-0) = E(+0) \Rightarrow \frac{qN_A d_\mathrm{p}}{\epsilon} = \frac{qN_D d_\mathrm{n}}{\epsilon} \equiv E_0 \tag{1.93}$$

$$\Rightarrow N_A d_\mathrm{p} = N_D d_\mathrm{n} \tag{1.94}$$

and

$$\phi(-0) = \phi(+0) \Rightarrow \phi(-\infty) + \frac{qN_A d_\mathrm{p}^2}{2\epsilon} = \phi(\infty) - \frac{qN_D d_\mathrm{n}^2}{2\epsilon} \tag{1.95}$$

$$\Rightarrow \frac{q}{2\epsilon}(N_A d_\mathrm{p}^2 + N_D d_\mathrm{n}^2) = \phi(\infty) - \phi(-\infty) = V_\mathrm{bi} \tag{1.96}$$

where E_0 is the absolute value of the electric field at $x = 0$. Equation (1.94) means that charge neutrality in the entire device is maintained. Choosing $\phi(0) = 0$ makes $\phi(\infty) = qN_D d_\mathrm{n}^2/(2\epsilon)$ and $\phi(-\infty) = -qN_A d_\mathrm{p}^2/(2\epsilon)$. Equation (1.96) together with Eq. (1.94) yields

$$V_\mathrm{bi} = \frac{q}{2\epsilon} N_A d_\mathrm{p}(d_\mathrm{p} + d_\mathrm{n}) \tag{1.97}$$

$$= \frac{1}{2} E_0 W \qquad (W = d_\mathrm{p} + d_\mathrm{n}), \tag{1.98}$$

and the same time

$$V_{bi} = \frac{q}{2\epsilon}(N_A d_p^2 + N_D d_n^2) \tag{1.99}$$

$$= \frac{\epsilon}{2q}\left[\frac{1}{N_A}\left(\frac{qN_A d_p}{\epsilon}\right)^2 + \frac{1}{N_D}\left(\frac{qN_D d_n}{\epsilon}\right)^2\right] \tag{1.100}$$

$$= \frac{\epsilon}{2qE_0^2}\frac{N_A + N_D}{N_A N_D}. \tag{1.101}$$

The depletion width W is then determined by eliminating E_0 from Eqs. (1.98) and (1.101). This results in the relation

$$W = d_p + d_n = \sqrt{\frac{2\epsilon}{q}\frac{N_A + N_D}{N_A N_D}V_{bi}} \tag{1.102}$$

where the built-in potential V_{bi} can be calculated by Eq. (1.83). The depletion lengths in p and n regions can be derived from Eqs. (1.93), (1.98) and (1.102)

$$d_p = \frac{2\epsilon}{qN_A}\frac{V_{bi}}{W} = \sqrt{\frac{2\epsilon}{q}\frac{N_D/N_A}{N_A + N_D}V_{bi}}, \tag{1.103}$$

$$d_n = \frac{2\epsilon}{qN_D}\frac{V_{bi}}{W} = \sqrt{\frac{2\epsilon}{q}\frac{N_A/N_D}{N_A + N_D}V_{bi}}. \tag{1.104}$$

Figure 1.21 shows the depletion layer widths of an abrupt p-n junction formed in silicon as a function of N_A for fixed $N_D = 10^{16}\text{cm}^{-3}$. For the case of $N_A = N_D(= 10^{16}\text{cm}^{-3})$, depletion layer sizes of both the p and n sides are the same, which is a fairly trivial case. If N_A is different from N_D, the depletion layer width shrinks on the higher impurity density side and grows on the lower impurity density side, because the charge neutrality (Eq. (1.94)) should be satisfied in the entire devices. If we look at the points, for example, where the impurity density of one side is 10 times larger than that of the other side, the complete depletion layer width is already almost identical to the depletion width in the lower impurity density region. Such a situation is often called a one-sided abrupt junction, for which Eq. (1.102) reduces to

$$W = \sqrt{\frac{2\epsilon}{qN_B}}, \tag{1.105}$$

$$N_B = \begin{cases} N_A & (N_D \gg N_A) \\ N_D & (N_A \gg N_D). \end{cases} \tag{1.106}$$

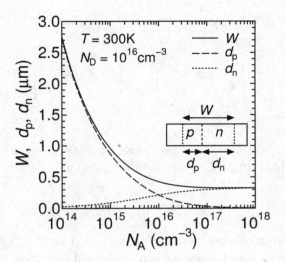

Fig. 1.21   Calculated results of the depletion layer widths of an abrupt silicon p-n junctions as a function of acceptor concentration for fixed donor concentration. The solid line shows the complete depletion layer width on both sides of the junction. Broken and dotted lines depict the depletion widths of p and n regions, respectively.

The potential profile and the electric field across an abrupt p-n junction in silicon are additionally shown in Figs. 1.22 and 1.23. The maximum values of the electric field, which occur at the position of the p-n junction, increase with doped impurity densities as shown in Fig. 1.23, because of a tighter curvature in the potential profile for higher impurity density (Eq. (1.92)).

1.3.2   P-N Junction with External Voltages

In this section we consider a p-n junction with an externally applied voltage V. We define the external voltage V as positive if it elevates the potential of the p-side of the junction with respect to the n-side. The voltage V can be assumed to change the potential profile only in the depletion layer, because the depletion layer has much higher resistance than the homogeneous p- and n-region due to its very low carrier densities. Therefore, the change in potential across the depletion region can be written as

$$\Delta\phi(V) = V_{\text{bi}} - V. \tag{1.107}$$

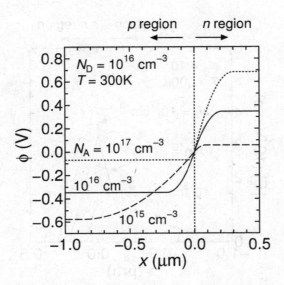

Fig. 1.22 The potential profiles of the silicon abrupt p-n junctions for various acceptor concentrations.

The application of the voltage V also modifies the thickness of depletion layer through the potential change. Replacing V_{bi} in Eqs. (1.102), (1.103) and (1.104) by Eq. (1.107) yields

$$W(V) = \sqrt{\frac{2\epsilon}{q}\frac{N_A + N_D}{N_A N_D}(V_{bi} - V)} = W(0)\sqrt{1 - \frac{V}{V_{bi}}}, \qquad (1.108)$$

$$d_p(V) = \sqrt{\frac{2\epsilon}{q}\frac{N_D/N_A}{N_A + N_D}(V_{bi} - V)} = d_p(0)\sqrt{1 - \frac{V}{V_{bi}}}, \qquad (1.109)$$

$$d_n(V) = \sqrt{\frac{2\epsilon}{q}\frac{N_A/N_D}{N_A + N_D}(V_{bi} - V)} = d_n(V)\sqrt{1 - \frac{V}{V_{bi}}}. \qquad (1.110)$$

Figure 1.24 illustrates schematically the behavior of the potential profile and the depletion layer width with respect to externally applied voltages. For the case of a silicon abrupt p-n junction, the width of the depletion layer is shown quantitatively in Fig. 1.25 for a negative, or reverse, external voltages.

When $V \neq 0$, hole and electron currents flow through the p-n junction.

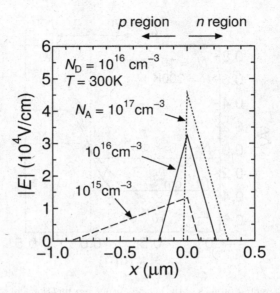

Fig. 1.23   The absolute values of electric fields across the silicon abrupt *p-n* junctions for various acceptor concentrations.

The total current can be expressed as

$$J = J_{\mathrm{p}} + J_{\mathrm{n}} = J_{\mathrm{s}} \left[\exp\left(\frac{qV}{kT} \right) - 1 \right] \qquad (1.111)$$

where J_{p} and J_{n} are the hole and electron currents, respectively. The prefactor J_{s} in Eq. (1.111) is given within the Shockley approach by [5, 7]

$$J_{\mathrm{s}} = \frac{qD_{\mathrm{p}}p_{\mathrm{n0}}}{L_{\mathrm{p}}} + \frac{qD_{\mathrm{n}}n_{\mathrm{p0}}}{L_{\mathrm{n}}} \qquad (1.112)$$

where D and L denote the diffusion constant and the diffusion length of the minority carriers in the respective region (see pages from 430 to 432 of reference [5] for more detailed explanation and derivation of the Shockley diode equation). Figure 1.26 shows the current-voltage characteristics of a *p-n* junction.

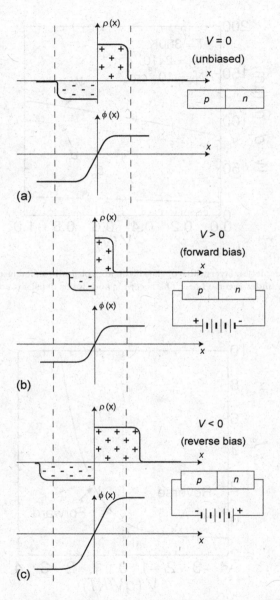

Fig. 1.24   The charge and the potential profile of the depletion layer across a *p-n* junction under (a) the unbiased, (b) the forward biased ($V > 0$) and (c) reverse biased ($V < 0$) conditions [8].

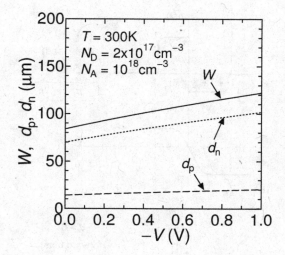

Fig. 1.25   The depletion layer width of the silicon abrupt *p-n* junction as a function of the external voltage under reverse bias. The acceptor and donor densities are $1 \times 10^{18} \text{cm}^{-3}$ and $2 \times 10^{17} \text{cm}^{-3}$, respectively.

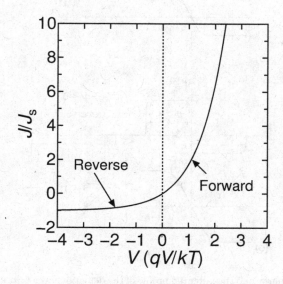

Fig. 1.26   Current flowing through a *p-n* junction as a function of the externally applied voltage. The voltage is normalized by q/kT.

1.4 Device Simulation

In this section we derive the basic equations used in semiconductor device simulations by describing carrier motion using a fluid-flow model, and give some brief explanations of a typical numerical method for solving these basic equations.

1.4.1 *Basic Equations*

The Maxwell equations for an electromagnetic wave in materials are the starting point and are expressed as

$$\nabla \cdot \boldsymbol{D} = \rho, \tag{1.113}$$

$$\nabla \cdot \boldsymbol{B} = 0, \tag{1.114}$$

$$\nabla \times \boldsymbol{E} = -\frac{\partial \boldsymbol{B}}{\partial t}, \tag{1.115}$$

$$\nabla \times \boldsymbol{H} = \boldsymbol{J} + \frac{\partial \boldsymbol{D}}{\partial t}, \tag{1.116}$$

$$\boldsymbol{D} = \epsilon \boldsymbol{E}, \tag{1.117}$$

$$\boldsymbol{B} = \mu \boldsymbol{H} \tag{1.118}$$

where ρ is the charge density and $\boldsymbol{J}$ is the current density. $\boldsymbol{E}$ is the electric field and $\boldsymbol{D}$ is the displacement vector, while $\boldsymbol{H}$ and $\boldsymbol{B}$ are the magnetic field and the induction vector, respectively. In the Eqs. (1.117) and (1.118), ϵ is the permittivity and μ is the permeability in the materials under consideration. Note that Eq. (1.117) is valid only if the electric field is static or of changes with a very low frequency. Taking the divergence of Eq. (1.116) and exploiting the relation of $\nabla \cdot (\nabla \times \boldsymbol{X}) = 0$ for an arbitrary vector $\boldsymbol{X}$ leads to

$$\nabla \cdot (\nabla \times \boldsymbol{H}) = 0 = \nabla \cdot \boldsymbol{J} + \frac{\partial \nabla \cdot \boldsymbol{D}}{\partial t}. \tag{1.119}$$

When applying Eq. (1.113) to the above equation, we obtain

$$\nabla \cdot \boldsymbol{J} + \frac{\partial \rho}{\partial t} = 0. \tag{1.120}$$

If furthermore, as in the electrostatic approximation, $\boldsymbol{B}$ and its change with respect to time are assumed to be sufficiently small, then the Maxwell equations from Eq. (1.113) to Eq. (1.118) reduce to the following 3 equations:

$$\nabla \cdot \epsilon \boldsymbol{E} \doteq \rho, \tag{1.121}$$

$$\nabla \times \boldsymbol{E} = \boldsymbol{0}, \tag{1.122}$$

$$\nabla \cdot \boldsymbol{J} + \frac{\partial \rho}{\partial t} = 0. \tag{1.123}$$

We can now apply a general theorem from vector analysis, which states that for an arbitrary vector $\boldsymbol{X}$ satisfying the condition

$$\nabla \times \boldsymbol{X} = \boldsymbol{0}, \tag{1.124}$$

there exists a scalar field ϕ from which $\boldsymbol{X}$ is determined by

$$\nabla \phi = -\boldsymbol{X}. \tag{1.125}$$

Therefore, the validity of Eq. (1.122) assures that a scalar field ϕ, or an electric potential, exists for each electric field $\boldsymbol{E}$. Using this electric potential ϕ yields

$$- \nabla \cdot (\epsilon \nabla \phi) = \rho, \tag{1.126}$$

$$\nabla \cdot \boldsymbol{J} + \frac{\partial \rho}{\partial t} = 0. \tag{1.127}$$

In semiconducting materials the charge density ρ can be expressed by using the sum of donor density N_D, acceptor density N_A, electron density n and hole density p, which is then multiplied with the elementary charge q. Assuming there is no charged generation-recombination-trapping centers and only the donor and acceptor dopant impurities which are all ionized, we have

$$\rho \doteq q(N_\mathrm{D} - N_\mathrm{A} + p - n). \tag{1.128}$$

Furthermore, the current density $\boldsymbol{J}$ is known to consist of the electron current $\boldsymbol{J}_\mathrm{n}$ and hole current $\boldsymbol{J}_\mathrm{p}$. When we consequently rewrite the current density as $\boldsymbol{J} = \boldsymbol{J}_\mathrm{n} + \boldsymbol{J}_\mathrm{p}$, Eq. (1.127) becomes

$$\nabla \cdot \boldsymbol{J}_\mathrm{n} - q\frac{\partial n}{\partial t} = - \left(\nabla \cdot \boldsymbol{J}_\mathrm{p} + q\frac{\partial p}{\partial t} \right) \equiv -q(g - r). \tag{1.129}$$

Here the terms g and r have been introduced to include the special phenomena of carrier creation and anihilation, respectively. g and r mean the number of electrons or holes generated and recombined per unit time, respectively. Note that g and r contain only electron-hole generation-recombination between an electron in the conduction bands and a hole in the valence bands, and carrire trapping is not taken into account in Eq. (1.129). When we define the generation rates of electrons g_n and holes g_p, and the recombination rates of electrons r_n and holes r_p, Eq. (1.129) has to be valid separately for electrons and holes so that the basic equations for semiconductors derived from the Maxwell equations are expressed as

$$\nabla \cdot (\epsilon \nabla \phi) = -q(N_D - N_A + p - n), \tag{1.130}$$

$$\nabla \cdot \boldsymbol{J}_n - q\frac{\partial n}{\partial t} = -q(g_n - r_n), \tag{1.131}$$

$$\nabla \cdot \boldsymbol{J}_p + q\frac{\partial p}{\partial t} = q(g_p - r_p). \tag{1.132}$$

Equation (1.130) is the well-known trap-free Poisson equation which describes the relation between the electric potential and the space charge density. Equations (1.131) and (1.132) are the continuity equations representing the conservation law of the carrier number.

The currents in semiconductors are the results of the combined motion of the electrons and holes. Therefore, these currents under an external electric field can be obtained using a statistical density function, $f(\boldsymbol{v}, \boldsymbol{r}, t)$, of the velocity $\boldsymbol{v}$ and position $\boldsymbol{r}$ of these carriers at time t in a volume element $d\boldsymbol{v}$ in the velocity space and $d\boldsymbol{r}$ in the position space, at a time t. The number of carriers with position from $\boldsymbol{r}$ to $\boldsymbol{r} + d\boldsymbol{r}$ and with velocity from $\boldsymbol{v}$ to $\boldsymbol{v} + d\boldsymbol{v}$ at time t is given by

$$f(\boldsymbol{r}, \boldsymbol{v}, t)\, d\boldsymbol{v}\, d\boldsymbol{r}. \tag{1.133}$$

Consider now the temporal change of this distribution function between $t - dt$ and t where dt is an incremental time step. There are two possible histories which carriers can have during their travel in the time step dt: one is an undisturbed flight without collision, and the other is a collision with a scatterer such as another carrier or the lattice of the semiconductor crystal. For the undisturbed flight, initial position and velocity of a carrier before dt are given by $\boldsymbol{r} - \boldsymbol{v}dt$ and $\boldsymbol{v} - (\boldsymbol{F}/m)dt$, respectively, where $\boldsymbol{F}$ is the force acting on the carrier and m is the carrier mass. Therefore, the number of carriers flowing into the phase-space element $(\boldsymbol{r}, \boldsymbol{r} + d\boldsymbol{r})$, $(\boldsymbol{v}, \boldsymbol{v} + d\boldsymbol{v})$ at time t through an undisturbed flight is

$$f(\boldsymbol{r} - \boldsymbol{v}dt, \boldsymbol{v} - (\boldsymbol{F}/m)dt, t - dt)\, d\boldsymbol{v}\, d\boldsymbol{r}. \tag{1.134}$$

For the case of a disturbed flight, since the number of collisions may be propotional to dt, the number of carriers entering into the phase-space element $(r, r + dr)$, $(v, v + dv)$ at time t is given by

$$f_{\text{coll}} \, dr \, dv \, dt \qquad (1.135)$$

where f_{coll} is a factor related to the scattering rates. In consequence, Eq. (1.133) is expressed by the sum of Eqs. (1.134) and (1.135):

$$f(r, v, t) \, dv \, dr = f(r - vdt, v - (F/m)dt, t - dt) \, dv \, dr + f_{\text{coll}} \, dr \, dv \, dt. \qquad (1.136)$$

Expanding the distribution function $f(r - vdt, v - (F/m)dt, t - dt)$ on the right hand side of Eq. (1.136) into a Taylor series and neglecting the second or higher order terms with respect to dt, leads to

$$f(r, v, t) = f(r, v, t) - vdt \cdot \nabla_r f - \frac{F}{m} dt \cdot \nabla_v f - \frac{\partial f}{\partial t} dt + f_{\text{coll}} \, dt, \quad (1.137)$$

$$\Rightarrow v \cdot \nabla_r f + \frac{F}{m} \cdot \nabla_v f + \frac{\partial f}{\partial t} = f_{\text{coll}}. \qquad (1.138)$$

where ∇_r and ∇_v are the gradient operators for the r and v dimensions, respectively. Equation (1.138) is the Boltzmann transport equation describing the behavior of the introduced distribution function [11].

Next we solve the Boltzmann transport equation under the relaxation-time approximation where the collisional term f_{coll} is approximated by the relation

$$f_{\text{coll}} = -\frac{f - f_0}{\tau}. \qquad (1.139)$$

Here f_0 and τ are the distribution function under thermal equilibrium and the relaxation time required to recover the thermal equilibrium from an off-equilibrium state, respectively. Furthermore, we assume that the change in the distribution function from the equlibrium state is sufficiently small. Thus the off-equilibrium distribution function f can be written as

$$f = f_0 + \delta f, \qquad (1.140)$$

where δf is the small amount of change. Substituting Eqs. (1.139) and (1.140) to Eq. (1.138) yields

$$f = f_0 - \tau \left[v \cdot \nabla_r f_0 + \frac{F}{m} \cdot \nabla_v f_0 \right]. \qquad (1.141)$$

In the algebraic calculation for deriving the above equation both $\nabla_r \delta f = 0$ and $\nabla_v \delta f = 0$ are approximated because δf is a very small quantity. Using the fact that the distribution function is a function of $\xi = (E - E_{\mathrm{f}})/kT$ where E is the kinetic energy of carriers, $\nabla_r f_0$ in Eq. (1.141) becomes

$$\nabla_r f_0 = \frac{df_0}{d\xi} \nabla_r \left(\frac{E - E_{\mathrm{f}}}{kT} \right) \tag{1.142}$$

$$= kT \frac{\partial f_0}{\partial E} \left(\frac{\nabla_r E - \nabla_r E_{\mathrm{f}}}{kT} - \frac{E - E_{\mathrm{f}}}{kT^2} \nabla_r T \right) \tag{1.143}$$

$$= \frac{\partial f_0}{\partial E} \left(\nabla_r E - \nabla_r E_{\mathrm{f}} - \frac{E - E_{\mathrm{f}}}{T} \nabla_r T \right). \tag{1.144}$$

The gradient of the carrier energy $\nabla_r E$ is given by $q\boldsymbol{E}$, because $\nabla_r E + q\nabla_r \phi = 0$ must be valid when the total energy is constant ($E + q\phi = \text{const.}$). Applying this relation to Eq. (1.144) yields

$$\nabla_r f_0 = \frac{\partial f_0}{\partial E} \left(q\boldsymbol{E} - \nabla_r E_{\mathrm{f}} - \frac{E - E_{\mathrm{f}}}{T} \nabla_r T \right). \tag{1.145}$$

In a similar way $\nabla_v f_0$ is derived as

$$\nabla_v f_0 = \frac{\partial f_0}{\partial E} \nabla_v E = \frac{\partial f_0}{\partial E} m\boldsymbol{v}. \tag{1.146}$$

Combining Eqs. (1.141), (1.145) and (1.146) gives

$$f = f_0 - \tau\boldsymbol{v} \left(q\boldsymbol{E} - \nabla_r E_{\mathrm{f}} - \frac{E - E_{\mathrm{f}}}{T} \nabla_r T + \boldsymbol{F} \right) \frac{\partial f_0}{\partial E} \tag{1.147}$$

$$= f_0 + \tau\boldsymbol{v} \left(\nabla_r E_{\mathrm{f}} + \frac{E - E_{\mathrm{f}}}{T} \nabla_r T \right) \frac{\partial f_0}{\partial E}. \tag{1.148}$$

Here we employed the relation $\boldsymbol{F} = -q\boldsymbol{E}$ to obtain the simplification from Eq. (1.147) to Eq. (1.148). The term relating to $\nabla_r T$ describes the carrier motion arising from the spatial gradient of the temperature, which is the origin of thermoelectric forces. In the following discussion we only consider the system under uniform temperature condition, so that we can set $\nabla_r T$ to zero. The distribution function based on the relaxation time approximation in a semiconductor with uniform temperature is then given by

$$f = f_0 + \tau\boldsymbol{v} \cdot (\nabla_r E_{\mathrm{f}}) \frac{\partial f_0}{\partial E}. \tag{1.149}$$

Once the distribution function is obtained, we can calculate the expectation values of physical quantities. The current density $\boldsymbol{J}$, which is the quantity we are interested in, is expressed as

$$J = q \int v f_0 N(E) \, dE + q \left(\nabla_r E_f \right) \int \tau v^2 \frac{\partial f_0}{\partial E} N(E) \, dE \qquad (1.150)$$

where $N(E)$ is the carrier density at energy E. The first term of the right hand side indicates the current under the thermal equilibrium which should obviously be zero. Thus, using the relation $E_f = -q\phi_f$, we obtain

$$J = -q\rho\mu \left(\nabla_r \phi_F \right), \qquad (1.151)$$

$$\mu \equiv \frac{q}{\rho} \int \tau v^2 \frac{\partial f_0}{\partial E} N(E) \, dE \qquad (1.152)$$

where ϕ_F is the electric potential at the Fermi energy and μ is the carrier mobility. Equation (1.151) reveals that the current density is linearly proportional to the gradient of the Fermi energy.

We further proceed to the formulation of the current density for the case of electrons. We rewrite the Fermi energy by using the conduction band edge E_c:

$$E_f = E_f - E_c + E_c = E_c - (E_c - E_f). \qquad (1.153)$$

Employing Eq. (1.62) for rewriting the $E_c - E_f$ term, the Fermi energy is expressed as

$$E_f = E_c - kT \log \left(\frac{N_c}{n} \right). \qquad (1.154)$$

Using the relations $E_f = -q\phi_F$ and $E_c = -q\phi_C$, Eq. (1.154) becomes

$$\phi_F = \phi_C + \frac{kT}{q} \log \left(\frac{N_c}{n} \right). \qquad (1.155)$$

The gradient of the Fermi potential ϕ_F is then given by

$$\nabla \phi_F = \nabla \phi_C + \frac{kT}{q} \nabla \log \left(\frac{N_c}{n} \right) \qquad (1.156)$$

$$= -E - \frac{kT}{qn} \nabla n. \qquad (1.157)$$

From Eq. (1.156) onwards, the subscript r of the gradient symbol is removed for simplicity since the gradient symbol will be only used with this meaning. Substituting Eq. (1.157) into Eq. (1.151) yields

$$J_n = qn\mu_n E + kT\mu_n \nabla n \qquad (1.158)$$

where μ_n is the electron mobility. The hole current density $\boldsymbol{J}_p$ is also obtained by applying the same derivation procedure as above

$$\boldsymbol{J}_p = qp\mu_p\boldsymbol{E} - kT\mu_p\nabla p. \tag{1.159}$$

Introducing carrier diffusion constants D_n for electrons and D_p for holes, and incorporating the Einstein relationship $D = (kT/q)\mu$, the current densities of electrons and holes are expressed as

$$\boldsymbol{J}_n = qn\mu_n\boldsymbol{E} + qD_n\nabla n, \tag{1.160}$$

$$\boldsymbol{J}_p = qp\mu_p\boldsymbol{E} - qD_p\nabla p, \tag{1.161}$$

which are two the basic equations for the current densities derived from the Boltzmann transport equation. Equations (1.160) and (1.161) mean that the currents consist of a drift component caused by the applied electric field (the first term of the right hand side) and a diffusion component caused by the carrier density gradient (the second term of the right hand side).

We have finished the derevation of the basic equations for semiconductor devices and can conclude by summarizing these equations as follows:

$$\nabla \cdot (\epsilon\nabla\phi) = -q(N_D - N_A + p - n), \tag{1.162}$$

$$\nabla \cdot \boldsymbol{J}_n - q\frac{\partial n}{\partial t} = -q(G_n - R_n), \tag{1.163}$$

$$\nabla \cdot \boldsymbol{J}_p + q\frac{\partial p}{\partial t} = q(G_p - R_p), \tag{1.164}$$

$$\boldsymbol{J}_n = qn\mu_n\boldsymbol{E} + qD_n\nabla n, \tag{1.165}$$

$$\boldsymbol{J}_p = qp\mu_p\boldsymbol{E} - qD_p\nabla p. \tag{1.166}$$

These equations are the Shockley equations for a two-carrier species such as a semiconductor including only generation-recombination of the two carrier pairs. Trapping of electrons is excluded. The Shcokley equations containing a trapping/detrapping process of electrons are explained in the reference [5] (see Eqs. (350.1) to (350.6) in page 268). Generally, the Shockley equations from Eq. (1.162) to Eq. (1.166) are coupled, and are solved self-consistently by applying numerical solution methodologies in semiconductor-device simulators.

1.4.2 *Linearization and Discretization of Poisson Equation*

In this section linearization and discretization methods for the Poisson equation without trapped electrons are briefly explained [11], which are the key techniques in today's numerical semiconductor device simulators for solving these basic equations.

Device simulators have to solve the basic equations by using iterative techniques because they are implicit. When solving the Poisson equation, for example, we start with an initial guess of the potential ϕ_0. Then an update value $\delta\phi$ of the potential is calculated by substituting ϕ_0 into the Poisson equation. The solution is considered to be achieved when the amount of the update becomes less than an allowable value. In the Poisson equation the carrier densities n and p also contain the potential ϕ in exponential terms, which makes the Poisson equation a nonlinear differential equation. Therefore, linearization techniques are important for achieving fast convergence, i.e. for obtaining a low computational burden of each iteration step as well as a small number of iteration steps.

Suppose that the solution ϕ of the next iteration step is represented by the sum of the initial guess ϕ_0 and the small displacement $\delta\phi$:

$$\phi = \phi_0 + \delta\phi. \tag{1.167}$$

The electron and hole densities in the next iteration step are then given by

$$n = N_\mathrm{c} \exp\left(\frac{q[\phi_0 - \phi_\mathrm{F} + \delta\phi]}{kT}\right) = n_0 \exp\left(\frac{q\delta\phi}{kT}\right) \approx n_0 \left(1 + \frac{q\delta\phi}{kT}\right), \tag{1.168}$$

$$p = N_\mathrm{v} \exp\left(\frac{q[\phi_\mathrm{F} - \phi_0 - \delta\phi]}{kT}\right) = p_0 \exp\left(-\frac{q\delta\phi}{kT}\right) \approx p_0 \left(1 - \frac{q\delta\phi}{kT}\right) \tag{1.169}$$

where n_0 and p_0 are the carrier densities for the initial potential ϕ_0. Substituting Eqs. (1.167), (1.168) and (1.169) into the Poisson equation (Eq. (1.162)) leads to

$$\nabla \cdot (\epsilon\nabla\delta\phi) - \frac{q(n_0 + p_0)}{kT}\delta\phi = -\nabla \cdot (\epsilon\nabla\phi_0) - q(p_0 - n_0 + N_\mathrm{D} - N_\mathrm{A}). \tag{1.170}$$

Equation (1.170) is now a more simple linear differential equation with respect to the small displacement $\delta\phi$.

In principle the basic equations should be solved in a continuous space, however, digital computers can only perform a finite number of calculations. Therefore, discretization with respect to the spatial dimension is necessary to numerically solve these equations. Here we explain the control volume technique, which we obtain by discretizing the linearized Poisson equation (Eq. (1.170)), as an example.

Figure 1.27 shows a mesh in an orthogonal grid pattern in 2 dimensions for discretizing a simulation domain and the corresponding control volume around one grid point indicated by the box with a dashed line [11]. First

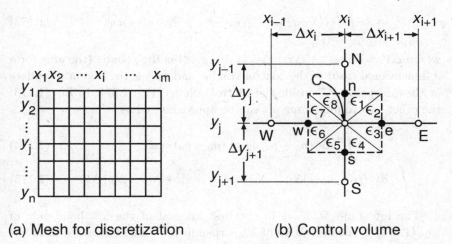

(a) Mesh for discretization   (b) Control volume

Fig. 1.27   Schematic of (a) a mesh for discretizing a simulation domain and (b) a closeup view of a grid point and the control volume CV around this grid point [11].

we integrate the linearlized Poisson equation Eq. (1.170) over the control volume CV.

$$\int_{CV} dV \, \nabla \cdot (\epsilon \nabla \delta \phi) - \frac{q}{kT} \int_{CV} dV \, (n_0 + p_0) \delta \phi$$
$$= - \int_{CV} dV \, \nabla \cdot (\epsilon \nabla \phi_0) - q \int_{CV} dV \, (p_0 - n_0 + N_D - N_A). \qquad (1.171)$$

According to the Gauss theorem the following relation for a vector X is always satisfied if this vector X and its first derivative are continuous over the region of interest V

$$\int_V dV \, \nabla \cdot X = \int_S ds \, X \cdot n. \qquad (1.172)$$

Here n is the unit normal vector outward from the volume V and S is the surface of the volume V. In other words, the volume integral of the divergence of a vector over the domain volume surrounded by a closed surface is identical to the surface integral of the vector over this closed surface. Therefore Eq. (1.171) becomes

$$\int_S ds\,\epsilon(\nabla\delta\phi\cdot\boldsymbol{n}) - \frac{q(n_0+p_0)}{kT}\delta\phi_C\Delta V$$

$$= -\int_S ds\,\epsilon(\nabla\phi_0\cdot\boldsymbol{n}) - q(p_0-n_0+N_D-N_A)\Delta V \qquad (1.173)$$

where $\Delta V = (\Delta x_{i+1}+\Delta x_i)(\Delta y_{j+1}+\Delta y_j)/4$ is the volume (the area for a 2 dimensional case) of the control volume and $\delta\phi_C$ is the potential update at the grid point "C" inside the control volume (Fig. 1.27). In Eq. (1.173) the other two volume integrals can be approximated as

$$\int_{CV} dV\,(n_0+p_0)\,\delta\phi \approx (n_0+p_0)\,\delta\phi dV, \qquad (1.174)$$

$$\int_{CV} dV\,(p_0-n_0+N_D-N_A) \approx (p_0-n_0+N_D-N_A)\Delta V. \qquad (1.175)$$

The integrand $\nabla\delta\phi\cdot\boldsymbol{n}$ in the first integral of the left hand side of Eq. (1.173) is the derivative of $\delta\phi$ perpendicular to and outward from the control volume boundary. When performing the surface integrals, the control volume is divided into the 8 triangular areas, and we assign permittivity values from ϵ_1 to ϵ_8 to each area. Hence

$$\int_S ds\,\epsilon(\nabla\delta\phi\cdot\boldsymbol{n}) \approx \left(\epsilon_1\frac{\Delta x_{i+1}}{2}+\epsilon_8\frac{\Delta x_i}{2}\right)\frac{\delta\phi_N-\delta\phi_C}{\Delta y_j}$$

$$+\left(\epsilon_7\frac{\Delta y_j}{2}+\epsilon_6\frac{\Delta y_{j+1}}{2}\right)\frac{\delta\phi_W-\delta\phi_C}{\Delta x_i}$$

$$+\left(\epsilon_4\frac{\Delta x_{i+1}}{2}+\epsilon_5\frac{\Delta x_i}{2}\right)\frac{\delta\phi_S-\delta\phi_C}{\Delta y_{j+1}}$$

$$+\left(\epsilon_3\frac{\Delta y_{j+1}}{2}+\epsilon_2\frac{\Delta y_j}{2}\right)\frac{\delta\phi_E-\delta\phi_C}{\Delta x_{i+1}} \qquad (1.176)$$

where subscripts N, W, S and E are the labels of the 4 nearest neighbor grid points of the point C [11]. Finally the discretized Poisson equation is expressed as

$$A\delta\phi_N + B\delta\phi_W - [E + \frac{q(n_0+p_0)}{kT}\Delta V]\delta\phi_C + C\delta\phi_E + D\delta\phi_S$$

$$= -A\phi_N - B\phi_W + E\phi_C - C\phi_E - D\phi_S - q(p_0-n_0+N_D-N_A)\Delta V$$

$$(1.177)$$

where

$$A = \frac{\epsilon_1 \Delta x_{i+1} + \epsilon_8 \Delta x_i}{2\Delta y_j}, \qquad (1.178)$$

$$B = \frac{\epsilon_7 \Delta y_j + \epsilon_6 \Delta y_{j+1}}{2\Delta x_i}, \qquad (1.179)$$

$$C = \frac{\epsilon_3 \Delta y_{j+1} + \epsilon_2 \Delta y_j}{2\Delta x_{i+1}}, \qquad (1.180)$$

$$D = \frac{\epsilon_4 \Delta x_{i+1} + \epsilon_5 \Delta x_i}{2\Delta y_{j+1}}, \qquad (1.181)$$

$$E = A + B + C + D. \qquad (1.182)$$

1.4.3 Device Simulation of MOSFETs

In this section we explain the basic method for performing device simulations of the MOSFET (Metal Oxide Semiconductor Field Effect Transistors) to obtain the MOSFET's characteristics. Figure 1.28 shows a 3-D view of an n-MOSFET, where electrons flow through the channel when a positive voltage is applied to the gate terminal. In this structure, an ox-

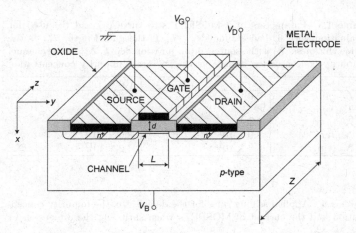

Fig. 1.28 Three dimensional illustration of the structure of an n-MOSFET [12].

ide layer with width d is sandwiched between the p-type silicon substrate and the gate electrode. Figure 1.29 gives a schematic explanation of the operation mechanisms of MOSFETs with the help of 2-dimensional band structure plots in a vertically cut plan (x-y plane) through source, gate and drain of the MOSFET structure. As illustrated in Fig. 1.29 (b) application of a positive gate voltage lowers the potential barrier at both p-n junctions, which turns on the electron diffusion into the p region below the gate oxide. In conjunction with a positive drain voltage, an electron current starts to flow from the source to the drain (Fig. 1.29 (c)). Since the gate voltage mainly controls the diffusive current from the n region (the source) to the p region (the channel), the gate can turn the MOSFET's current on and off.

Table 1.3 lists the device parameters of 2 simulated MOSFET structures. These parameters are determined by employing the MOSFET scaling rule [7] shown in Tab. 1.4, which is useful for designing the MOSFET structures down to $L_\text{g} \approx 1\mu$m. Figure 1.30 shows a calculated result of the potential profile for the MOSFET structure (a) in Tab. 1.3 with the help of a 2-dimensional device simulators. As it becomes clear from Fig. 1.30, the potential contour lines are tightly-packed in the vicinity of the interface between the gate oxide and p-doped substrate, which is an indication that the electronic characteristics of MOSFETs are predominantly determined by the potential just below the gate-oxide interface, known as the surface potential.

Table 1.3   Parameters of 2 MOSFET structures (a) and (b) used for evaluation with a device simulator. L_g is the gate length, T_ox is the oxide thickness, X_j is the source/drain junction depth, N_ch is the channel impurity concentration and N_s is the substrate impurity concentration

	(a)	(b)
L_g (μm)	2.0	1.0
T_ox (nm)	8	4
X_j (μm)	0.4	0.2
N_ch(cm^{-3})	2.0×10^{17}	8.0×10^{17}
N_s (cm^{-3})	8.0×10^{16}	3.2×10^{17}

Table 1.4   Applied scaling rules for the device size, the impurity concentration and the current of MOSFETs when shrinking the dimension by $1/k$.

	Constant voltage	Constant field
Length	$1/k$	$1/k$
Impurity concentration	k^2	k
Current	k	$1/k$

Figures 1.31 and 1.32 show the calculated threshold voltages as a func-

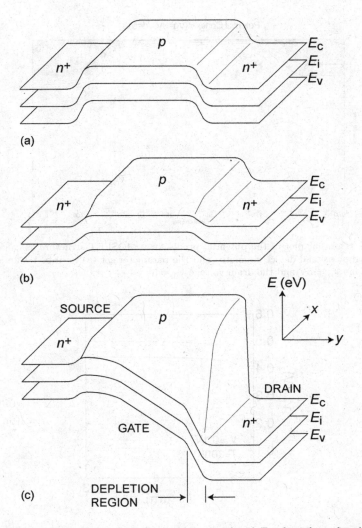

Fig. 1.29 Two dimensional views of the conduction band E_c, the valence band E_v and the intrinsic Fermi level E_i of an n-MOSFET [13]. (a) Band diagram under the flat-band condition. (b) Equilibrium condition with a positive gate voltage as compared to source and drain. (c) Nonequilibrium condition under both positive gate and drain voltages.

tion of gate length and the normalized drain currents as a function of drain voltage. The threshold voltages for the MOSFET parameters (a) and (b) in Tab. 1.3 are in good agreement when the difference of the built-in potentials arising from the difference between the respective impurity concentrations

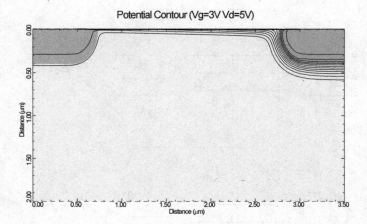

Fig. 1.30 Contour plot of the potential profile for a MOSFET structure as calculated with a 2-dimensional device simulator and the parameter set (a) in Tab. 1.3 under the gate voltage $V_g = 3$V and the drain voltage $V_d = 5$V.

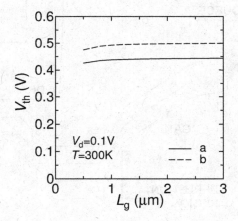

Fig. 1.31 Threshold voltages as calculated with a 2-dimensional device simulator as a function of gate length for the MOSFET parameter sets (a) and (b) in Tab. 1.3. The drain voltage $V_d = 0.1$V and the temperature $T = 300$K are fixed.

are included. Indeed the difference of built-in potentials of the silicon p-type substrate, which is identical to the difference of Fermi levels in this case, is calculated to be 36mV by using Eq. (1.63). This value is not far from 56mV estimated from Fig. 1.31. The normalized drain currents of these two structures calculated by the 2D-device simulator shown in Fig. 1.32

are almost identical to each other, which means the scaling rule of current (Tab. 1.4) is valid for the case between the structures (a) and (b) shown in Tab. 1.3.

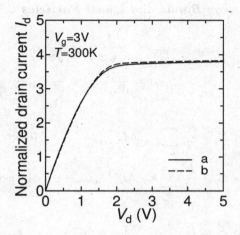

Fig. 1.32 Normalized drain currents as calculated with the 2D-device simulator as a function of drain voltage for the MOSFET parameter sets (a) and (b) in Tab. 1.3. The gate voltage $V_g = 3$V and the temperature $T = 300$K are fixed.

1.5   Summary of Equations and Symbols Presented in Chapter 1 for Semiconductor Device Physics

Section 1.1: Energy Bands and Quasi Particles

$$\psi_{\text{anti-bonding}} = +\frac{1}{\sqrt{2}}\psi_1 - \frac{1}{\sqrt{2}}\psi_2$$

$$\psi_{\text{bonding}} = +\frac{1}{\sqrt{2}}\psi_1 + \frac{1}{\sqrt{2}}\psi_2.$$

$$E(\boldsymbol{k}) = E_{\text{c}} + \frac{\hbar^2}{2}\sum_{\mu\nu} k_\mu (\boldsymbol{M}^{-1})_{\mu\nu} k_\nu \qquad \text{(electrons)},$$

$$E(\boldsymbol{k}) = E_{\text{v}} - \frac{\hbar^2}{2}\sum_{\mu\nu} k_\mu (\boldsymbol{M}^{-1})_{\mu\nu} k_\nu \qquad \text{(holes)}.$$

$$E(\boldsymbol{k}) = E_{\text{c}} + \hbar^2 \left(\frac{k_1^2}{2m_1^*} + \frac{k_2^2}{2m_2^*} + \frac{k_3^2}{2m_3^*} \right) \qquad \text{(electrons)},$$

$$E(\boldsymbol{k}) = E_{\text{v}} - \hbar^2 \left(\frac{k_1^2}{2m_1^*} + \frac{k_2^2}{2m_2^*} + \frac{k_3^2}{2m_3^*} \right) \qquad \text{(holes)}.$$

Table 1.5   Symbols presented in section 1.1.

$\psi_{\text{anti-bonding}}$	wavefunction of anti-bonding orbital
ψ_{bonding}	wavefunction of bonding orbital
$\boldsymbol{k}$	wave number
$E(\boldsymbol{k})$	energy dispersion
E_{c}	energy of conduction band minima
E_{v}	energy of valence band maxima
$\boldsymbol{M}^{-1}$	inverse effective mass tensor
m_{x}^*	effective mass ($x = 1, 2, 3$)

Section 1.2: Carrier Density and Fermi Level in Semiconductors

$$a_0 = \frac{4\pi\epsilon_0 \hbar^2}{m_e q^2} = 0.529\text{Å},$$

$$E_0 = \frac{m_e q^4}{(4\pi\epsilon_0)^2 2\hbar^2} = 13.6\text{eV}.$$

$$r = \frac{4\pi\epsilon\hbar^2}{m^*q^2} = \kappa\frac{m_e}{m^*} \cdot \frac{4\pi\epsilon_0\hbar^2}{m_eq^2} = \kappa\frac{m_e}{m^*} \cdot a_0 = \kappa\frac{m_e}{m^*} \times 0.529\text{Å},$$

$$E = \frac{m^*q^4}{(4\pi\epsilon)^22\hbar^2} = \frac{1}{\kappa^2}\frac{m^*}{m_e} \cdot \frac{m_eq^4}{(4\pi\epsilon_0)^22\hbar^2} = \frac{1}{\kappa^2}\frac{m^*}{m_e} \cdot E_0 = \frac{1}{\kappa^2}\frac{m^*}{m_e} \times 13.6\text{eV}.$$

$$E(\boldsymbol{k}) = E_c + \frac{\hbar^2k^2}{2m_e} \qquad \text{(electrons)},$$

$$E(\boldsymbol{k}) = E_v - \frac{\hbar^2k^2}{2m_p} \qquad \text{(holes)}.$$

$$N(\boldsymbol{k}) = 2\iiint_V dx\,dy\,dz\frac{\hbar^3d^3\boldsymbol{k}}{(2\pi\hbar)^3} = 2\frac{V}{8\pi^3}d^3\boldsymbol{k},$$

$$N(\boldsymbol{k}) = \frac{V}{4\pi^3}k^2\,dk\sin\theta\,d\theta\,d\phi.$$

$$N(E)\,dE = \frac{V}{4\pi^3}\int_0^{2\pi}d\phi\int_0^\pi d\theta\sin\theta\,k^2\,dk = \frac{V}{4\pi^3} \cdot 2\pi \cdot 2\,k^2\,dk = \frac{V}{\pi^2}\,k^2\,dk,$$

$$g(E) = \frac{k^2}{\pi^2}\left(\frac{dE}{dk}\right)^{-1},$$

$$k^2\left(\frac{dE}{dk}\right)^{-1} = k^2\left(\frac{\hbar^2k}{m_e}\right)^{-1} = \frac{m_ek}{\hbar^2} = \frac{1}{2}\left(\frac{2m_e}{\hbar^2}\right)^{3/2}\sqrt{E - E_c},$$

$$g(E) = \begin{cases} \dfrac{1}{2\pi^2}\left(\dfrac{2m_e}{\hbar^2}\right)^{3/2}\sqrt{E - E_c} & \text{(for electrons)} \\[3mm] \dfrac{1}{2\pi^2}\left(\dfrac{2m_p}{\hbar^2}\right)^{3/2}\sqrt{E_v - E} & \text{(for holes)}. \end{cases}$$

$$f_{\text{FD}}(E) = \frac{1}{\exp[(E - E_f)/kT] + 1}.$$

$$n(T) = \int_{E_{\rm c}}^{\infty} dE\, g(E) f_{\rm FD}(E)$$

$$= \frac{1}{2\pi^2} \left(\frac{2m_{\rm e}}{\hbar^2}\right)^{3/2} \int_{E_{\rm c}}^{\infty} dE \frac{\sqrt{E - E_{\rm c}}}{\exp[(E - E_{\rm f})/kT] + 1},$$

$$p(T) = \int_{-\infty}^{E_{\rm v}} dE\, g(E)[1 - f_{\rm FD}(E)]$$

$$= \frac{1}{2\pi^2} \left(\frac{2m_{\rm p}}{\hbar^2}\right)^{3/2} \int_{-\infty}^{E_{\rm v}} dE \sqrt{E_{\rm v} - E} \left(1 - \frac{1}{\exp[(E - E_{\rm f})/kT] + 1}\right)$$

$$= \frac{1}{2\pi^2} \left(\frac{2m_{\rm p}}{\hbar^2}\right)^{3/2} \int_{-\infty}^{E_{\rm v}} dE \frac{\sqrt{E_{\rm v} - E}}{\exp[(E_{\rm f} - E)/kT] + 1}.$$

$$E_{\rm c} - E_{\rm f} \gg kT,$$
$$E_{\rm f} - E_{\rm v} \gg kT,$$

$$f_{\rm FD}(E) = \frac{1}{\exp[(E - E_{\rm f})/kT] + 1} \approx \exp\left(-\frac{E - E_{\rm f}}{kT}\right).$$

$$n(T) = \frac{1}{2\pi^2} \left(\frac{2m_{\rm e}}{\hbar^2}\right)^{3/2} \int_{E_{\rm c}}^{\infty} dE \sqrt{E - E_{\rm c}}\, \exp\left(-\frac{E - E_{\rm f}}{kT}\right),$$

$$p(T) = \frac{1}{2\pi^2} \left(\frac{2m_{\rm p}}{\hbar^2}\right)^{3/2} \int_{-\infty}^{E_{\rm v}} dE \sqrt{E_{\rm v} - E}\, \exp\left(-\frac{E_{\rm f} - E}{kT}\right).$$

$$\int_{0}^{\infty} dx\, \sqrt{x}\, \exp(x) = \frac{\sqrt{\pi}}{2}.$$

$$n(T) = N_{\rm c}(T) \exp\left(-\frac{E_{\rm c} - E_{\rm f}}{kT}\right),$$

$$p(T) = N_{\rm v}(T) \exp\left(-\frac{E_{\rm f} - E_{\rm v}}{kT}\right).$$

$$N_{\rm c}(T) = \frac{1}{4}\left(\frac{2m_{\rm e}kT}{\pi\hbar^2}\right)^{3/2}$$

$$= 2.509\times10^{19}\left(\frac{m_{\rm e}}{m}\right)^{3/2}\left(\frac{T}{300}\right)^{3/2}{\rm cm}^{-3},$$

$$N_{\rm v}(T) = \frac{1}{4}\left(\frac{2m_{\rm p}kT}{\pi\hbar^2}\right)^{3/2}$$

$$= 2.509\times10^{19}\left(\frac{m_{\rm p}}{m}\right)^{3/2}\left(\frac{T}{300}\right)^{3/2}{\rm cm}^{-3}.$$

$$n(T)p(T) = N_{\rm c}(T)N_{\rm v}(T)\exp\left(-\frac{E_{\rm c}-E_{\rm v}}{kT}\right) = N_{\rm c}(T)N_{\rm v}(T)\exp\left(-\frac{E_{\rm g}}{kT}\right).$$

$$n(T) = p(T) \equiv n_{\rm i}(T).$$

$$n_{\rm i}(T) = [N_{\rm c}(T)N_{\rm v}(T)]^{1/2}\exp\left(-\frac{E_{\rm g}}{2kT}\right)$$

$$= \frac{1}{4}\left(\frac{2kT}{\pi\hbar^2}\right)^{3/2}(m_{\rm e}m_{\rm p})^{3/4}\exp\left(-\frac{E_{\rm g}}{2kT}\right).$$

$$n(T)p(T) = n_{\rm i}^2(T).$$

$$E_{\rm i} = E_{\rm v} + \frac{1}{2}E_{\rm g} + \frac{1}{2}kT\log\left(\frac{N_{\rm v}}{N_{\rm c}}\right),$$

$$E_{\rm i} = E_{\rm v} + \frac{1}{2}E_{\rm g} + \frac{3}{4}kT\log\left(\frac{m_{\rm p}}{m_{\rm e}}\right).$$

$$p + N_{\rm D}^+ = n + N_{\rm A}^-.$$

$$N_{\rm D}^+ = n + N_{\rm A},$$

$$N_{\text{D}}^+ = N_{\text{D}} \left\{ 1 + \left[1 + \frac{1}{g_{\text{D}}} \exp\left(\frac{E_{\text{D}} - E_{\text{f}}}{kT} \right) \right]^{-1} \right\}.$$

$$N_{\text{D}}^+ = N_{\text{D}} \left[1 + \frac{g_{\text{D}} n}{N_{\text{c}}} \exp\left(\frac{\Delta E_{\text{D}}}{kT} \right) \right]^{-1} = N_{\text{D}} \left[\frac{n_{1\text{D}}}{n_{1\text{D}} + n} \right].$$

$$n = N_{\text{D}}^+ \approx \frac{N_{\text{D}}}{\dfrac{g_{\text{D}} n}{N_{\text{c}}} \exp\left(\dfrac{\Delta E_{\text{D}}}{kT} \right)},$$

$$n \approx \sqrt{\frac{N_{\text{D}} N_{\text{c}}}{g_{\text{D}}}} \exp\left(\frac{\Delta E_{\text{D}}}{2kT} \right).$$

$$N_{\text{A}} = N_{\text{D}}^+ = N_{\text{D}} \left[1 + \frac{g_{\text{D}} n}{N_{\text{c}}} \exp\left(\frac{\Delta E_{\text{D}}}{kT} \right) \right]^{-1},$$

$$n = \frac{N_{\text{c}}}{g_{\text{D}}} \frac{N_{\text{D}} - N_{\text{A}}}{N_{\text{A}}} \exp\left(\frac{\Delta E_{\text{D}}}{kT} \right).$$

$$N_{\text{A}}^+ = N_{\text{A}} \left[1 + \frac{g_{\text{A}} n}{N_{\text{v}}} \exp\left(\frac{\Delta E_{\text{A}}}{kT} \right) \right]^{-1},$$

$$p \approx \sqrt{\frac{N_{\text{A}} N_{\text{v}}}{g_{\text{A}}}} \exp\left(\frac{\Delta E_{\text{A}}}{2kT} \right) \qquad \text{(nearly uncompensated } p\text{-type region)},$$

$$p = \frac{N_{\text{v}}}{g_{\text{A}}} \frac{N_{\text{A}} - N_{\text{D}}}{N_{\text{D}}} \exp\left(\frac{\Delta E_{\text{A}}}{kT} \right) \qquad \text{(freeze-out region)}.$$

$$n + N_{\text{A}}^- = p + N_{\text{D}}^+ = \frac{n_{\text{i}}^2}{n} + N_{\text{D}}^+,$$

$$\Rightarrow n^2 - (N_{\text{D}}^+ - N_{\text{A}}^-)n - n_{\text{i}}^2 = 0,$$

$$p + N_{\text{D}}^+ = n + N_{\text{A}}^- = \frac{n_{\text{i}}^2}{p} + N_{\text{D}}^+,$$

$$\Rightarrow p^2 - (N_{\text{A}}^- - N_{\text{D}}^+)p - n_{\text{i}}^2 = 0.$$

$$n = \frac{1}{2}\left[(N_D - N_A) + \sqrt{(N_D - N_A)^2 + 4n_i^2}\right]$$
$$p = \frac{1}{2}\left[(N_A - N_D) + \sqrt{(N_A - N_D)^2 + 4n_i^2}\right].$$

$$n \approx n_i + \frac{1}{2}(N_D - N_A),$$
$$p \approx n_i + \frac{1}{2}(N_A - N_D).$$

$$\left. \begin{array}{l} n \approx N_D - N_A \\[2mm] p \approx \dfrac{n_i^2}{N_D - N_A} \end{array} \right\} \quad N_D > N_A,$$

$$\left. \begin{array}{l} n \approx \dfrac{n_i^2}{N_A - N_D} \\[2mm] p \approx N_A - N_D \end{array} \right\} \quad N_A > N_D.$$

$$E_c - E_f = kT \log\left(\frac{N_c}{n}\right),$$
$$E_f - E_v = kT \log\left(\frac{N_v}{p}\right).$$

$$E_f - E_i = E_f - E_v - \frac{1}{2}E_g - \frac{1}{2}kT \log\left(\frac{N_v}{N_c}\right)$$
$$= E_f - E_v - \frac{E_c - E_v}{2} - \frac{1}{2}kT \log\left(\frac{N_v}{N_c}\right)$$
$$= E_f - \frac{E_c + E_v}{2} - \frac{1}{2}kT \log\left(\frac{N_v}{N_c}\right)$$
$$= -\frac{E_c - E_f}{2} + \frac{E_f - E_v}{2} - \frac{1}{2}kT \log\left(\frac{N_v}{N_c}\right).$$

$$E_f - E_i = \frac{1}{2}kT\left[-\log\left(\frac{N_c}{n}\right) + \log\left(\frac{N_v}{p}\right) - \log\left(\frac{N_v}{N_c}\right)\right]$$
$$= \frac{1}{2}kT \log\left(\frac{n}{p}\right).$$

$$E_\text{f} - E_\text{i} = \frac{1}{2}kT \log\left(\frac{n^2}{np}\right) = \frac{1}{2}kT \log\left(\frac{n^2}{n_\text{i}^2}\right) = kT \log\left(\frac{n}{n_\text{i}}\right).$$

$$E_\text{i} - E_\text{f} = kT \log\left(\frac{p}{n_\text{i}}\right).$$

Table 1.6   Symbols presented in section 1.2.

a_0	radius of the first Bohr orbit of a hydrogen atom
E_0	ground state energy of a hydrogen atom
r	Bohr radius of impurity level
E	ground state energy of impurity level
m_e	effective mass of electron
m_p	effective mass of hole
$N(\boldsymbol{k})$, $N(E)$	density of the electronic state in the elemental volume
$g(E)$	density of state
f_FD	Fermi-Dirac statistics function
n	electron concentration
p	hole concentration
E_f	Fermi level
E_i	intrinsic Fermi level
E_g	band gap
E_D	donor level
E_A	acceptor level
N_c	effective densities of states in the conduction band
N_v	effective densities of states in the valence band
N_D	donor density
N_A	acceptor density
N_D^+	ionized donor density
N_A^-	ionized acceptor density
g_D	number of degeneracy of bound orbitals for electrons
g_{A^-}	number of degeneracy of bound orbitals for holes
ΔE_D	binding energy of the electron trapped at a donor ion
ΔE_A	binding energy of the hole trapped at an acceptor ion
n_i	intrinsic carrier concentration

Section 1.3: P-N Junction

$$-q\phi(x) \equiv E_\text{c}(x) - E_\text{c}(0) = E_\text{v}(x) - E_\text{v}(0).$$

$$n(x) = N_c(T) \exp\left(-\frac{E_c(x) - E_f}{kT}\right)$$
$$= N_c(T) \exp\left(-\frac{-q\phi(x) + E_c(0) - E_f}{kT}\right),$$
$$p(x) = N_v \exp\left(-\frac{E_f - E_v(x)}{kT}\right)$$
$$= N_v(T) \exp\left(-\frac{E_f + q\phi(x) - E_v(0)}{kT}\right).$$

$$n_{n0} = n(\infty) = N_c \exp\left(-\frac{-q\phi(\infty) + E_c(0) - E_f}{kT}\right),$$
$$p_{p0} = p(-\infty) = N_v \exp\left(-\frac{E_f + q\phi(-\infty) - E_v(0)}{kT}\right).$$

$$n_{n0}\,p_{p0} = N_c N_v \exp\left(-\frac{-q[\phi(\infty) - \phi(-\infty)] + E_c(0) - E_v(0)}{kT}\right),$$
$$= N_c N_v \exp\left(-\frac{-qV_{bi} + E_g}{kT}\right),$$
$$qV_{bi} = E_g + kT \log\left(\frac{n_{n0}\,p_{p0}}{N_c N_v}\right).$$

$$qV_{bi} = -kT \log\left(\frac{n_i^2}{N_c N_v}\right) + kT \log\left(\frac{n_{n0}\,p_{p0}}{N_c N_v}\right),$$
$$= kT \log\left(\frac{n_{n0}\,p_{p0}}{n_i^2}\right).$$

$$-\nabla^2\phi = -\frac{d^2\phi(x)}{dx^2} = \frac{\rho(x)}{\epsilon},$$
$$\rho(x) = q[N_D(x) - N_A(x) - n(x) + p(x)].$$

$$n(x) = n_{n0} \exp\left(-\frac{q[\phi(\infty) - \phi(x)]}{kT}\right).$$

$$p(x) = p_{p0} \exp\left(-\frac{q[\phi(x) - \phi(-\infty)]}{kT}\right).$$

$$N_D(x) = \begin{cases} N_D, & (x > 0) \\ 0, & (x < 0) \end{cases}$$

$$N_A(x) = \begin{cases} 0, & (x > 0) \\ N_A, & (x < 0) \end{cases}$$

$$\frac{d^2\phi(x)}{dx^2} = -\frac{\rho(x)}{\epsilon} = \begin{cases} 0, & x < -d_p \\ \dfrac{qN_A}{\epsilon}, & -d_p < x < 0, \\ -\dfrac{qN_D}{\epsilon}, & 0 < x < d_n, \\ 0, & x > d_n. \end{cases}$$

$$E(x) = -\frac{d\phi(x)}{dx} = \begin{cases} 0, & x < -d_p \\ -\dfrac{qN_A}{\epsilon}(x + d_p), & -d_p < x < 0, \\ \dfrac{qN_D}{\epsilon}(x - d_n), & 0 < x < d_n, \\ 0, & x > d_n, \end{cases}$$

$$\phi(x) = \begin{cases} \phi(-\infty), & x < -d_p \\ \phi(-\infty) + \dfrac{qN_A}{2\epsilon}(x + d_p)^2, & -d_p < x < 0, \\ \phi(\infty) - \dfrac{qN_D}{2\epsilon}(x - d_n)^2, & 0 < x < d_n, \\ \phi(\infty), & x > d_n \end{cases}$$

$$E(-0) = E(+0) \Rightarrow \frac{qN_A d_p}{\epsilon} = \frac{qN_D d_n}{\epsilon} \equiv E_0$$

$$\Rightarrow N_A d_p = N_D d_n,$$

$$\phi(-0) = \phi(+0) \Rightarrow \phi(-\infty) + \frac{qN_A d_p^2}{2\epsilon} = \phi(\infty) - \frac{qN_D d_n^2}{2\epsilon}$$

$$\Rightarrow \frac{q}{2\epsilon}(N_A d_p^2 + N_D d_n^2) = \phi(\infty) - \phi(-\infty) = V_{bi}.$$

$$V_{bi} = \frac{q}{2\epsilon} N_A d_p (d_p + d_n)$$

$$= \frac{1}{2} E_0 W \qquad (W = d_p + d_n),$$

$$V_{bi} = \frac{q}{2\epsilon}(N_A d_p^2 + N_D d_n^2)$$

$$= \frac{\epsilon}{2q}\left[\frac{1}{N_A}\left(\frac{qN_A d_p}{\epsilon}\right)^2 + \frac{1}{N_D}\left(\frac{qN_D d_n}{\epsilon}\right)^2\right]$$

$$= \frac{\epsilon}{2qE_0^2}\frac{N_A + N_D}{N_A N_D}.$$

$$W = d_p + d_n = \sqrt{\frac{2\epsilon}{q}\frac{N_A + N_D}{N_A N_D}V_{bi}}$$

$$d_p = \frac{2\epsilon}{qN_A}\frac{V_{bi}}{W} = \sqrt{\frac{2\epsilon}{q}\frac{N_D/N_A}{N_A + N_D}V_{bi}},$$

$$d_n = \frac{2\epsilon}{qN_D}\frac{V_{bi}}{W} = \sqrt{\frac{2\epsilon}{q}\frac{N_A/N_D}{N_A + N_D}V_{bi}}.$$

$$W = \sqrt{\frac{2\epsilon}{qN_B}},$$

$$N_B = \begin{cases} N_A & (N_D \gg N_A) \\ N_D & (N_A \gg N_D). \end{cases}$$

$$\Delta\phi(V) = V_{bi} - V.$$

$$W(V) = \sqrt{\frac{2\epsilon}{q}\frac{N_A + N_D}{N_A N_D}(V_{bi} - V)} = W(0)\sqrt{1 - \frac{V}{V_{bi}}},$$

$$d_p(V) = \sqrt{\frac{2\epsilon}{q}\frac{N_D/N_A}{N_A + N_D}(V_{bi} - V)} = d_p(0)\sqrt{1 - \frac{V}{V_{bi}}},$$

$$d_n(V) = \sqrt{\frac{2\epsilon}{q}\frac{N_A/N_D}{N_A + N_D}(V_{bi} - V)} = d_n(V)\sqrt{1 - \frac{V}{V_{bi}}}.$$

$$J = J_p + J_n = J_s\left[\exp\left(\frac{qV}{kT}\right) - 1\right],$$

$$J_s = \frac{q D_p p_{n0}}{L_p} + \frac{q D_n n_{p0}}{L_n}.$$

Table 1.7   Symbols presented in section 1.3.

x	position perpendicular to the p/n junction
ϕ	potential
n_{n0}	electron concentration in n-type semiconductors under the thermal equilibrium
n_{p0}	electron concentration in p-type semiconductors under the thermal equilibrium
p_{p0}	hole concentration in p-type semiconductors under the thermal equilibrium
p_{n0}	hole concentration in n-type semiconductors under the thermal equilibrium
V_{bi}	built-in potential across the p/n junction
ρ	charge density
d_n	depletion layer thickness on the n side
d_p	depletion layer thickness on the p side
W	depletion layer thickness across the p/n junction
J	total current density across the p/n junction
J_n	electron current density across the p/n junction
J_p	hole current density across the p/n junction
J_s	prefactor given within the Shockley approach [5, 7]
D_n	diffusion constant of electron
D_p	diffusion constant of hole
L_n	diffusion length of electron
L_p	diffusion length of hole

Section 1.4: Device Simulation

$$\nabla \cdot \boldsymbol{D} = \rho,$$

$$\nabla \cdot \boldsymbol{B} = 0,$$

$$\nabla \times \boldsymbol{E} = -\frac{\partial \boldsymbol{B}}{\partial t},$$

$$\nabla \times \boldsymbol{H} = \boldsymbol{J} + \frac{\partial \boldsymbol{D}}{\partial t},$$

$$\boldsymbol{D} = \epsilon \boldsymbol{E},$$

$$\boldsymbol{B} = \mu \boldsymbol{H}.$$

$$\nabla \cdot (\nabla \times \boldsymbol{H}) = 0 = \nabla \cdot \boldsymbol{J} + \frac{\partial \nabla \cdot \boldsymbol{D}}{\partial t}.$$

$$\nabla \cdot \boldsymbol{J} + \frac{\partial \rho}{\partial t} = 0.$$

$$\nabla \cdot \epsilon \boldsymbol{E} = \rho,$$

$$\nabla \times \boldsymbol{E} = \boldsymbol{0},$$

$$\nabla \cdot \boldsymbol{J} + \frac{\partial \rho}{\partial t} = 0.$$

An arbitrary vector $\boldsymbol{X}$ satisfying the condition:

$$\nabla \times \boldsymbol{X} = \boldsymbol{0}.$$

$$\nabla \phi = -\boldsymbol{X}.$$

$$-\nabla \cdot (\epsilon \nabla \phi) = \rho,$$

$$\nabla \cdot \boldsymbol{J} + \frac{\partial \rho}{\partial t} = 0.$$

$$\rho = q(N_{\mathrm{D}} - N_{\mathrm{A}} + p - n).$$

$$\nabla \cdot \boldsymbol{J}_{\mathrm{n}} - q\frac{\partial n}{\partial t} = -\left(\nabla \cdot \boldsymbol{J}_{\mathrm{p}} + q\frac{\partial p}{\partial t}\right) \equiv -q(G - R).$$

$$\nabla \cdot (\epsilon \nabla \phi) = -q(N_{\mathrm{D}} - N_{\mathrm{A}} + p - n),$$

$$\nabla \cdot \boldsymbol{J}_{\mathrm{n}} - q\frac{\partial n}{\partial t} = -q(G_{\mathrm{n}} - R_{\mathrm{n}}),$$

$$\nabla \cdot \boldsymbol{J}_{\mathrm{p}} + q\frac{\partial p}{\partial t} = q(G_{\mathrm{p}} - R_{\mathrm{p}}).$$

$$f(\boldsymbol{r}, \boldsymbol{v}, t)\, d\boldsymbol{v}\, d\boldsymbol{r},$$

$$f(\boldsymbol{r} - \boldsymbol{v}dt, \boldsymbol{v} - (\boldsymbol{F}/m)dt, t - dt)\, d\boldsymbol{v}\, d\boldsymbol{r}.$$

$$f_{\mathrm{coll}}\, d\boldsymbol{r}\, d\boldsymbol{v}\, dt,$$

$$f(\boldsymbol{r}, \boldsymbol{v}, t)\, d\boldsymbol{v}\, d\boldsymbol{r} = f(\boldsymbol{r} - \boldsymbol{v}dt, \boldsymbol{v} - (\boldsymbol{F}/m)dt, t - dt)\, d\boldsymbol{v}\, d\boldsymbol{r} + f_{\mathrm{coll}}\, d\boldsymbol{r}\, d\boldsymbol{v}\, dt.$$

$$f(\boldsymbol{r}, \boldsymbol{v}, t) = f(\boldsymbol{r}, \boldsymbol{v}, t) - \boldsymbol{v}dt \cdot \nabla_{\boldsymbol{r}} f - \frac{\boldsymbol{F}}{m} dt \cdot \nabla_{\boldsymbol{v}} f - \frac{\partial f}{\partial t} dt + f_{\text{coll}} dt,$$

$$\Rightarrow \boldsymbol{v} \cdot \nabla_{\boldsymbol{r}} f + \frac{\boldsymbol{F}}{m} \cdot \nabla_{\boldsymbol{v}} f + \frac{\partial f}{\partial t} = f_{\text{coll}}.$$

$$f_{\text{coll}} = -\frac{f - f_0}{\tau}.$$

$$f = f_0 + \delta f,$$

$$f = f_0 - \tau \left[\boldsymbol{v} \cdot \nabla_{\boldsymbol{r}} f_0 + \frac{\boldsymbol{F}}{m} \cdot \nabla_{\boldsymbol{v}} f_0 \right].$$

$$\nabla_{\boldsymbol{r}} f_0 = \frac{df_0}{d\xi} \nabla_{\boldsymbol{r}} \left(\frac{E - E_{\text{f}}}{kT} \right)$$

$$= kT \frac{\partial f_0}{\partial E} \left(\frac{\nabla_{\boldsymbol{r}} E - \nabla_{\boldsymbol{r}} E_{\text{f}}}{kT} - \frac{E - E_{\text{f}}}{kT^2} \nabla_{\boldsymbol{r}} T \right)$$

$$= \frac{\partial f_0}{\partial E} \left(\nabla_{\boldsymbol{r}} E - \nabla_{\boldsymbol{r}} E_{\text{f}} - \frac{E - E_{\text{f}}}{T} \nabla_{\boldsymbol{r}} T \right).$$

$$\nabla_{\boldsymbol{r}} f_0 = \frac{\partial f_0}{\partial E} \left(q\boldsymbol{E} - \nabla_{\boldsymbol{r}} E_{\text{f}} - \frac{E - E_{\text{f}}}{T} \nabla_{\boldsymbol{r}} T \right).$$

$$\nabla_{\boldsymbol{v}} f_0 = \frac{\partial f_0}{\partial E} \nabla_{\boldsymbol{v}} E = \frac{\partial f_0}{\partial E} m\boldsymbol{v}.$$

$$f = f_0 - \tau \boldsymbol{v} \left(q\boldsymbol{E} - \nabla_{\boldsymbol{r}} E_{\text{f}} - \frac{E - E_{\text{f}}}{T} \nabla_{\boldsymbol{r}} T + \boldsymbol{F} \right) \frac{\partial f_0}{\partial E}$$

$$= f_0 + \tau \boldsymbol{v} \left(\nabla_{\boldsymbol{r}} E_{\text{f}} + \frac{E - E_{\text{f}}}{T} \nabla_{\boldsymbol{r}} T \right) \frac{\partial f_0}{\partial E}.$$

$$f = f_0 + \tau \boldsymbol{v} \cdot (\nabla_{\boldsymbol{r}} E_{\text{f}}) \frac{\partial f_0}{\partial E}.$$

$$\boldsymbol{J} = q \int \boldsymbol{v} f_0 N(E) \, dE + q \left(\nabla_{\boldsymbol{r}} E_{\text{f}} \right) \int \tau v^2 \frac{\partial f_0}{\partial E} N(E) \, dE,$$

$$\boldsymbol{J} = -q\rho\mu\left(\nabla_r \phi_F\right),$$

$$\mu \equiv \frac{q}{\rho} \int \tau v^2 \frac{\partial f_0}{\partial E} N(E)\, dE.$$

$$E_f = E_f - E_c + E_c = E_c - (E_c - E_f).$$

$$E_f = E_c - kT \log\left(\frac{N_c}{n}\right).$$

$$\phi_F = \phi_C + \frac{kT}{q} \log\left(\frac{N_c}{n}\right).$$

$$\nabla\phi_F = \nabla\phi_C + \frac{kT}{q}\nabla \log\left(\frac{N_c}{n}\right)$$

$$= -\boldsymbol{E} - \frac{kT}{qn}\nabla n.$$

$$\boldsymbol{J}_n = qn\mu_n \boldsymbol{E} + kT\mu_n \nabla n,$$

$$\boldsymbol{J}_p = qp\mu_p \boldsymbol{E} - kT\mu_p \nabla p.$$

$$\boldsymbol{J}_n = qn\mu_n \boldsymbol{E} + qD_n \nabla n,$$

$$\boldsymbol{J}_p = qp\mu_p \boldsymbol{E} - qD_p \nabla p.$$

$$\nabla \cdot (\epsilon\nabla\phi) = -q(N_D - N_A + p - n),$$

$$\nabla \cdot \boldsymbol{J}_n - q\frac{\partial n}{\partial t} = -q(G_n - R_n),$$

$$\nabla \cdot \boldsymbol{J}_p + q\frac{\partial p}{\partial t} = q(G_p - R_p),$$

$$\boldsymbol{J}_n = qn\mu_n \boldsymbol{E} + qD_n \nabla n,$$

$$\boldsymbol{J}_p = qp\mu_p \boldsymbol{E} - qD_p \nabla p.$$

$$\phi = \phi_0 + \delta\phi.$$

$$n = N_{\mathrm{c}} \exp\left(\frac{q[\phi_0 - \phi_{\mathrm{F}} + \delta\phi]}{kT} \right) = n_0 \exp\left(\frac{q\delta\phi}{kT} \right) \approx n_0 \left(1 + \frac{q\delta\phi}{kT} \right),$$

$$p = N_{\mathrm{v}} \exp\left(\frac{q[\phi_{\mathrm{F}} - \phi_0 - \delta\phi]}{kT} \right) = p_0 \exp\left(-\frac{q\delta\phi}{kT} \right) \approx p_0 \left(1 - \frac{q\delta\phi}{kT} \right),$$

$$\nabla \cdot (\epsilon \nabla \delta\phi) - \frac{q(n_0 + p_0)}{kT} \delta\phi = -\nabla \cdot (\epsilon \nabla \phi_0) - q(p_0 - n_0 + N_{\mathrm{D}} - N_{\mathrm{A}}).$$

$$\int_{\mathrm{CV}} dV \, \nabla \cdot (\epsilon \nabla \delta\phi) - \frac{q}{kT} \int_{\mathrm{CV}} dV \, (n_0 + p_0)\delta\phi$$

$$= -\int_{\mathrm{CV}} dV \, \nabla \cdot (\epsilon \nabla \phi_0) - q \int_{\mathrm{CV}} dV \, (p_0 - n_0 + N_{\mathrm{D}} - N_{\mathrm{A}}).$$

$$\int_V dV \, \nabla \cdot \boldsymbol{X} = \int_S ds \, \boldsymbol{X} \cdot \boldsymbol{n}.$$

$$\int_S ds \, \epsilon(\nabla \delta\phi \cdot \boldsymbol{n}) - \frac{q(n_0 + p_0)}{kT} \delta\phi_C \Delta V$$

$$= -\int_A ds \, \epsilon(\nabla \phi_0 \cdot \boldsymbol{n}) - q(p_0 - n_0 + N_{\mathrm{D}} - N_{\mathrm{A}})\Delta V,$$

$$\int_{\mathrm{CV}} dV \, (n_0 + p_0) \, \delta\phi \approx (n_0 + p_0) \, \delta\phi dV,$$

$$\int_{\mathrm{CV}} dV \, (p_0 - n_0 + N_{\mathrm{D}} - N_{\mathrm{A}}) \approx (p_0 - n_0 + N_{\mathrm{D}} - N_{\mathrm{A}})\Delta V.$$

$$\int_S ds \, \epsilon(\nabla \delta\phi \cdot \boldsymbol{n}) \approx \left(\epsilon_1 \frac{\Delta x_{i+1}}{2} + \epsilon_8 \frac{\Delta x_i}{2} \right) \frac{\delta\phi_N - \delta\phi_C}{\Delta y_j}$$

$$+ \left(\epsilon_7 \frac{\Delta y_j}{2} + \epsilon_6 \frac{\Delta y_{j+1}}{2} \right) \frac{\delta\phi_W - \delta\phi_C}{\Delta x_i}$$

$$+ \left(\epsilon_4 \frac{\Delta x_{i+1}}{2} + \epsilon_5 \frac{\Delta x_i}{2} \right) \frac{\delta\phi_S - \delta\phi_C}{\Delta y_{j+1}}$$

$$+ \left(\epsilon_3 \frac{\Delta y_{j+1}}{2} + \epsilon_2 \frac{\Delta y_j}{2} \right) \frac{\delta\phi_E - \delta\phi_C}{\Delta x_{i+1}}.$$

$$A\delta\phi_N + B\delta\phi_W - [E + \frac{q(n_0 + p_0)}{kT}\Delta V]\delta\phi_C + C\delta\phi_E + D\delta\phi_S$$
$$= -A\phi_N - B\phi_W + E\phi_C - C\phi_E - D\phi_S - q(p_0 - n_0 + N_{\rm D} - N_{\rm A})\Delta V,$$

$$A = \frac{\epsilon_1\Delta x_{i+1} + \epsilon_8\Delta x_i}{2\Delta y_j},$$

$$B = \frac{\epsilon_7\Delta y_j + \epsilon_6\Delta y_{j+1}}{2\Delta x_i},$$

$$C = \frac{\epsilon_3\Delta y_{j+1} + \epsilon_2\Delta y_j}{2\Delta x_{i+1}},$$

$$D = \frac{\epsilon_4\Delta x_{i+1} + \epsilon_5\Delta x_i}{2\Delta y_{j+1}},$$

$$E = A + B + C + D.$$

Table 1.8 Symbols presented in section 1.4.

$\boldsymbol{D}$	displacement vector
$\boldsymbol{B}$	induction vector
$\boldsymbol{E}$	electric field
$\boldsymbol{H}$	magnetic field
$\boldsymbol{J}$	current density
$\boldsymbol{J}_{\rm n}$	electron current density
$\boldsymbol{J}_{\rm p}$	hole current density
G	generation rate
$G_{\rm n}$	generation rate of electron
$G_{\rm p}$	generation rate of hole
R	recombination rate
$R_{\rm n}$	recombination rate of electron
$R_{\rm p}$	recombination rate of hole
f	distribution function
$f_{\rm coll}$	factor related to the scattering rates
f_0	distribution function under the thermal equilibrium
τ	relaxation time
μ	mobility
$\mu_{\rm n}$	electron mobility
$\mu_{\rm p}$	hole mobility

Bibliography

[1] W. A. Harrison, "Electronic Structure and the Properties of Solid: The Physics of the Chemical Bond", W. H. Freeman and Company, 1980.

[2]   A. Szabo and N. S. Ostlund, "Modern Quantum Chemistry: Introduction to Advanced Electronic Structure Theory," Macmillan Publishing Co., Inc., 1982.

[3]   D. Bohm, "Quantum Theory," Dover Publications, Inc., New York, 1989.

[4]   L. I. Schiff, "Quantum Mechanics: International Edition," McGraw-Hill Book Company, 1968.

[5]   C. T. Sah, "FUNDAMENTALS OF SOLID-STATE ELECTRONICS", World Scientific Publishing Co. Pte. Ltd., 1991.

[6]   K. F. Komatsubara, "Valence Electron Theory for Solid State Physics (in Japanese)", Recent Engineering Co., 1982.

[7]   S. M. Sze, "Physics of Semiconductor Device," New York, John Wiley & Sons, Inc., 1981.

[8]   K. Seeger, "Semiconductor Physics An Introduction," Springer.

[9]   N. W. Ashcroft and N. D. Mermin, "Solid State Physics," Saunders College Publishing, 1976.

[10] G. B. Arfken and H. J. Weber, "Mathematical Methods for Physicists," Academic Press, 1995.

[11] T. Sugano, R. Dan, K. Taniguchi, T. Wada, K. Asada, J. Ueda, and K. Kato, "Process and Device Simulation Technology (in Japanese)", Sangyo-Tosho Co. Ltd., 1988.

[12] D. Kahng and M. M. Atalla, "Silicon-Silicon Dioxide Field Induced Surface Devices", IRE Solid-State Device Res. Conf., Carnegie Institute of Technology, Pittsburgh, Pa., 1960.

[13] H. C. Pao and C. T. Sah, "Effects of Diffusion Current on Characteristics of Metal-Oxide (Insulator)-Semiconductor Transistors (MOST)", Solid State Electron., **9**, pp. 927-937 (1966).

Chapter 2

Basic Compact Surface-Potential Model of the MOSFET

2.1 Compact Modeling Concept

To describe the operational features of a MOSFET device based on the device physics three basic sets of equations have to be solved simultaneously. From Section 1.1, these are

- Poisson equation

$$\nabla^2 \phi = -\frac{q}{\epsilon_{Si}} (N_D - N_A + p - n) \tag{2.1}$$

$$n = n_i \exp \frac{q(\phi - \phi_{fn})}{kT}$$

$$p = n_i \exp \frac{q(\phi_{fp} - \phi)}{kT}$$

- Current-density equations

$$j_n = -q\mu_n n \frac{\partial \phi}{\partial y} + q D_n \nabla n \tag{2.2}$$

$$j_p = -q\mu_p p \frac{\partial \phi}{\partial y} - q D_p \nabla p$$

- Continuity equations

$$\frac{\partial n}{\partial t} = \frac{1}{q} \nabla j_n \tag{2.3}$$

$$\frac{\partial p}{\partial t} = -\frac{1}{q} \nabla j_p$$

where ϕ, q, ϵ_{Si}, k, T are potential, electron charge, silicon permittivity, Boltzmann constant, and lattice temperature in Kelvin, respectively. N_D, N_A, p, n, and n_i are donor, acceptor, hole, electron and intrinsic carrier concentrations. ϕ_{fn} and ϕ_{fp} are quasi-Fermi potentials of electron and hole, respectively. The carrier mobility and the carrier diffusivity, usually treated by the diffusion constant, are denoted by μ and D. In contrast to the numerical 2D-device simulator solving all equations simultaneously, the coupling among these equations is neglected and they are solved independently in compact modeling. The validity of this treatment is demonstrated in Fig. 2.1, where calculated potential distributions along the channel with a 2D-device simulator are compared with and without current flow. For the calculation of the case without current flow the carrier mobility μ is reduced in the 2D-device simulator to a level low enough so that no obvious current flow is observed. The difference between the results with and without current flow is small in comparison to the potentail value itself. However, it has to be noticed that the validity is limited to the region where the potential increases only gradually, which is refered as the gradual-channel approximation in the MOSFET, and where no strong field increase and thus no carrier generation occur. These conditions are generally fullfilled for the devices and their corresponding operating conditions, for which compact modeling is applied to obtain accurate models for the purpose of circuit simulation. In 2D-device simulation, on the other hand, the generation and recombination terms are included in Eq. (2.3).

The Poisson equation describes the relation between potential and carrier concentrations, and the current equations under the drift-diffusion theory, first employed by Sah and Pao, describe the mechanism of current flow [1, 2]. The diffusion term $(qD_n\nabla n, qD_p\nabla p)$ dominates before the inversion channel is formed, and the drift term $(q\mu_n n\partial\phi/\partial y, q\mu_p p\partial\phi/\partial y)$ dominates beyond the inversion channel formation as schematically shown in Fig. 2.2

Circuit simulators follow the charge-control approximation to give the time-dependent current-voltage characteristics. The basic equation is obtained by integrating Eq. (2.3) along the channel under the assumption that the potential responds instantaneously to the terminal voltage change and is given by [3]

$$I_x(t) = I_{x0}(V(t)) + \frac{dQ_x(t)}{dt}; \quad x = \text{S, D, B, or G} \qquad (2.4)$$

Here S, D, B and G stand for the four MOSFET terminals(see Fig. (2.3)), namely source, drain, bulk and gate, respectively. The assumption of an instantaneous potential response to terminal voltages has to be improved in the modeling of Non-Quasi-Static (NQS) effects (see Section 6.2). Equation (3.154) shows that the total drain/source terminal current in transient operation can be derived from the superposition of two components, the

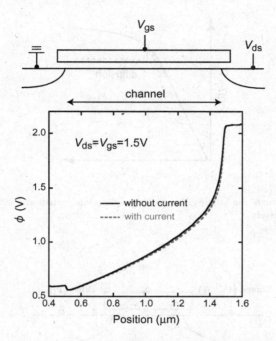

Fig. 2.1 Simulated potential distributions along the channel with current flow in the MOSFET and without. For the simulation the gate oxide thickness T_{ox} and the substrate impurity concentration are fixed to 3nm and $1 \times 10^{17} \text{cm}^{-3}$ respectively. The gate voltage V_{gs} and the drain voltage V_{ds} are both set to 1.5V.

conductive current I_{x0} and the charging curren dQ_x/dt [3]. I_{x0} is a function of the instantaneous terminal voltages and is approximated by the steady-state solution. The source and drain charging currents are the time derivatives of the charges Q_S and Q_D, induced in these terminals by the charges in the channel due to the voltage changes at the respective terminals. In quasi-static (QS) approximation, all charges are again approximated by the steady-state solutions. As a result, the QS approximation becomes valid only for constant and slowly time-varying input signals. In practical circuit simulation Eq. (3.154) has to be solved for an ensemble of many MOSFETs, interconnected to form a circuit, at the same time [4]. Each of these MOSFETs has four terminals, as depicted in Fig. 2.3, which maybe connected to different circuit nodes.

The compact circuit-simulation models usually solve the two remaining sets, namely Eqs. (2.1) and (2.2) of the basic device equations [5,6]. Applied

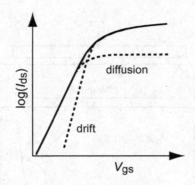

Fig. 2.2   Schematic of the MOSFET operating regions where drift and diffusion currents dominate, respectively.

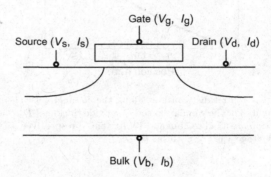

Fig. 2.3   MOSFET terminals with their basic characteristic terminal conditions Important MOSFET parameters, namely the gate length L_{gate}, the gate-oxide thickness T_{ox}, and the substrate impurity concentration N_{sub} are also indicated.

voltages on each node of the MOSFET induce, as already mentioned the device-internal potential distribution. On the other hand, its operational features, including the potential distribution, are also determined by the MOSFET's geometrical structure parameters such as the oxide thickness T_{ox} and its impurity concentrations in the semiconducting areas such as the substrate-impurity concentration. Here different dopant types N_{D} and N_{A} are not distinguished but the impurity concentration is identified by the notation N_{sub}. The quantitative values of these structural MOSFET parameters influence strongly the potential distribution in the MOSFET, and therefore all MOSFET operational features which are determined by the potential distribution.

A further schematic of the MOSFET cross-section along the channel with typical terminal connections is shown in Fig. 2.4 for the n-channel case together with energy diagrams near source and drain. The voltages at the four MOSFET terminals are normally referred to as gate voltage V_g, drain voltage V_d, bulk voltage V_b, and source voltage V_s. Usually, the so-called source-reference scheme is applied where V_s is taken as the reference potential so that applied voltages are determined as gate-source voltage V_{gs}, drain-source voltage V_{ds}, and bulk-source voltage V_{bs}, respectively. The MOSFET is mainly controlled by the two applied voltages, V_{gs} and V_{ds}. V_{gs} controls the number of carriers and in particular the formation of the inversion layer or inversion channel, when $E_f > E_i$ (see Fig. 2.4 [A]), where E_f is the Fermi level and E_i is the intrinsic Fermi level. V_{ds} controls mainly the carrier flow from source to drain. If V_{ds} is larger than V_{gs}, namely the V_{ds} control becomes stronger than that of V_{gs}, which occurs usually near the drain end of the channel, the energy diagram becomes more complicated (see Fig. 2.4 [B]). An important feature of the MOSFET is that carriers can easily move in the inversion channel formed underneath the gate oxide when $E_f > E_i$. The thickness of the inversion layer is very thin in comparison with the channel length, so that neglection of the carrier distribution transverse to the surface is possible. Consequently, a compact modeling approach, based on the surface potential along the channel as the main measure for modeling, becomes conceivable. HiSIM (Hiroshima-university STARC IGFET Model) is the first complete surface-potential-based MOSFET model actually applied for real circuit simulation based on the drift-diffusion model [7]. The most important advantage of the surface-potential-based modeling is the unified description of the MOSFET characteristics for all bias conditions. The physical reliability of the drift-diffusion theory has been proved by 2D-device simulations with channel lengths even down to below $0.1\mu m$ [8].

Conventional models based on the threshold voltage V_{th}, often called also piece-wise or regional models [9], require different equations in specific operating regions to cover all bias conditions. The reason for this necessity is that a V_{th}-based model is valid only under the strong-inversion condition because it considers only the drift contribution, so that the subthreshold and saturation conditions have to be modeled separately by different sets of equations. In particular, neglection of the diffusion contribution leads to simplified equations, which are only functions of applied voltages [10] and are given by the well-known threshold-voltage-based formulation of the drain current I_{ds}

$$I_{ds} = \frac{W}{L}\mu C_{ox}\left[(V_G - V_{th})V_{ds} - \left(\frac{1}{2} + \frac{1}{4C_{ox}}\sqrt{\frac{2\epsilon_{Si}qN_{sub}}{2\Phi_B}}\right)V_{ds}^2\right] \quad (2.5)$$

$$V_G = V_{gs} - V_{fb} \quad (2.6)$$

where V_{th} is the threshold voltage. W, L, C_{ox}, N_{sub}, and V_{fb} denote MOS-FET width, MOSFET length, the oxide capacitance, the substrate impurity concentration, and the flat-band voltage, respectively. The form of Eq. (2.5) has a big advantage because applied voltages can be directly measured and many MOSFET characteristics can be estimated by simplified hand calculations. A serious and unavoidable disadvantage is that the validity of Eq. (2.5) is the restricted to only one of the special operating regions and that discontinuities occur at the boundaries to the other specific operating regions which have to be described by different equations.

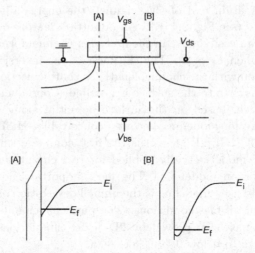

Fig. 2.4   Schematic of a MOSFET cross-section and its energy diagrams at the positions [A] and [B] perpendicular to the channel for the n-channel case. E_i and E_f are the intrinsic Fermi energy and the Fermi energy of the p-type silicon substrate.

In the surface potential-based drift-diffusion model, two further simplifications are made, namely the constant transverse electric field approximation, often referred as the charge-sheet approximation, and the gradual-channel approximation. All device characteristics are described in HiSIM analytically by the channel-surface potentials at the source end (ϕ_{S0}) and at the drain end ($\phi_{\rm SL}$) as schematically shown in Fig. 2.5 [11–13]. This is the long-channel basis of the HiSIM model, and extensions of these model approximations are done for advanced technologies. All newly appearing phenomena such as short-channel and reverse-short-channel effects are included in the surface potential calculations causing modifications which result from the features of these advanced technologies [14] (see Section 2.3).

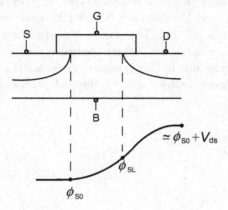

Fig. 2.5   Schematic of the surface potential distribution in the channel.

The surface potentials are calculated from the Poisson equation (Eq. (2.1)) together with the Gauss law. The final equations for the surface potential calculations are implicit functions of the applied voltages. Thus model-internal iteration procedures are introduced in addition to the global iteration of the circuit simulator [6]. By choosing the iterative solution for HiSIM, we preserve the MOSFET physics, which is built into the set of implicit functions. Such an iterative solution is commonly believed to result in an execution time penalty for the compact model [15]. Therefore, specific attention is directed towards calculating the surface potentials with enough accuracy and also with small CPU run-time. The number of HiSIM-internal iteration steps could be reduced down to two steps. Good initial guesses for the iteration and a good algorithm for solving the Poisson equation realize the fast simulation time.

In addition to the reduction of the calculation time for surface potentials, HiSIM includes many efforts to take into account the impurity profiles as explicitly and as realistic as possible. The aim here is predicting the scaling features of the MOSFET as accurately as possible even for new technologies by undertaking intensive channel engineering, which exploits impurity profile adjustments to suppress the disadvantages of device-size minimization.

In this book the focus is given on bulk MOSFETs, where the validity has been tested for channel lengths down to the 65nm technology node. The basic concepts used in HiSIM are, however, applicable for any future MOSFET generation. Extensions to other types of MOSFETs such as SOI-MOSFETs [16–18], and Double-Gate-MOSFETs have consequently been developed successfully [19–23]. This is the big advantage of following the basic device physics, which is not different between the various device types.

Though all descriptions in this book are given for the n-channel MOS-FET, they are analogously also valid for the p-channel case.

2.2 Device Structure Parameters of the MOSFET

Fig. 2.6 shows a cross section of the MOSFET, which schematically explains important structure parameters. The effective channel length L_{eff} and width W_{eff} are calculated from the geometrical gate length L_{gate} and width W_{gate}. L_{gate} and W_{gate} are equal to the gate nominal length and width, drawn in the layout for specifying the optical masks (L_{drawn} and W_{drawn}) for the photo-lithography step of the fabrication process:

$$L_{\text{gate}} = L_{\text{drawn}} \tag{2.7}$$

$$W_{\text{gate}} = W_{\text{drawn}} \tag{2.8}$$

$$L_{\text{poly}} = L_{\text{gate}} - 2 \cdot dL \tag{2.9}$$

$$W_{\text{poly}} = W_{\text{gate}} - 2 \cdot dW \tag{2.10}$$

In Eqs. (2.9) and (2.10) dL and dW include the size dependences due to lithography effects, modeled as

$$dL = \frac{LL}{(L_{\text{gate}} + LLD)^{LLN}} \tag{2.11}$$

$$dW = \frac{WL}{(W_{\text{gate}} + WLD)^{WLN}} \tag{2.12}$$

where LL, LLD, LLN, WL, WLD, and WLN are model parameters for including L_{gate} or W_{gate} dependencies in L_{poly} and W_{poly}. These parameters are becoming more important as the device size approaches the critical dimension of the lithography.

The channel length and width used in the actual electrical MOSFET modeling are

$$L_{\text{eff}} = L_{\text{poly}} - 2 \cdot XLD \tag{2.13}$$

$$W_{\text{eff}} = W_{\text{poly}} - 2 \cdot XWD \tag{2.14}$$

where XLD and XWD account for the overlaps between source/drain contacts and the gate oxide.

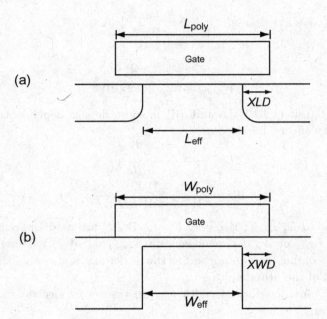

Fig. 2.6 Cross section of the MOSFET schematically explaining important structure parameters, (a) along the channel direction and (b) along the width direction.

2.3   Surface Potentials

The Gauss law describes the boundary condition for a given V_{gs} value of the MOSFET [43]

$$\epsilon_{ox} \cdot E_{ox} = \epsilon_{Si} \cdot E_s \tag{2.15}$$

$$E_{ox} = \frac{(V_G' - \phi_S(y))}{T_{ox}} \tag{2.16}$$

$$V_G' = V_{gs} - V_{fb} - \Delta V_{th} \tag{2.17}$$

$$= V_G - \Delta V_{th} \tag{2.18}$$

where ϵ_{ox} is the permittivity in the oxide, and E_{ox} and E_s are the vertical electric field in the gate oxide and at the surface of the substrate, respectively. V_G' is the effective gate voltage, which determines the carrier dynamics, and $\phi_S(y)$ is the potential at the surface of the MOSFET channel, which is called the surface potential (see Fig. 2.7). ΔV_{th} is the

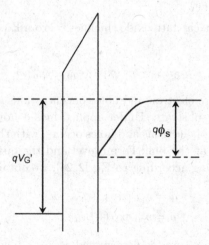

Fig. 2.7   Schematic band diagram vertical to the surface, showing the effective gate voltage V_G' and the surface potential ϕ_S.

threshold voltage shift from the long-channel V_{th} value (see Section 3.1), and V_{fb} gives the flat-band voltage. If no short-channel effect is observable, ΔV_{th} reduces to zero.

The equation for $E_s(y)$ is obtained by integrating the Poisson equation vertical to the channel at position y along the channel, leading to an implicit equation for the surface potential ϕ_S in the form

$$C_{ox}(V'_G - \phi_S(y)) = \epsilon_{Si} \cdot E_s$$

$$= \sqrt{\frac{2\epsilon_{Si}qN_{sub}}{\beta}}$$

$$\cdot \left[\exp\{-\beta(\phi_S(y) - V_{bs})\} + \beta(\phi_S(y) - V_{bs}) - 1 \right.$$

$$\left. + \frac{n_{p0}}{p_{p0}} \left\{ \exp(\beta(\phi_S(y) - \phi_f(y))) - \exp(\beta(V_{bs} - \phi_f(y))) \right\} \right]^{\frac{1}{2}}$$

$$(2.19)$$

$$C_{ox} = \frac{\epsilon_{ox}}{T_{ox}} \tag{2.20}$$

$$\beta = \frac{q}{kT} \tag{2.21}$$

$$n_{p0} = \frac{n_i^2}{p_{p0}} \tag{2.22}$$

where T_{ox} is the physical gate-oxide thickness. From Eq. (2.1) the *pn* product is written

$$p_{p0}n_{p0} = n_i^2 \exp \beta(\phi_{fp} - \phi_{fn}) \tag{2.23}$$

where n_{p0} and p_{p0} are electron and hole concentrations at equilibrium condition in the p-type substrate. Under applied bias condition, separation of the Fermi potentials of electrons and holes occurs, with the hole quasi-Fermi potential remaining at the bulk Fermi level and the quasi-Fermi potential for electrons increasing according to V_{ds} [2, 24], we obtain

$$n = n_{p0} \exp[\beta(\phi_s - \phi_f)] \tag{2.24}$$

$$p = p_{p0} \exp(-\beta\phi_s) \tag{2.25}$$

where ϕ_f is equal to ϕ_{fn}, and is determined as

$$\phi_f(L_{eff}) - \phi_f(0) = V_{ds} \tag{2.26}$$

The hole concentration at equilibrium condition p_{p0} is further approximated to be equal to the substrate impurity concentration N_{sub}. The intrinsic carrier concentration n_i is (see Eq. (1.35) in Chapter 1)

$$n_i = n_{i0}T^{\frac{3}{2}} \exp\left(-\frac{E_g}{2q}\beta\right), \tag{2.27}$$

where E_g represents the temperature dependent bandgap. The Eqs. from (2.15) to (2.26) are generally valid for MOSFETs fabricated with any technology generation.

To account also for different gate-oxide materials, which are increasingly applied to prevent gate leakage currents [25], a model parameter $KAPPA$ is introduced for the permittivity of the gate dielectric κ so that Eq. (2.20) is rewritten as

$$C_{\text{ox}} = \frac{\kappa}{T_{\text{ox}}} = \frac{KAPPA}{T_{\text{ox}}} \qquad (2.28)$$

If $KAPPA$ is not specified otherwise, the permittivity of the silicon oxide ϵ_{ox} is taken by default. Under the condition that the quantum mechanical effects cannot be ignored, the oxide thickness is replaced with the effecttive one as

$$T_{\text{ox}} = T_{\text{ox}} + \Delta T_{\text{ox}} \qquad (2.29)$$

where ΔT_{ox} is given in Eq. (3.69).

The Poisson equation Eq. (2.1) is in principle valid for 3 dimensions (3D), which means

$$\nabla^2 \phi = \frac{\partial^2 \phi}{\partial x^2} + \frac{\partial^2 \phi}{\partial y^2} + \frac{\partial^2 \phi}{\partial z^2} \qquad (2.30)$$

where x and y give the directions vertical and parallel to the surface, and z is the device width direction as shown in Fig. 2.8.

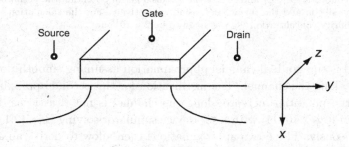

Fig. 2.8   Definition of the notation for the directions relative to the MOSFET channel, as applied in this book.

The drain current flows in y direction near the surface and has relatively large width in the z direction. Thus it is sufficient to solve the Poisson equation in 2 dimensions (2D), namely in the x and y directions. In stead of actually solving the Poisson equation in 2D, which is possible only numerically, we follow an analytical approach by introducing approximations not seriously affecting the validity of the device physics. This approach obtains analytical solutions for describing MOSFET performances by introducing the charge-sheet approximation of an inversion layer with zero thickness [11–13]. The justification for introducing this approximation is that the channel thickness is at maximum a few nm as seen in Fig. 2.9, which is much smaller than the gate length L_{gate}. A second approximation in-

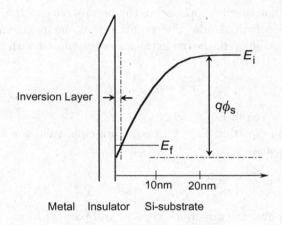

Fig. 2.9 Typical potential distribution perpendicular to the channel under the strong-inversion condition. E_f, E_f and ϕ_S are the intrinsic Fermi potential, the Fermi potential of a p-type silicon and the surface potential, respectively. For the calculation the substrate impurity concentration N_{sub} is fixed to be $2 \times 10^{18} \text{cm}^3$.

troduced is the gradual-channel approximation assuming smooth potential increase along the channel. The justification for this second approximation is that the potential increase along the channel is not drastic as demonstrated in Fig. 2.10 [11] with a 2D-device simulator solving Eqs. (2.1)- (2.3) simultaneously. These two approximations then allow to derive an analytical formulation for all MOSFET performances as a function of only the surface potentials at the source end ϕ_{S0} and the drain end ϕ_{SL}. It has to be noticed that all device equations described in this text are valid for device sizes larger than electron wave length which is about 10nm [26].

Calculated ϕ_S characteristics at source and drain are depicted schematically in Fig. 2.11 as a function of V_{gs}, calculated from Eq. (2.38) derived

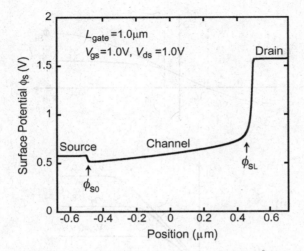

Fig. 2.10 Surface potential simulation result as a function of the position along the MOSFET channel from source to drain with a 2D-device simulator.

later in Section 2.4. Under the stronginversion condition with large V_{gs} and relatively small V_{ds}, the relation

$$\phi_{SL} = \phi_{S0} + V_{ds} \tag{2.31}$$

is preserved as can be seen in Fig. 2.11b. On the other hand, ϕ_{SL} does not reach $\phi_{S0} + V_{ds}$ for larger V_{ds} but saturates, reflecting the fact that the MOSFET operates under the saturation condition. For smaller V_{gs} the difference between ϕ_{SL} and ϕ_{S0} reduces, and this difference becomes negligible below threshold voltage, as can be seen in Fig. 2.11a, resulting in practically no current flow. At the flat-band condition ϕ_{S0} and ϕ_{SL} become zero, and their sign becomes negative for further reduced V_{gs} causing the accumulation condition. Calculated surface potentials are compared with 2D-device simulation results in Fig. 2.12 [27, 28]. ϕ_{SL} in Fig. 2.12 gives the surface-potential value at the end of the gradual-channel approximation also called pinch-off point. Beyond the pinch-off point steep potential increase is observed. A potential increase of the magnitude $V_{ds} - \phi_{SL}$ occurs both in the pinch-off region and the overlap region as seen in the figure.

HiSIM solves the implicit surface-potential equations (see Eq. (2.38)) iteratively with the Newton method in a similar way as numerical device

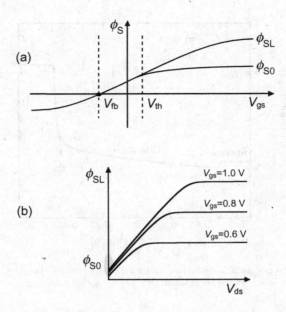

Fig. 2.11 Surface potentials at source and drain of the MOSFET as a function of (a) the gate voltage V_{gs}, and (b) the drain voltage V_{ds}.

simulators. Calculated surface-potential values of HiSIM are compared in more detail with 2D-device simulation results in Fig. 2.13 for various conditions. Good agreement is verified. Since the impurity profile used for the 2D-device simulation is not calibrated to the studied devices, slight differences can also be seen. It is important to note that the accuracy of the surface-potential calculation must be smaller than 10^{-12}V to obtain sufficiently accurate MOSFET capacitance calculations, both in signs and in magnitude [6].

Derivatives of all MOSFET characteristics influence on analog as well as RF-circuit performances [29]. The features of all these derivatives are mainly determined by the surface-potential derivatives with respect to all possible applied voltages. Calculated derivatives are compared with 2D numerical device-simulation results in Fig. 2.14. HiSIM results show again the same characteristics as the 2D-device simulation results for all applied bias conditions. In particular all complicated detailed features, which are not observable in usual surface potential versus voltage plots, are seen to be accurately preserved in the HiSIM results [30].

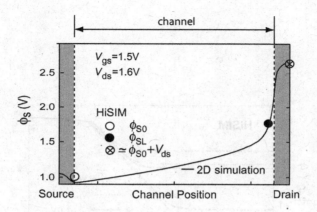

Fig. 2.12   Calculated surface potential by HiSIM at source end ϕ_{S0} and at drain end ϕ_{SL} under the saturation condition. The simulated surface-potential distribution along the channel with a 2D-device simulator is depicted for comparison. The shadowed regions denote the overlaps between source/drain contacts and the gate oxide.

Instead of solving the Poisson equation iteratively, the recently developed compact MOSFET model PSP approximates the surface potentials by explicit mathematical functions with applied voltages as variables, determined as 1st order perturbation theory results after a slight modification of the original implicit equation [31]. The purpose of the description with closed form equations is to reduce the simulation time by eliminating otherwise unavoidable iterations. However, the simulation time of the iterative HiSIM approach turns out to be nevertheless faster than with the analytical approach of PSP [6, 32, 33] due to the simpler structure of HiSIM's equations. A big advantage of HiSIM's approach, namely to keep the basic physical device equations as much as possible, is to preserve device physics even in the compact model and thereby to maintain a better extendability to future developments of MOSFET technology such as the SOI-MOSFET or the double-gate MOSFET.

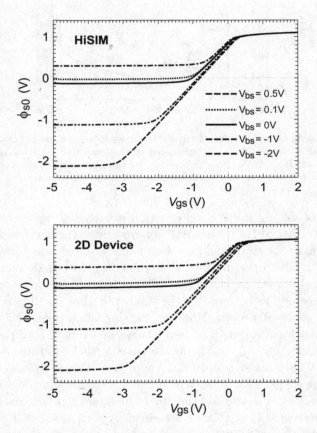

Fig. 2.13   Comparison of calculated surface potential at source end ϕ_{S0} by HiSIM and a 2D-device simulator for various bulk voltages V_{bs} and at fixed drain voltage of V_{ds}=0.1V. Gate length is fixed to 2μm.

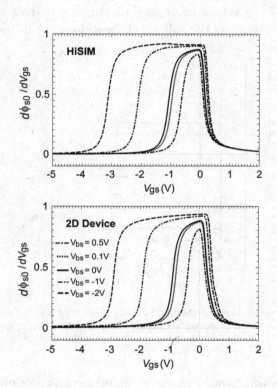

Fig. 2.14 Comparison of calculated derivatives of the surface potential at source end ϕ_{S0} by HiSIM and a 2D-device simulator. Bias conditions are the same as given in Fig. 2.13.

2.4   Charge Densities

Applied voltages cause a band structure bending within the MOSFET as schematically shown in Fig. 2.15, where the vertical profile at a certain position along the channel is depicted. In the substrate the depletion charge Q_b and the inversion charge Q_i are induced, and the same amount of charge with an opposite sign is induced as the gate charge Q_g in the gate at the gate insulator. Fig. 2.16 shows the induced charges along the channel direction.

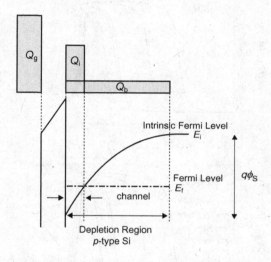

Fig. 2.15   Relationship between the band structure bending and the space charge induced in the *p*-type silicon substrate, consisting of the depletion charge and the inversion charge.

These induced charges can be viewed as the origin of all operational features of the MOSFET device. Integration of the inversion charge multiplied with the carrier velocity along the channel allows to derive the equation for the current flow, and the derivatives of all induced MOSFET charges with respect to the terminal potentials allow to derive the equations for all internal capacitances which appear in the MOSFET [29]. Thus it is easily said that the MOSFET charges are determined by the potential distribution along the channel which is induced by the applied terminal voltages of the MOSFET.

Under the charge-sheet approximation [34–36] the charges on the four MOSFET terminals Q_G(gate), Q_B(bulk), Q_D(drain), and Q_S(source), are obtained by integrating charge distributions along the channel direction y from source to drain as [3]:

$$Q_{\mathrm{G}} = -(Q_{\mathrm{B}} + Q_{\mathrm{I}}) = -Q_{\mathrm{SP}} \qquad (2.32)$$

$$Q_{\mathrm{B}} = W_{\mathrm{eff}} \int_0^{L_{\mathrm{eff}}} Q_{\mathrm{b}}(y)\,dy \qquad (2.33)$$

$$Q_{\mathrm{I}} = W_{\mathrm{eff}} \int_0^{L_{\mathrm{eff}}} Q_{\mathrm{i}}(y)\,dy \qquad (2.34)$$

$$Q_{\mathrm{D}} = W_{\mathrm{eff}} \int_0^{L_{\mathrm{eff}}} \frac{y}{L_{\mathrm{eff}}} Q_{\mathrm{i}}(y)\,dy \qquad (2.35)$$

$$Q_{\mathrm{S}} = Q_{\mathrm{I}} - Q_{\mathrm{D}} \qquad (2.36)$$

where Q_{I} is the inversion charge and Q_{SP} is the space charge. $y = 0$ and $y = L_{\mathrm{eff}}$ are the channel-end positions at the source end and the drain end, respectively.

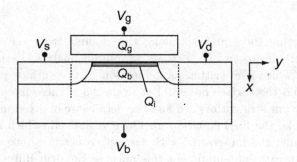

Fig. 2.16   Charges induced in the MOSFET by its terminal voltages, viewed along the channel direction.

By applying the Gauss law, the position dependent space-charge density

$Q_{\text{sp}}(y)$ is derived from the Poisson equation as:

$$-Q_{\text{sp}}(y) = \epsilon_{\text{Si}} E_{\text{s}}(y)$$

$$= \sqrt{\frac{2\epsilon_{\text{Si}} q N_{\text{sub}}}{\beta} \left[\exp\{-\beta(\phi_{\text{S}}(y) - V_{\text{bs}})\} + \beta(\phi_{\text{S}}(y) - V_{\text{bs}}) - 1 \right.}$$

$$\left. + \frac{n_{\text{p0}}}{p_{\text{p0}}} \left\{ \exp\big(\beta(\phi_{\text{S}}(y) - \phi_{\text{f}}(y))\big) - \exp\big(\beta(V_{\text{bs}} - \phi_{\text{f}}(y))\big) \right\} \right]^{\frac{1}{2}}$$

$$\tag{2.37}$$

$$= \epsilon_{\text{ox}} E_{\text{ox}}(y) = C_{\text{ox}}\,(V_{\text{G}}' - \phi_{\text{S}}(y)) \tag{2.38}$$

The Poisson equation and the Gauss law are also used to derive the separate inversion-charge density and depletion-charge density equations under the assumption of a homogeneous substrate impurity concentration as

$$Q_{\text{b}}(y) = -\sqrt{\frac{2\epsilon_{\text{Si}} q N_{\text{sub}}}{\beta} \left[\exp\{-\beta(\phi_{\text{S}}(y) - V_{\text{bs}})\} + \beta(\phi_{\text{S}}(y) - V_{\text{bs}}) - 1 \right]^{\frac{1}{2}}}$$

$$\tag{2.39}$$

$$Q_{\text{i}}(y) = -C_{\text{ox}}(V_{\text{G}}' - \phi_{\text{S}}(y))$$

$$+ \sqrt{\frac{2\epsilon_{\text{Si}} q N_{\text{sub}}}{\beta} \left[\exp\{-\beta(\phi_{\text{S}}(y) - V_{\text{bs}})\} + \beta(\phi_{\text{S}}(y) - V_{\text{bs}}) - 1 \right]^{\frac{1}{2}}}$$

$$\tag{2.40}$$

After integrating the equations along the channel from the source end ($y = 0$) to the drain end ($y = L_{\text{eff}}$), we obtain analytical equations for Q_{B} and Q_{I}, which are written as a function of the surface potential at source ϕ_{S0} and the surface potential at drain ϕ_{SL}. These integrations are quite lengthy but straightforward and the details are omitted in this book. As an example, the final equation for Q_{B} is explained, which is obtained by transforming the integration with respect to position into an integration with respect to potential. For this purpose the drift-diffusion-current equation (see Eq. (2.51)) is used in the form

$$dy = \frac{W_{\text{eff}} \mu Q_{\text{b}}(y) \left\{ Q_{\text{i}}(y) \beta d\phi_{\text{S}} - dQ_{\text{i}}(y) \right\}}{\beta I_{\text{ds}}} \tag{2.41}$$

where I_{ds} is the drain current and μ is the carrier mobility. The integration is done by substituting Eq. (3.19) into Eq. (2.33), resulting in

$$
\begin{aligned}
Q_{\mathrm{B}} &= W \int_0^L Q_{\mathrm{b}}(y)\,dy \\
&= -\frac{kT}{q}\frac{\mu W_{\mathrm{eff}}^2}{I_{\mathrm{ds}}}\int \left\{ Q_{\mathrm{b}}(y)Q_{\mathrm{i}}(y)\beta d\phi_{\mathrm{S}} - Q_{\mathrm{b}}(y)dQ_{\mathrm{i}}(y) \right\} \\
&= -\frac{\mu W_{\mathrm{eff}}^2}{I_{\mathrm{ds}}}\left[const0\, C_{\mathrm{ox}}(V_{\mathrm{G}}')\frac{1}{\beta}\frac{2}{3}\left[\{\beta(\phi_{\mathrm{S}} - V_{\mathrm{bs}}) - 1\}^{\frac{3}{2}} \right]_{\phi_{\mathrm{SO}}}^{\phi_{\mathrm{SL}}} \right. \\
&\quad - const0\, C_{\mathrm{ox}}\frac{1}{\beta}\frac{2}{3}\left[\phi_{\mathrm{S}}\{\beta(\phi_{\mathrm{S}} - V_{\mathrm{bs}}) - 1\}^{\frac{3}{2}} \right]_{\phi_{\mathrm{SO}}}^{\phi_{\mathrm{SL}}} \\
&\quad + const0\, C_{\mathrm{ox}}\frac{1}{\beta}\frac{2}{3}\frac{1}{\beta}\frac{2}{5}\left[\{\beta(\phi_{\mathrm{S}} - V_{\mathrm{bs}}) - 1\}^{\frac{5}{2}} \right]_{\phi_{\mathrm{SO}}}^{\phi_{\mathrm{SL}}} \\
&\quad \left. - const0^2\frac{1}{\beta}\frac{1}{2}\left[\beta^2(\phi_{\mathrm{SL}} - V_{\mathrm{bs}})^2 - 2\beta(\phi_{\mathrm{SL}} - V_{\mathrm{bs}}) + 1 \right]_{\phi_{\mathrm{SO}}}^{\phi_{\mathrm{SL}}} \right] \\
&\quad - \frac{1}{\beta}\frac{\mu W_{\mathrm{eff}}^2}{I_{\mathrm{ds}}}\left[const0\, C_{\mathrm{ox}}\frac{1}{\beta}\frac{2}{3}\{\beta(\phi_{\mathrm{S}} - V_{\mathrm{bs}}) - 1\}^{\frac{3}{2}} + \frac{1}{2}const0^2\beta\phi_{\mathrm{S}} \right]_{\phi_{\mathrm{SO}}}^{\phi_{\mathrm{SL}}}
\end{aligned}
\tag{2.42}
$$

Here $const0$ is defined as

$$
const0 = \sqrt{\frac{2\epsilon_{\mathrm{Si}}qN_{\mathrm{sub}}}{\beta}}
\tag{2.43}
$$

To reduce simulation time without sacrificing accuracy, a major effort in HiSIM is directed towards the efficient calculation of the three independent charges Q_{B}, Q_{I} and Q_{D}. However, accurate calculation of these charges for all bias conditions is indispensable not only within the regime of normal bias conditions but also in a wide range outside of this regime. The purpose of this requirement is to achieve stable circuit simulation by providing stable Newton iteration. Fig. 2.17 shows a schematic plot of these independent charges (a) as a function of V_{gs} for two fixed V_{ds} values and (b) as a function of V_{ds} for two fixed V_{gs} values. Around threshold condition the inversion charge Q_{I} starts to increase rapidly, whereas the depletion charge Q_{B} enters a saturation like condition (see Fig. 2.17a). The observed characteristics of the independent charges are directly correlated with the surface potential characteristics shown in Fig. 2.11. Further Fig. 2.17 shows that an increase of V_{ds} causes a reduction of Q_{I}. This effect is schematically illustrated in Fig. 2.18 and more quantitatively depicted in the 2D-device simulation results of Fig. 2.19. The explanation for the charge reduction with increased V_{ds} comes from the increase of the carrier-velocity v, which

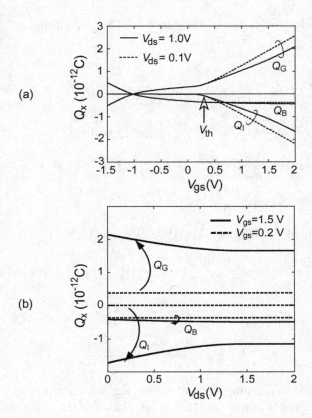

Fig. 2.17   Induced charges of the MOSFET as a function of (a) V_{gs} and (b) V_{ds}.

suppresses the carrier concentration n in order to preserve the continuity of the drain current, given in Eq. (2.3). Under the DC condition $\partial n/\partial t$ reduces to zero and thus the continuity equation is simplified as

$$\frac{1}{q}\nabla j_{\mathrm{n}} = 0 \qquad (2.44)$$

Under the strong inversion condition the current-density equation is simplified as

$$j_{\mathrm{n}} \simeq -q\mu_n n \frac{\partial \phi}{\partial y} = -q\mu_n n E_{\mathrm{y}} \qquad (2.45)$$

If we approximate that the carrier mobility μ is independent of the field

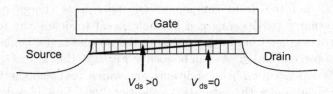

Fig. 2.18   Schematic illustration of the inversion charge reduction with increasing volt-age V_{ds} at the drain terminal. The charge distribution for $V_{ds} = 0V$ is hatched by vertical lines and that for $V_{ds} > 0V$ is marked by thick lines.

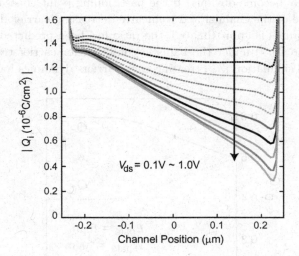

Fig. 2.19   Quantitative 2D-device simulation of the inversion-charge density from source to drain as a function of the drain voltage V_{ds}. The inversion charge Q_i is the integrated value along the inversion layer thickness. The position scale is chosen to have the zero value at the mid-point of the channel.

E_y, Eq. (2.45) is simply written as

$$j_n \propto Q_i \cdot v \tag{2.46}$$

where $Q_i = qn$, and q is the magnitude of the electron charge. Equa-tion (2.46) verifies that increasing carrier velocity v requires Q_i to decrease to preserve a constant current density j_n. The relation between Q_i and Q_I is seen in Eq. (2.34).

The inversion charge Q_I is normally partitioned into the source component Q_S and the drain component Q_D [3]. This charge partition-ing—indexcharge partitioning can be understood to define the terminals where carriers come from or go to during the processes of channel forma-tion or channel extinction. As can be seen in Fig. 2.20 the ratio of Q_S to Q_D ($Q_S{:}Q_D$) is dependent on bias conditions, and varies from 50:50 in the linear region to 60:40 under the saturation condition [28]. This bias-dependent dynamic partitioning; which results from the carrier-distribution change along the channel becomes important for accurate prediction of circuit per-formances under high frequencies. This partitioning scheme is expected to be changed for ultra short-channel MOSFETs, where the ballistic carrier transport may become obvious. If the partitioning point is assumed to be in the middle of the channel and a linearly decreasing carrier distribution along the channel is approximated, the ratio has been predicted to change from 60:40 to 70:30 under the ballistic condition for carrier transport by Monte Carlo simulation [37], treating each carriers dynamics individually.

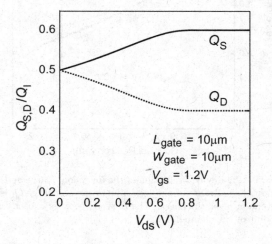

Fig. 2.20   Dynamic partitioning of the inversion charge Q_I into source charge Q_S and drain charge Q_D as a function of drain voltage V_{ds}.

Since the MOSFET charges are the origin of currents and capaci-tances, compact models based on these charges have also been developed. Inversion-charge-controlled models are a recent type of such models, de-scribing MOSFET features by the inversion charge at the source end and at the drain end [38–41] instead of using origin of the charges, namely the surface potentials. The purpose of the inversion-charge-controlled modeling approach is to get a smooth transition from the subthreshold condition to the inversion condition, aiming at improving the shortcomings of the V_{th}-based models for analog applications. Closed-form equations for the charges have been derived by approximating a linear dependence of the inversion charge on V_{gs} [42].

2.5   Drain Current

The drift-diffusion model describes the current density j_n for nMOSFET as (see Eq. (2.2)) [2]

$$j_n = -q\mu_n n \frac{\partial \phi}{\partial y} + qD_n \nabla n \tag{2.47}$$

The equation is rewritten using the Einstein relation ($D_n = \mu_n/\beta$), approximating no correlation among carriers, which is the case for usual semiconductors

$$j_n = -q\mu_n n \left(\frac{\partial \phi}{\partial y} - \frac{1}{\beta} \frac{\nabla n}{n} \right) \tag{2.48}$$

$$= -q\mu_n n \left(\frac{d\phi}{dy} - \frac{1}{\beta} \frac{dn}{n} \right) \tag{2.49}$$

$$= -q\mu_n n \left(\frac{d\phi}{dy} - \frac{1}{\beta} \frac{d\ln n}{dy} \right) \tag{2.50}$$

Under the charge-sheet approximation the drain current I_{ds} is written

$$I_{ds} = W_{eff} \cdot q \cdot \mu \cdot n(y) \cdot \left(\frac{d\phi_S(y)}{dy} - \frac{1}{\beta} \cdot \frac{d\ln n(y)}{dy} \right) \tag{2.51}$$

where $n(y)$ is the carrier density at the channel position y, and is the integrated value with negligible thickness of the inversion layer x_i

$$n(y) = \int_0^{x_i} n(y)dx \tag{2.52}$$

Eq. (2.51) is rewritten with the quasi-Fermi level ϕ_f measured from the Fermi level of the substrate as [43]

$$I_{ds} = W_{eff} \cdot q \cdot \mu \cdot n(y) \cdot \frac{d\phi_f}{dy} \tag{2.53}$$

where

$$\frac{d\phi_f}{dy} = \frac{d\phi_S(y)}{dy} - \frac{1}{\beta} \cdot \frac{d\ln n(y)}{dy} \tag{2.54}$$

The carrier density at the position y in the channel $n(y)$ is derived with Eq. (2.54) as

$$n(y) = n(0) \cdot \exp\left\{ \beta\left(\phi_S(y) - \phi_S(0)\right) - \beta\left(\phi_f(y) - \phi_f(0)\right) \right\} \tag{2.55}$$

The derivative of Eq. (2.55) is substituted into Eq. (2.49), and we obtain the relationship

$$n(y) = n(0) \exp\{\phi_S(y) - \phi_S(0)\} \tag{2.56}$$
$$- \frac{\beta I_{ds}}{q\mu W_{eff}} \exp\{\beta\phi_S(y)\} \int_0^L \exp\{-\beta\phi_S(y')\}dy'$$

The above equation together with the Gauss law

$$C_{ox}[V_G' - \phi_S(y)] = -[Q_b(y) + Q_i(y)] \tag{2.57}$$

leads to a closed form equation for drain current I_{ds} after the integration along the channel [6,11], which is shown in Eqs. (2.58) to (2.61).

$$I_{ds} = \frac{W_{eff}}{L_{eff}} \cdot \mu \cdot \frac{I_{dd}}{\beta} \tag{2.58}$$

$$I_{dd} = C_{ox}(\beta V_G' + 1)(\phi_{SL} - \phi_{S0}) - \frac{\beta}{2}C_{ox}(\phi_{SL}^2 - \phi_{S0}^2)$$
$$- \frac{2}{3}const0\left[\{\beta(\phi_{SL} - V_{bs}) - 1\}^{\frac{3}{2}} - \{\beta(\phi_{S0} - V_{bs}) - 1\}^{\frac{3}{2}}\right] \tag{2.59}$$
$$+ const0\left[\{\beta(\phi_{SL} - V_{bs}) - 1\}^{\frac{1}{2}} - \{\beta(\phi_{S0} - V_{bs}) - 1\}^{\frac{1}{2}}\right]$$

$$V_G' = V_{gs} - V_{fb} - \Delta V_{th} \tag{2.60}$$

$$const0 = \sqrt{\frac{2\epsilon_{Si}qN_{sub}}{\beta}} \tag{2.61}$$

Here ΔV_{th} is the threshold voltage shift from the long-channel MOSFET threshold-voltage value, and describes the short-channel effect discussed in detail in Section 3.1. The relationship with the charge density $Q_i(y)$ of Eq. (2.40) is established by

$$Q_i(y) = q \cdot n(y) \tag{2.62}$$

The gradual-channel approximation together with further approximations of an idealized gate structure and uniform channel doping were applied for the derivatives. The gradual-channel approximation allows to integrate the current density equation simply along the y direction, independent of the vertical direction x. The above description includes also the approximation that the mobility μ is independent of the position along the channel y. A constant mobility approximation along the channel has been estimated to cause only a few percent of inaccuracy for a $0.3\mu m$ technology [6], and the inaccuracy due to the approximation can be absorbed by the mobility fitting. However, the position dependent mobility has strong influence on higher-order phenomena related to distributed effects along the channel, which become obvious under high-frequency operation. Therefore, the potential distribution along the channel, which is exactly the origin of the position dependence of the mobility, has to be considered to derive analytical equations for these higher-order phenomena (see Chapter 4 and Section 6.1).

By fixing the surface potentials to

$$\phi_{S0} = 2\Phi_B \ , \ 2\Phi_B = \frac{2}{\beta} \ln\left(\frac{N_{sub}}{n_i}\right) \tag{2.63}$$

$$\phi_{SL} = 2\Phi_B + V_{ds} \tag{2.64}$$

in Eqs. (2.58) and (2.59), the well-known Sah equation for the long-channel case [10] is obtained as

$$I_{ds} = \frac{W_{eff}}{L_{eff}}\mu C_{ox}\left[(V_G' - V_{th})V_{ds} - \left(\frac{1}{2} + \frac{1}{4C_{ox}}\sqrt{\frac{2\epsilon_{Si}qN_{sub}}{2\Phi_B}}\right)V_{ds}^2\right] \tag{2.65}$$

$$V_{th} = V_{fb} + 2\Phi_B + \frac{\sqrt{2\epsilon_{Si}qN_{sub}\cdot 2\Phi_B}}{C_{ox}} \tag{2.66}$$

where V_{th} is the threshold voltage. Equations (2.65) and (2.66) are applied in the V_{th}-based models and are very advantageous for providing an explicit dependence of I_{ds} on applied voltages. Consequently, pinning of the surface potentials to the values given in Eqs. (2.63) and (2.64) is equivalent to the conventional drift approximation with V_{th} as a model parameter, as applied in the V_{th}-based models [9]. The meaning of the pinning, which is a very severe approximation, is shown by dashed lines in Fig. 2.21. In reality, the surface potentials are strongly dependent on applied voltages. The strong surface potential increase at the source end as a function of V_{gs} is largely suppressed after the formation of the inversion layer. Although, ϕ_{S0} still increases continuously with increasing V_{gs}, as can be been in Fig. 2.21a, this increase is limited by the fact that the surface potential value cannot exceed the energy gap value, namely approximately 1.1V for the Silicon substrate.

The surface potential at the drain end ϕ_{SL} enters the saturation condition after the so-called the pinch-off condition occurs, and never reaches $\phi_{S0} + V_{ds}$ shown by a dashed line in Fig. 2.21b. The potential ϕ_{SL} defines the value at the drain end of the gradual-channel approximation as can be seen in Fig. 2.12. Beyond the end of point of the gradual channel, where the surface-potential value ϕ_{SL} is defined, steep potential increase occurs and the carrier velocity reaches its maximum. In the V_{th}-based modeling this type modeling ϕ_{SL} under the saturation condition is often referred as the saturation voltage V_{dsat} [1, 2], and this type of modeling was employed by the threshold-voltage-based compact models, such as BSIM [9]

$$V_{\mathrm{dsat}} = V_G' + \frac{\epsilon_{\mathrm{Si}} q N_{\mathrm{sub}}}{C_{\mathrm{ox}}^2} \left[1 - \sqrt{1 + \frac{2C_{\mathrm{ox}}^2}{\beta \epsilon_{\mathrm{Si}} q N_{\mathrm{sub}}} \{\beta(V_G' - V_{\mathrm{bs}}) - 1\}} \right] \quad (2.67)$$

The potential increase from ϕ_{SL} up to $\phi_{\mathrm{S0}} + V_{\mathrm{ds}}$ occurs in the drain collector region, often called the pinch-off region, as well as in the drain contact region. However, it is difficult to obtain information about the exact position where the gradual-channel approximation terminates and where the pinch-off region starts. In HiSIM the problem of potential description beyond the carrier depletion line, often referred to as pinch-off point, is solved by introducing a steep increase of the surface potential with a model parameter named CLM (see section 3.5).

The gradual-channel approximation, being introduced to derive closed-form equations, would in principle limit the validity of the model description to the non-saturated condition. The saturation condition results in a weakened inversion condition at the drain end in the channel in the pinch-off region, as shown in Fig. 2.4 [B], which is mainly controlled by the longitudinal electric field instead of the vertical one [14]. In HiSIM this saturated condition is treated as a channel-length-modulation effect. Thus L_{eff} in the I_{ds} equation is replaced with the length including the modulation effect [44, 45].

$$L_{\mathrm{eff}} = L_{\mathrm{eff}} - \Delta L \quad (2.68)$$

The modeling details are explained in the section on channel-length-modulation modeling (Section 3.5). For long-channel transistors the pinch-off region is practically negligible and the equation derived under the gradual-channel approximation ignoring the pinch-off effect reproduces measurements with very good accuracy. This is verified in Fig. 2.22, where (a) depicts the I_{ds} versus V_{gs} characteristics and (b) depicts the $g_{\mathrm{m}}/I_{\mathrm{ds}}$ versus I_{ds} characteristics. g_{m} is the transconductance defined as

$$g_{\mathrm{m}} = \frac{dI_{\mathrm{ds}}}{dV_{\mathrm{gs}}} \quad (2.69)$$

MOSFETs for analog applications intensively utilize the region where

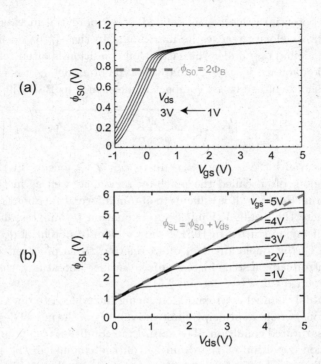

Fig. 2.21   Calculated surface potentials (a) at source end as a function of V_{gs}, and (b) at drain end as a function of V_{ds}. The gate length is fixed to 0.3μm, the minimum for the applied technology. The dashed lines show approximations which are effectively made in threshold voltage based models like those of the BSIM series.

g_m/I_{ds} varies most strongly. This strong variation occurs normally for current a bit above the threshold condition, where both the diffusion and the drift contributions are comparable, and give a threshold currents, which is small compared with the saturation current $I_{ds,sat}$.

If MOSFETs respond to input signals linearly, the output signals are a linear function of the input signals. In reality this is not the case but MOSFET characteristics are nonlinear functions of input signals, so that higher-order terms contribute to the output characteristics. This is usually analyzed with a harmonic distortion analysis, and is discussed in section 3.10.

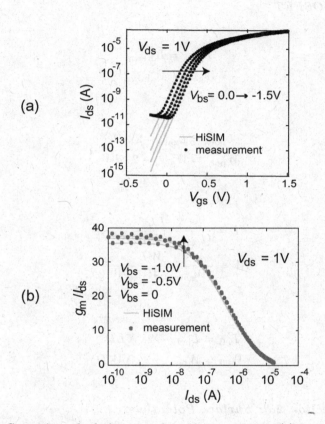

Fig. 2.22 Comparison of calculation results with measurements (a) I_{ds} as a function V_{gs} and (b) g_m/I_{ds} as a function of I_{ds}. The gate length is long and fixed to 10μm.

2.6   Summary of Equations and Model Parameters Presented in Chapter 2 for Basic Compact Surface-Potential Model of the MOSFET

2.6.1   *Section 2.2: Device Structure Parameters of the MOSFET*

$$L_{\text{gate}} = L_{\text{drawn}}$$
$$W_{\text{gate}} = W_{\text{drawn}}$$

$$L_{\text{poly}} = L_{\text{gate}} - 2 \cdot dL$$
$$W_{\text{poly}} = W_{\text{gate}} - 2 \cdot dW$$

$$dL = \frac{LL}{(L_{\text{gate}} + LLD)^{LLN}}$$
$$dW = \frac{WL}{(W_{\text{gate}} + WLD)^{WLN}}$$

$$L_{\text{eff}} = L_{\text{poly}} - 2 \cdot XLD$$
$$W_{\text{eff}} = W_{\text{poly}} - 2 \cdot XWD$$

2.6.2   *Section 2.3: Surface Potentials*

$$C_{\text{ox}}\left(V_G' - \phi_S(y)\right) = \sqrt{\frac{2\epsilon_{\text{Si}} q N_{\text{sub}}}{\beta}}$$

$$\cdot \left[\exp\{-\beta(\phi_S(y) - V_{\text{bs}})\} + \beta(\phi_S(y) - V_{\text{bs}}) - 1 \right.$$

$$\left. + \frac{n_{\text{p0}}}{p_{\text{p0}}} \left\{ \exp\left(\beta(\phi_S(y) - \phi_f(y))\right) - \exp\left(\beta(V_{\text{bs}} - \phi_f(y))\right) \right\} \right]^{\frac{1}{2}}$$

2.6.3 Section 2.4: Charge Densities

$$Q_B = W \int_0^L Q_b(y)dy$$

$$= -\frac{kT}{q}\frac{\mu W_{\text{eff}}^2}{I_{\text{ds}}} \int \{Q_b(y)Q_i(y)\beta d\phi_S - Q_b(y)dQ_i(y))\}$$

$$= -\frac{\mu W_{\text{eff}}^2}{I_{\text{ds}}} \left[const0\, C_{\text{ox}}(V_G')\frac{1}{\beta}\frac{2}{3}\Big[\{\beta(\phi_S - V_{\text{bs}}) - 1\}^{\frac{3}{2}}\Big]_{\phi_{\text{S0}}}^{\phi_{\text{SL}}}\right.$$

$$- const0\, C_{\text{ox}}\frac{1}{\beta}\frac{2}{3}\Big[\phi_S\{\beta(\phi_S - V_{\text{bs}}) - 1\}^{\frac{3}{2}}\Big]_{\phi_{\text{S0}}}^{\phi_{\text{SL}}}$$

$$+ const0\, C_{\text{ox}}\frac{1}{\beta}\frac{2}{3}\frac{1}{\beta}\frac{2}{5}\Big[\{\beta(\phi_S - V_{\text{bs}}) - 1\}^{\frac{5}{2}}\Big]_{\phi_{\text{S0}}}^{\phi_{\text{SL}}}$$

$$\left.- const0^2\frac{1}{\beta}\frac{1}{2}\Big[\beta^2(\phi_{\text{SL}} - V_{\text{bs}})^2 - 2\beta(\phi_{\text{SL}} - V_{\text{bs}}) + 1\Big]_{\phi_{\text{S0}}}^{\phi_{\text{SL}}}\right]$$

$$- \frac{1}{\beta}\frac{\mu W_{\text{eff}}^2}{I_{\text{ds}}}\left[const0\, C_{\text{ox}}\frac{1}{\beta}\frac{2}{3}\{\beta(\phi_S - V_{\text{bs}}) - 1\}^{\frac{3}{2}} + \frac{1}{2}const0^2\beta\phi_S\right]_{\phi_{\text{S0}}}^{\phi_{\text{SL}}}$$

2.6.4 Section 2.5: Drain Current

$$I_{\text{ds}} = \frac{W_{\text{eff}}}{(L_{\text{eff}} - \Delta L)} \cdot \mu \cdot \frac{I_{\text{dd}}}{\beta}$$

$$I_{\text{dd}} = C_{\text{ox}}(\beta V_G' + 1)(\phi_{\text{SL}} - \phi_{\text{S0}}) - \frac{\beta}{2}C_{\text{ox}}(\phi_{\text{SL}}^2 - \phi_{\text{S0}}^2)$$

$$- \frac{2}{3}const0\Big[\{\beta(\phi_{\text{SL}} - V_{\text{bs}}) - 1\}^{\frac{3}{2}} - \{\beta(\phi_{\text{S0}} - V_{\text{bs}}) - 1\}^{\frac{3}{2}}\Big]$$

$$+ const0\Big[\{\beta(\phi_{\text{SL}} - V_{\text{bs}}) - 1\}^{\frac{1}{2}} - \{\beta(\phi_{\text{S0}} - V_{\text{bs}}) - 1\}^{\frac{1}{2}}\Big]$$

$$V_G' = V_{\text{gs}} - V_{\text{fb}} + \Delta V_{\text{th}}$$

$$const0 = \sqrt{\frac{2\epsilon_{\text{Si}}qN_{\text{sub}}}{\beta}}$$

$$C_{\text{ox}} = \frac{\epsilon_{\text{ox}}}{T_{\text{ox}}} = \frac{\kappa}{T_{\text{ox}}}$$

$$\beta = \frac{q}{kT}$$

N_{sub} is determined in Eqs. (3.39) and (3.40), and ΔV_{th} is determined in Eq. (3.111). ΔT_{ox} is given in Eq. (3.69), and ΔL and μ are determined in Eq. (3.101) and Eq. (3.84), respectively.

Table 2.1   HiSIM model parameters for basic surface-potentail model.

XLD	gate-overlap length
LL	coefficient of gate-length modulation
LLD	coefficient of gate-length modulation
LLN	coefficient of gate-length modulation
XWD	gate-overlap width
WL	coefficient of gate-width modulation
WLD	coefficient of gate-width modulation
WLN	coefficient of gate-width modulation
TOX	physical oxide thickness T_{ox}
VFB	flat-band voltage V_{fb}
$KAPPA$	permittivity of gate dielectric ϵ

Bibliography

[1]   C. T. Sah and H. C. Pao, "The effects of fixed bulk charge on the characteristics of metal-oxide-semiconductor transistors," *IEEE Trans. Electron Devices*, vol. ED-13, no. 4, pp. 393–409, 1966.

[2]   H. C. Pao and C. T. Sah, "Effects of diffusion current on characteristics of metal-oxide (insulator)-semiconductor transistors," *Solid-State Electron.*, vol. 9, no. 10, pp. 927–937, 1966.

[3]   S.-Y. Oh, D. E. Ward, and R. W. Dutton, "Transient Analysis of MOS Transistors," *IEEE J. Solid-State Circ.*, vol. SC–15, no. 8, pp. 636–643, 1980.

[4]   L. W. Nagel and D. O. Pederson, "Simulation program with integrated circuit emphasis (SPICE)," *Proc. 16th Midwest Symp. on Circuit Theory*, Waterloo, April, 1973.

[5]   A. R. Boothroyd, S. W. Tarasewicz, and C. Slaby, "MISNAN–A physically based continuous MOSFET model for CAD applications," *IEEE Trans. Electron Devices*, vol. 10, no. 12, pp. 1512–1529, 1991.

[6]   M. Miura-Mattausch, U. Feldmann, A. Rahm, M. Bollu, and D. Savignac, "Unified complete MOSFET model for analysis of digital and analog circuits," *IEEE Trans. Computer-Aided Design*, vol. 15, no. 1, pp. 1–7, 1996.

[7]   M. Miura-Mattausch, U. Feldmann, A. Rahm, M. Bollu, and D. Savignac, "Unified complete MOSFET model for analysis of digital and analog circuits," *ICCAD Dig. Tech. Papers*, pp. 264–267, San Jose, Nov. 1994.

[8]   J. D. Bude, "MOSFET modeling into the ballistic regime," *Proc. SISPAD*, pp. 23–26, Seattle, Sept., 2000.

[9]   *BSIM4.0.0 MOSFET Model, User's Manual*, Department of Electrical En-

gineering and Computer Science, University of California, Berkeley CA, 2000.

[10] C. T. Sah, "Characteristics of the metal-oxide-semiconductor transistors," *IEEE Trans. Electron Devices*, vol. ED-11, no. 7, pp. 324–345, 1964.

[11] J. R. Brews, "A charge-sheet model of the MOSFET," *Solid-State Electron.*,vol. 21, no. 2, pp. 345–355, 1978.

[12] G. Baccarani, M. Rudan, and G. Spadini, "Analytical i.g.f.e.t model including drift and diffusion currents," *Solid-State Electron Devices*, vol: 2, no. 2, pp. 62–68, 1978.

[13] F. van der Wiele, "A long-channel MOSFET model," *Solid-State Electron.*, vol. 22, no. 12, pp. 991–997, 1979.

[14] *HiSIM1.1.0, User's Manual*, STARC & Hiroshima University, 2002.

[15] H. J. Park, P. K. Ko, and C. Hu, "A charge sheet capacitance model of short channel MOSFET's for SPICE," *IEEE Trans. Computer-Aided Design*, vol. 10, no. 3, pp. 376–389, 1991.

[16] D. Kitamaru, Y. Uetsuji, N. Sadachika, and M. Miura-Mattausch, "Complete surface-potential-based fully-depleted silicon-on-insulator metal-oxide-semiconductor field-effect-transistor model for circuit simulation," *Jpn. J. Appl. Phys.*, vol. 43, no. 4B, pp. 2166–2169, 2004.

[17] N. Sadatika, Y. Uetsuji, D. Kitamaru, H.J. Mattausch, M. Miura-Mattausch, L. Weiss, U. Feldmann, and S. Baba, "Fully-Depleted SOI-MOSFET Model for Circuit Simulation and Its Application to $1/f$ Noise Analysis," *Proc. SISPAD*, pp. 251-254, Munich, Sept., 2004.

[18] N. Sadachika, D. Kitamaru, Y. Uetsuji, D. Navarro, M. M. Yusoff, T. Ezaki, H. J. Mattausch, and M. Miura-Mattausch, "Completely Surface-Potential-Based Compact Model of the Fully Depleted SOI-MOSFET Including Short-Channel Effects," *IEEE Trans. Electron Devices*, vol. 53, no. 9, pp. 2017–2024, 2006.

[19] Y. Tauer, "Analytic Solutions of Charge and Capacitance in Symmetric and Asymmetric Double-Gate MOSFETs," *IEEE Trans. Electron Devices*, vol. 48, no. 12, pp. 2861–2869, 2001.

[20] M. V. Dunga, C.-H. Lin, X. Xi, D. D. Lu, A. M. Niknejad, and C. Hu, "Modeling Advanced FET Technology in a Compact Model," *IEEE Trans. Electron Devices*, vol. 53, no. 9, pp. 1971–1978, 2006.

[21] G. D. J. Smit, A. J. Scholten, N. Serra, R. M. T. Pijper, R. van Langevelde, A. Mercha, G. Gildenblat, and D. B. Klaassen, "PSP-Based Compact Fin-FET Model Describing DC and RF Measurements," *IEDM Tech. Dig.*, pp. 175–178, San Francisco, Dec. 2006.

[22] N. Sadachika, "Analysis and Modeling of Double-Gate MOSFET," *Master Dissertation, Hiroshima University*, Aug. 2006.

[23] N. Sadachika, H. Oka, R. Tanabe, T. Murakami, H. J. Mattausch, M. Miura-Mattausch, "Compact Double-Gate MOSFET Model Correctly Predicting Volume-Inversion Effects," *Proc. SISPAD*, pp. 289–292, Vienna, Sept., 2007.

[24] A. S. Grove and D. J. Fitzgerald, "Surface Effects on *p-n* Junctions: Characteristics of Surface Space-Charge Regions under Non-Equilibrium Conditions," *Solid-State Electron.*, vol. 9, pp. 783–806, 1966.

[25] M. L. Green, E. P. Gusev, R. Degrave, and E. Garfunkel, "Ultra (<4nm) SiO$_2$ and Si-O-N Gate Dielectric Layers for Silicon Microelectronics: Understanding the Processing, Structure, and Physical and Electrical Limits," *J. Appl. Phys.*, vol. 90, no. 5, pp. 2057–2021, Sept., 2001.

[26] S. Datta, "The Non-Equilibrium Green's Function (NEGF) Formalism: An Elementary Introduction," *IEDM Tech. Dig.*, pp. 703–706, San Francisco, Dec. 2002.

[27] M. Miura-Mattausch, H. Ueno, M. Tanaka, H. J. Mattausch, S. Kumashiro, T. Yamaguchi, K. Yamashita, and N. Nakayama, "HiSIM: A MOSFET model for circuit simulation connecting circuit performance with technology," *IEDM Tech. Dig.*, pp. 109–112, San Francisco, Dec. 2002.

[28] M. Miura-Mattasuch, H. Ueno, H. J. Mattausch, K. Morikawa, S. Itoh, A. Kobayashi, and H. Masuda, "100nm-MOSFET model for circuit simulation: challenges and solutions," *IEICE Trans. Electron.*, vol. E86-C, No. 6, pp. 1009–1021, 2003.

[29] Y. P. Tsividis, "Operation and Modeling of the MOS Transistor (Second Edition)," *McGraw-Hill*, 1999.

[30] M. Miura-Mattausch, N. Sadachika, D. Navarro, G. Suzuki, Y. Takeda, M. Miyake, T. Warabino, Y. Mizukane, R. Inagaki, T. Ezaki, H. J. Mattausch, T. Ohguro, T. Iizuka, M. Taguchi, S. Kumashiro, and S. Miyamoto, "HiSIM2: Advanced MOSFET Model Valid for RF Circuit Simulation," *IEEE Trans. Electron Devices*, vol. 53, no. 9, pp. 1994–2007, 2006.

[31] G. Gildenblat, X. Li, H. Wang, W. Wu, R. van Langevelde, A. J. Scholten, G. D. J. Smit and D. B. M. Klaassen, "Introduction to PSP MOSFET model," *Tech. Proc. WCM*, pp. 19–24, Anaheim, May, 2005.

[32] R. Rios, S. Mudanai, W. -K. Shih and P. Packan, "An efficient surface potential solution algorithm for compact MOSFET models," *IEDM Tech. Dig.*, pp. 755–758, San Francisco, Dec., 2004.

[33] http://www.simucad.com/; H. J. Mattausch, M. Miyake, T. Yoshida, S. Hazama, D. Navarro, N. Sadachika, T. Ezaki, M. Miura-Mattausch, "HiSIM2 Circuit Simulation – Solving the Speed versus Accuracy Crisis," *IEEE Circuits and Devices Magazine*, vol. 22, no. 5, pp. 29–38, 2006.

[34] J. R. Brews, "A charge-sheet model of the MOSFET," *Solid-State Electron.*, vol. 21, no. 2, pp. 345–355, 1978.

[35] G. Baccarani, M. Rudan, and G. Spadini, "Analytical i.g.f.e.t model including drift and diffusion currents," *Solid-State Electron Devices*, vol. 2, no. 2, pp. 62–68, 1978.

[36] F. van der Wiele, "A long-channel MOSFET model," *Solid-State Electron.*, vol. 22, no. 12, pp. 991–997, 1979.

[37] T. Okagaki, M. Tanaka, H. Ueno, and M. Miura-Mattausch, "Importance of Ballistic Carriers for the Dynamic Response in Sub-100nm MOSFETs," *IEEE Electron Device Letters*, vol. 23, no. 3, pp. 154–156, 2002.

[38] C. C. Enz, F. Krummenacher, and E. A. Vittoz , "An analytical MOS transistor model valid in all regions of operation and dedicated to low-voltage and low-current applications," *Special Issue of the Analog Integrated Circuits and Signal Processing Journal on Low-Voltage and Low-Power Design,*

vol. 8, no., 7, pp. 83–114, 1995.

[39] C. Galup-Montoro, M. C. Schneider, and I. J. B. Loss, "Series-parallel association of FET's for high gain and high frequency applications," *IEEE J. Solid-State Circuits*, vol. 29, no. 9, pp. 1094–1101, 1994.

[40] H. K. Gummel and K. Singhal, "Inversion charge modeling," *IEEE Trans. Electron Devices*, vol. 48, no. 8, pp. 1585–1593, 2001.

[41] X. Xi, J. He, M. Dunga, C.–H. Lin, B. Heydari, H. Wam, M. Chan, A. M. Niknejad, and C. Hu, "The development of the next generation BSIM for sub-100nm mixed-signal circuit simulation," *Proc. Workshop on Compact Modeling*, pp. 70–73, Boston, March, 2004.

[42] J. Watts, C. McAndrew, C. Enz, C. Galup-Montoro, G. Gildenblat, C. Hu, R. van Langevelde, M. Miura-Mattausch, R. Rios, and C.–T. Sah, "Advanced Compact Models for MOSFETs," *Proc. Workshop on Compact Modeling*, pp. 3–12, Anaheim, May, 2005.

[43] S. M. Sze, "Physics of Semiconductor Device (Second Edition)", *New York, John Wiley & Sons, Inc.*, 1981.

[44] T. L. Chiu and C. T. Sah, "Correlation of experiments with a two-section model theory of the saturation drain conductance of MOS transistors," *Solid-State Electron.*, vol. 11, no. 11, pp. 1149–1163, 1968.

[45] D. Frohman-Bentchkowsky and A. S. Grove, "Conductance of MOS transistors in saturation," *IEEE Trans. Electron Devices*, vol. ED-16, no. 1, pp. 108–113, 1969.

Chapter 3

Advanced MOSFET Phenomena Modeling

3.1 Threshold Voltage Shift

Different from the drift approximation (see for example Eq. (2.65)), the drift-diffusion theory is exact thus it does not require the threshold voltage parameter V_{th} for describing device characteristics [1–4]. The MOSFET device parameters such as the oxide thickness T_{ox} and the substrate doping concentration N_{sub} determine the complete MOSFET behavior automatically and consistently as can be seen from Eqs. (2.58) [2] and (2.59) [3].

However, V_{th} provides as a good measure for characterizing MOSFET features [5], although consistent extraction of V_{th} for different technologies is not simple [6–11]. The most frequently used definition of V_{th} is the gate voltage inducing the normalized drain current of 1×10^{-7}A $(= I_{ds} \cdot L_{eff}/W_{eff})$. Measured V_{th} includes influences by various advanced MOSFET phenomena such as the short-channel effects, which cause a reduction of V_{th} for short-channel transistors in comparison to long-channel transistors as shown in Fig. 3.1. This so-called "V_{th} roll-off" (see also the curve in Fig. 3.2 depicted by solid triangles) is very much dependent on the technology used for MOSFET fabrication. Therefore, many detailed informations on the MOSFET fabrication technology, which are relevant for modeling device characteristics, are derived from the V_{th} shifts (ΔV_{th}) as a function of gate length (L_{gate}) [12]. The modeled ΔV_{th} is incorporated in the ϕ_S iteration through V_G' in Eq. (2.17).

The measured V_{th} can be viewed as consisting of two main effects or components [13, 14]:

(I) the short-channel effect: $\Delta V_{th,SC}$
(II) the reverse-short-channel effect: $\Delta V_{th,RSC}$

117

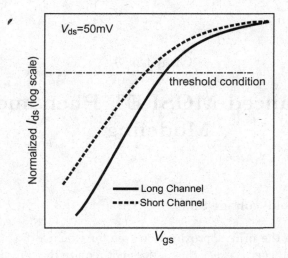

Fig. 3.1   Comparison of the normalized drain current for a long-channel and a short-channel MOSFET as a function of the gate voltage V_{gs}. The horizontal dotted-dashed line depicts the normalized current ($I_{ds} \cdot L_{eff}/W_{eff}$) of the threshold condition.

The separation of ΔV_{th} into these two components ($\Delta V_{th} = \Delta V_{th,SC} - \Delta V_{th,RSC}$) is schematically shown in Fig. 3.2.

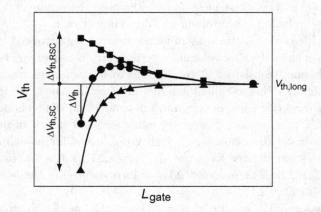

Fig. 3.2   Schematic plot of the separation of ΔV_{th} into the contributions of the short-channel (SC) and the reverse-short-channel (RSC) effect.

3.1.1 *(I) Short-Channel Effects*

As for the short-channel effects four important phenomena are observed: (i) reduction of V_{th} for reduced L_{gate} (V_{th} roll-off effect), (ii) V_{th} dependence on V_{ds}, (iii) reduction of the body effect coefficient (or body factor, see also Fig. 3.12), (iv) increase of the subthreshold swing. In particular (iv) is often not obvious for the usual fabrication technologies. (ii) is often also called the drain-induced barrier lowering (DIBL) a term coined by Troutman in 1979 [15]. Recent advanced technologies utilize aggressive geometrical scaling, which induces observable threshold degradation as shown in Fig. 3.3 for a MOSFET structure without suitable counter measures such as the pocket implant. The generally applied technology of a pocket implant, for example, reduces the threshold degradation because it causes a V_{th} increase, which is called reverse-short-channel effect, as schematically shown in Fig. 3.2 by solid squares, and thus shifts the threshold-degradation effect to shorter L_{gate}. To enable efficient modeling, the V_{th} variation as a

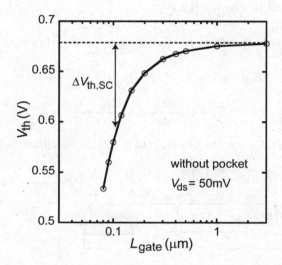

Fig. 3.3   2D-device simulation results of V_{th} as a function of gate length L_{gate} without pocket implant. No V_{th} increase is observed. $\Delta V_{th,SC}$ is the difference from the long L_{gate} value.

function of L_{gate} has to be separated according to the contributions of the short-channel (SC) and reverse-short-channel (RSC) effects as depicted in

Fig. 3.2, and is written as

$$V_{\text{th}} = V_{\text{th,long}} + \Delta V_{\text{th,RSC}} - \Delta V_{\text{th,SC}} \qquad (3.1)$$

where $V_{\text{th,long}}$ is the threshold voltage of a long channel transistor, $V_{\text{th,RSC}}$ describes the RSC effect, and $\Delta V_{\text{th,SC}}$ is the V_{th} shift due to the pure SC effect. We first concentrate on the modeling of $\Delta V_{\text{th,SC}}$.

All observed short-channel phenomena are caused by the longitudinal(lateral)-electric-field contribution in the MOSFET channel, which we subsequently refer to by the shorter term longitudinal electric field and which becomes important even at threshold condition with small V_{ds}. The contribution of the longitudinal electric field causes a 2D current flow in the channel. This effect was first modeled as the charge sharing by Yau [16]. Later on investigations have been undertaken to derive analytical solutions of the 2D Poisson equation [17, 18]. These solutions are rather complicated. A simple formulation of $\Delta V_{\text{th,SC}}$ is derived by applying the Gauss law to the square ABCD in the depletion region below the MOSFET gate as depicted in Fig. 3.4 [12, 13]

$$-\int_A^B E_{\text{y1}}(x)dx + \int_C^D E_{\text{y2}}(x)dx + \int_D^A E_{\text{x}}(x)dy = -\frac{Q_{\text{s}}}{\epsilon_{\text{Si}}} \qquad (3.2)$$

$$E_{\text{x}}dy + W_{\text{d}}(E_{\text{y2}} - E_{\text{y1}}) = -\frac{(Q_{\text{b}} + Q_{\text{i}})}{\epsilon_{\text{Si}}}dy \qquad (3.3)$$

$$E_{\text{x}} + W_{\text{d}}\frac{dE_{\text{y}}}{dy} = -\frac{(Q_{\text{b}} + Q_{\text{i}})}{\epsilon_{\text{Si}}} \qquad (3.4)$$

where E_{y} and E_{x} are the electric fields along the channel direction and

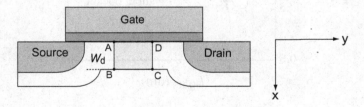

Fig. 3.4 Schematic illustration of the square ABCD in the MOSFET channel region to which the Gauss law is applied for deriving the equation for the short-channel threshold voltage shift.

perpendicular to the channel direction, respectively. The integration along B-C is zero, because the electric field at the boundary of the depletion region is zero. The depletion-layer thickness W_d is approximated to be position independent from source to drain and is written as

$$W_d = \sqrt{\frac{2\epsilon_{Si}(\phi_s - V_{bs})}{qN_{sub}}} \tag{3.5}$$

A parabolic potential distribution along the channel is also approximated, which results in a position independent gradient of the longitudinal-electric field dE_y/dy. If this longitudinal-field gradient is negligibly small, Eq. (3.4) reduces to the conventional equation for describing the relationship between the field at the substrate surface and the induced carrier density. Consequently, the second term on the left-hand side of Eq. (3.4) is exactly the reason for the threshold voltage reduction, i.e., the threshold voltage shift due to the SC effect.

To verify this conclusion, the electric field E_x at the surface, as simulated by a 2D-device simulator, is shown in Fig. 3.5 for different gate length under the threshold condition. Since the threshold condition requires the same amount of carriers, namely the same amount of potential bending, independent of the gate length L_{gate}, the contribution of the second term of the left-hand side compensates the reduction of E_x observed Fig. 3.5 [12].

Again with the Gauss law of $\epsilon_{Si}E_s = \epsilon_{ox}E_{ox}$, the relationship between the gate voltage and the induced field is written as

$$\left(E_x + W_d\frac{dE_y}{dy}\right)\epsilon_{Si} = (V_G' - \phi_S)\frac{\epsilon_{ox}}{T_{ox}}. \tag{3.6}$$

Since the threshold voltage is determined as the gate voltage V_{gs} for which ϕ_S is equal to $2\Phi_B$, given by

$$2\Phi_B = \frac{2}{\beta}\ln\left(\frac{N_{sub}}{n_i}\right) \tag{3.7}$$

it follows that the threshold voltage shift is induced by the additional contribution of the longitudinal field gradient. In this way, the final equation of $\Delta V_{th,SC}$ is derived as

$$\Delta V_{th,SC} = W_d\frac{\epsilon_{Si}}{C_{ox}}\frac{dE_y}{dy}. \tag{3.8}$$

Here the major concern for correct reproduction of the SC effects is an appropriate modeling of the gradient of the longitudinal-electric field. This is not an easy task, because a longitudinal impurity profile, determining

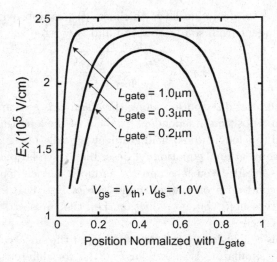

Fig. 3.5   Vertical electric field distribution at the surface as a function of channel position as simulated with a 2D-device simulator. Different gate lengths L_{gate} are compared with the gate voltage V_{gs} chosen at threshold condition.

the field gradient, is not known. Therefore, a simple phenomenological description is introduced in the form

$$\frac{dE_y}{dy} = \frac{2(V_{bi} - 2\Phi_B)}{(L_{gate} - PARL2)^2} \left(SC1 + SC2 \cdot V_{ds} + SC3 \cdot \frac{2\Phi_B - V_{bs}}{L_{gate}} \right) \quad (3.9)$$

where V_{bi} and $PARL2$ represent the built-in potential and the depletion width of the channel/contact junction vertical to the channel, respectively. The model parameter $SC1$ determines the threshold voltage shift for small V_{ds} and V_{bs}, and is expected to be unity for ideal cases. If the measured threshold voltage is plotted as a function of V_{ds}, it exhibits nearly a linear dependence as shown in Fig. 3.6. The V_{th} reduction as a function of V_{ds} is often called the drain-induced barrier lowering (DIBL) as we also mentioned before. The corresponding threshold-voltage shift ΔV_{th} in comparison to the V_{th}-value of a long-channel transistor is shown in Fig. 3.7. The gradient of this dependence is proportional to the parameter $SC2$. The parameter $SC3$ implements a correction of the charge-sheet approximation, which is expected to be small, and describes a modification of the V_{th} dependence on V_{bs} for an inhomogeneous substrate impurity profile.

The threshold voltage shift $\Delta V_{th,SC}$, which can be viewed as causing an

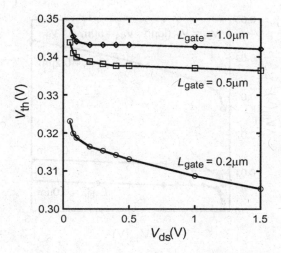

Fig. 3.6   Threshold voltage V_{th} as a function of V_{ds}, simulated with a 2D-device simulator for three different channel lengths.

additional contribution to vertical electric field E_x, is treated as an increase of V'_G in the overall compact modeling of the MOSFET

$$V'_G = V_{gs} - V_{fb} + \Delta V_{th,SC} \tag{3.10}$$

This modeling approach reproduces the parallel shift of the I_{ds}-V_{gs} characteristics in comparison to the long-channel case, which is usually observed in short-channel MOSFETs. However, a larger subthreshold swing S, which is defined as the gate voltage V_{gs} needed to increase the drain current by one decade as

$$S = \ln 10 \frac{dV_{gs}}{d(\ln I_{ds})} \tag{3.11}$$

is also observed for very short-channel lengths. This subthreshold-swing increase can be modeled with the field gradient in the form [12]

$$\frac{S}{\ln 10} = \left(1 + \frac{\gamma}{2\sqrt{\phi_S}} - \frac{A\sqrt{B}}{2\sqrt{\phi_S}}E_{yy} - A\sqrt{B\phi_S}\frac{dE_{yy}}{d\phi_S}\right)/\beta \tag{3.12}$$

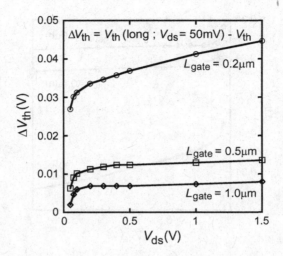

Fig. 3.7 Threshold voltage V_{th} as a function of V_{ds} simulated by a 2D-device simulator. The reduction for small V_{ds} is phenomenologically modeled separately with the model parameters $SCP21$ and $SCP22$ (see also Eq. (3.50)).

where

$$\gamma = \frac{\sqrt{2\epsilon_{Si} q N_{sub}}}{C_{ox}} \; ; \quad E_{yy} = \frac{dE_y}{dy}$$

$$A = \frac{\epsilon_{Si}}{C_{ox}} \; ; \quad B = \frac{2\epsilon_{Si}}{q N_{sub}}$$

The last two terms on the right-hand side of Eq. (3.12) have the opposite sign in comparison to the second term, which reduces the field-gradient contribution, resulting in a diminished degradation of the subthreshold swing. This is the reason why the swing increase is usually observed only for very short-channel cases. Equation (3.12) describes the features of the subthreshold swing, but is not suitable for compact modeling, because it requires an iteration process.

To obtain a more simple description, the V_{gs} dependence of the surface potential along the channel and vertical to the channel is studied in Fig. 3.8. Figure 3.8a shows the surface potential along the channel for different V_{gs} values. For small V_{gs} the built-in potential between the channel/contact junction determines the energy difference between the regions. By increasing V_{gs} the energy in the channel reduces and the potential difference due to the built-in potential diminishes. Figure 3.8b shows the band diagram in the vertical direction, where the well known band bending is seen with increased V_{gs} values. This strong V_{gs} dependence of ϕ_S in the subthreshold region has to be considered here. In the modeling of the short-channel effect, the surface potential ϕ_S is fixed to $2\Phi_B$, for modeling the threshold condition. Thus a simple formulation including the V_{gs} dependence of Φ_B, which is distinguished by $\Phi_B'(V_{gs})$, has to be developed to replace the constant Φ_B in Eq. (3.9). The desired simplified form of $\Phi_B'(V_{gs})$ is obtained by solving the Poisson equation analytically under the diffusion approximation without the drift component

$$\Phi_B'(V_{gs}) = V_G' + \left(\frac{const0}{C_{ox}}\frac{\beta}{2}\right)^2 \left[1 - \sqrt{1 - \frac{4\beta(V_G' - V_{bs} - 1)}{\beta^2 \left(\frac{const0}{C_{ox}}\right)^2}}\right] \qquad (3.13)$$

where $const0$ is written as

$$const0 = \sqrt{\frac{2\epsilon_{Si}qN_{sub}}{\beta}} \qquad (3.14)$$

This V_{gs} dependent $\Phi_B'(V_{gs})$ is used to modify Φ_B and to enable a correction of the subthreshold swing with a model parameter $PTHROU$

$$\Phi_{B,gs} = \Phi_B + PTHROU \cdot (\Phi_B'(V_{gs}) - \Phi_B) \qquad (3.15)$$

$\Phi_{B,gs}$ as defined in Eq. (3.15) replaces Φ_B in the term $(V_{bi} - 2\Phi_B)$ of Eq. (3.9). The model parameter $PTHROU$ enables the desired description of the increased subthreshold swing for short-channel transistors in HiSIM.

3.1.2 *(II) Reverse-Short-Channel Effects*

The reverse-short-channel (RSC) effect is categorized into resulting from two possible physical MOSFET properties:

(i) **Impurity concentration inhomogeneity perpendicular to the channel**
(obvious in the retrograded impurity implantation): $\Delta V_{th,R}$

(ii) **Impurity concentration inhomogeneity along the channel direction**
(obvious in the pocket implant technologies): $\Delta V_{th,P}$

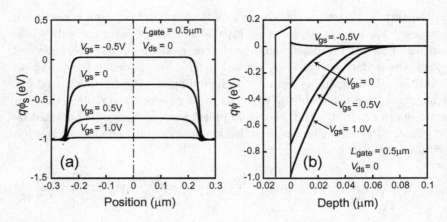

Fig. 3.8   Simulated energy diagram in the channel with a 2D-device simulator for different gate voltages V_{gs}: (a) The surface energy distribution along the channel. (b) The energy distribution in the depth direction at the channel middle. Position 0 in (a) shows the middle of the channel.

Since the channel impurity concentration determines not only V_{th} but also other important physical quantities such as the mobility, correct inclusion of the impurity concentration is a general key for accurate physical circuit simulation. This will be discussed further in Section 3.4.

(i) Impurity concentration inhomogeneity perpendicular to the channel

Fig. 3.9 shows 2D-device simulation results for the V_{th}-L_{gate} characteristics of a MOSFET technology without pocket implant. A slight V_{th} increase is observed. The origin of this type of reverse-short-channel (RSC) effect is a pileup of implanted impurities at the SiO_2/Si interface under the gate [19–21]. This pileup is due to interstitials created by the ion implantation for S/D contact formation. They diffuse out to the substrate and move to the surface underneath the gate in later fabrication steps and are accompanied by impurities in the substrate as shown in Fig. 3.10. As a result high impurity concentration regions are formed in the channel near S/D contacts. With reduced L_{gate} the two high concentration regions overlap, and thus the impurity concentration becomes L_{gate} dependent as can be seen in Fig. 3.11.

Figure 3.11 shows 2D-process simulation results of the N_{sub} profile at the channel middle into the channel depth direction [22, 23]. The profiles were used to simulate the V_{th}-L_{gate} characteristics shown in Fig. 3.9. The

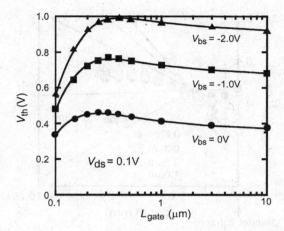

Fig. 3.9 Comparison of measured V_{th} (solid symbols) with model results (solid lines) for a MOSFET technology without pocket implants.

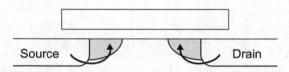

Fig. 3.10 Movement of interstitials, which are accompanied by impurities, to the bulk surface underneath the gate oxide.

fabrication process of these MOSFETs originally aimed at a profile with increasing impurity concentration along the depth direction, a so-called retrograded profile, as can be seen for the long L_{gate} data.

The substrate impurity pileup on top of the retrograded profile at the channel surface near the source/drain contacts is a cause for reverse-short-channel effects [21]. For inclusion of this physical cause in the compact model, the impurity profile $N_{\text{sub}}(x)$ is simplified as a linear function of the depth x [13]

$$N_{\text{sub}}(x) = NSUB0(L_{\text{gate}}) + NSUBG(L_{\text{gate}}) \times x \qquad (3.16)$$

where $NSUB0$ and $NSUBG$ are L_{gate} dependent parameters representing the impurity concentration at the surface of the substrate and its gradient in the vertical direction, respectively. The 1D Poisson equation under the

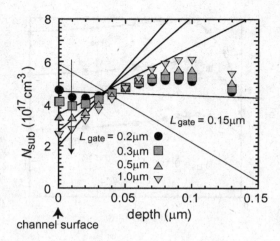

Fig. 3.11   Symbols are the impurity profiles used for the V_{th} simulation shown in Fig. 3.9 [22, 23]. Lines are extracted simplified linear profiles.

threshold condition

$$\frac{d^2\phi}{dx^2} = -\frac{q}{\epsilon_{\text{Si}}}\left[N_{\text{sub}}(x)\{\exp(\beta\phi) - 1\}\right] \tag{3.17}$$

is solved iteratively with Eq. (3.16) to calculate the depletion charge Q_{b}, and the threshold voltage is written as

$$V_{\text{th}} = V_{\text{fb}} + 2\Phi_{\text{B}} + \frac{Q_{\text{b}}}{C_{\text{ox}}} \tag{3.18}$$

For the homogeneous impurity case the closed form

$$Q_{\text{b}} = \sqrt{2\epsilon_{\text{Si}}qN_{\text{sub}}(2\Phi_{\text{B}} - V_{\text{bs}})} \tag{3.19}$$

is derived.

Figure 3.12 shows the same data as Fig. 3.9 but as a function of $\sqrt{\phi_{\text{S}} - V_{\text{bs}}}$, where ϕ_{S} is the surface potential fixed at threshold condition $2\Phi_{\text{B}}$ and V_{bs} is the bulk-source voltage. Here the ϕ_{S} value is taken from a long-channel case without the RSC effect. An interesting feature of the plot is that the linearly extrapolated V_{th} values at $\sqrt{\phi_{\text{S}} - V_{\text{bs}}}=0$ vary with L_{gate} and do not converge to the theoretically expected value of $\phi_{\text{S}} + V_{\text{fb}}$, where V_{fb} is the flat-band voltage.

The reason for the observed variation at $\sqrt{\phi_{\text{S}} - V_{\text{bs}}}=0$ can be attributed to an impurity (boron in our study) pileup beneath the MOSFET gate [24].

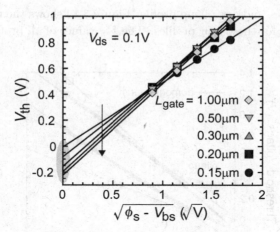

Fig. 3.12 Simulated V_{th}-$\sqrt{\phi_S - V_{bs}}$ characteristics. The gradient and the intersect at $\sqrt{\phi_S - V_{bs}}$=0 are dependent on the $N_{sub}(x)$ profile. ϕ_S is fixed to $2\Phi_B$ of a long-channel case.

This is verified by a simulation experiment with a 2D-device simulator. Figure 3.13 shows the four different vertical impurity profiles studied: Two ho-

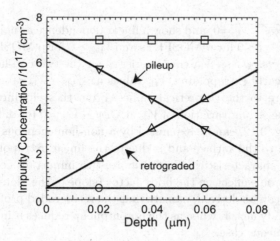

Fig. 3.13 Simulated V_{th}-$\sqrt{\phi_S - V_{bs}}$ characteristics. The gradient and the intersect are dependent on the $N_{sub}(x)$ profile.

mogeneous doping profiles with different impurity concentrations, one retrograded profile, and one pileup profile. Figure 3.14 shows the resulting V_{th} characteristics for these four profiles. The V_{th} values of all profiles meet, as

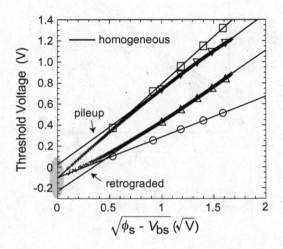

Fig. 3.14 Threshold voltage as a function of $\sqrt{\phi_S - V_{\text{bs}}}$ for pileup and retrograded impurity profiles in the channel, where ϕ_S is fixed to $2\phi_B$. The deviation from linearity for small V_{bs} is modeled with the parameters $BS1$ and $BS2$ (see Eq. (3.24)).

expected, at $\sqrt{\phi_S - V_{\text{bs}}}=0$, and show a linear dependence at normal operating conditions, $V_{\text{bs}} \leq 0$ for n-MOSFETs and $V_{\text{bs}} \geq 0$ for p-MOSFETs. However, if the V_{th} vs. $\sqrt{\phi_S - V_{\text{bs}}}$ characteristics for the two non-homogeneous profiles are linearly extrapolated, V_{th} values at $\sqrt{\phi_S - V_{\text{bs}}}=0$ converge obviously no more to the theoretical value. From these findings it is concluded, that the strong variation of V_{th} at $\sqrt{\phi_S - V_{\text{bs}}}=0$ for different L_{gate}, observed in Fig. 3.12, can be explained by a non-homogeneous N_{sub} distribution vertical to the surface and is due to the linear extrapolation of the V_{th}-$\sqrt{\phi_S - V_{\text{bs}}}$ characteristics. Furthermore, two important conclusions of the impurity-profile effect on the body factor (slope of the plots in Fig. 3.12 and Fig. 3.14) can be derived, namely that the retrograded profile results in an enhanced body factor, whereas the boron-pileup reduces it in comparison to the homogeneous case.

The parameters $NSUB0(L_{\text{gate}})$ and $NSUBG(L_{\text{gate}})$ are extracted so that calculated V_{th} vs. $\sqrt{\phi_S - V_{\text{bs}}}$ characteristics reproduce measured (or 2D numerical device-simulation results in our case) values.

With the depletion charge Q_{b}, the V_{th} shift in comparison to a long-channel transistor is written [13, 14] as

$$\Delta V_{\text{th,R}} = \frac{Q_{\text{b}}}{C_{\text{ox}}} - \frac{Q_{\text{b}}(\text{long})}{C_{\text{ox}}} \tag{3.20}$$

$$Q_{\text{b}} = q \int_0^{W_{\text{d}}} N_{\text{sub}}(x) dx \tag{3.21}$$

where $Q_{\text{b}}(\text{long})$ is the depletion charge of a long-channel transistor. Since a non-homogeneous impurity profile does not allow a description of the depletion width W_{d} in a closed form, Eq. (3.21) has to be solved iteratively with Eq. (3.17). The gradient of $N_{\text{sub}}(x)$ and its intersect at $x = 0$ are determined to reproduce measured (or simulated) V_{th}-$\sqrt{\phi_{\text{S}} - V_{\text{bs}}}$ characteristics. Figure 3.11 compares the extracted impurity profiles with the 2D-process simulation results [23]. Figure 3.9 compares simulated and measured V_{th} values as a function of L_{gate}.

An integrated form of the Q_{b} equation, represented by a polynomial function of L_{gate}, is implemented into the final compact model to eliminate the integration procedure during circuit simulation

$$Q_{\text{b}} = QDEPCC + \frac{QDEPCL}{L_{\text{gate}}^{QDEPCS}} + \left(QDEPBC + \frac{QDEPBL}{L_{\text{gate}}^{QDEPBS}} \right) \sqrt{\phi_{\text{S}} - V_{\text{bs}}}$$
$$\tag{3.22}$$

$QDEPCC$, $QDEPCL$, $QDEPCS$, $QDEPBC$, $QDEPBL$, and $QDEPBS$ are the resulting model parameters. The impurity concentration used for the surface-potential calculations is simplified by the value at the surface, $N_{\text{sub}}(0)$. The reason for this simplification is that the inversion-charge density Q_{i}, which mostly determines the MOSFET characteristics, extends only in a few nm into the vertical direction. The modeled threshold voltage is incorporated into V_{G}' as

$$V_{\text{G}}' = V_{\text{gs}} - V_{\text{fb}} + \Delta V_{\text{th,SC}} - \Delta V_{\text{th,R}} \tag{3.23}$$

Fig. 3.14 shows the V_{th} dependence as a function of $\sqrt{\phi_{\text{S}} - V_{\text{bs}}}$ for two different impurity profiles perpendicular to the channel. For cases where the inhomogeneity is large or where positive V_{bs} is applied, deviation from the linearity of V_{th} as a function of $\sqrt{\phi_{\text{S}} - V_{\text{bs}}}$ is modeled by a modification of Eq. (3.19) with two additional fitting parameters $BS1$ and $BS2$ as

$$Q_{\text{bmod}} = \sqrt{2q \cdot N_{\text{sub}} \cdot \epsilon_{Si} \cdot \left(\phi_{\text{S}} - V_{\text{bs}} - \frac{BS1}{BS2 - V_{\text{bs}}} \right)} \tag{3.24}$$

$BS1$ represents the strength of the deviation from linearity and $BS2$ is the starting value of $\sqrt{\phi_S - V_{bs}}$ where this deviation from linearity becomes visible.

It has to be noticed, however, that the model parameters $SC3$ (short-channel effect) and $SCP3$ (reverse-short-channel effect from pocket implant, see next subsection) can be successfully used to model the effects of an impurity concentration inhomogeneity perpendicular to the channel, if this inhomogeneity is not extremely large.

(ii) Impurity concentration inhomogeneity along the channel direction (Pocket Implantation)

The pocket (halo)-implant technology has been developed to further suppress the short-channel effects, and enables an L_{gate}-reduction of the MOSFETs down to the sub-100nm regime [25–27]. The reverse-short-channel effect (RSCE) due to the pocket implant is much stronger than that due to the unintended impurity pileup caused by interstitial movement [19, 21]. The impurity pileup mainly induces an impurity concentration inhomogeneity perpendicular to the channel surface which is located beneath the SiO_2/Si interface near source and drain, and the resulting RSCE is restricted to magnitudes up to 100mV as shown in Fig. 3.9. On the contrary, the RSCE, caused by the pocket implant, is normally clearly larger than 100mV as shown in Fig. 3.15. Two obvious new features are: (1) the V_{th} increase starts already from relatively long L_{gate} and (2) the other short-channel effects such as the V_{ds} dependence of V_{th} appear also even for long-channel transistors [28]. The observed abnormal V_{th}-characteristics depends in addition strongly on the pocket-implant condition which can result in different pocket-impurity profiles [29, 30] and inhomogeneity properties along the channel. However, in nearly every case the RSCE starts already at relatively long L_{gate} above $1\mu m$.

Fig. 3.16 shows the 2D pocket profile extracted from measured characteristics by inverse modeling with a 2D-process simulator. As can be seen from the simulation results of a 2D-device simulator with the 2D pocket profile, shown in Fig. 3.17, the pocket profile induces potential minima at the pocket concentration peaks. For such a potential-distribution case the threshold condition is determined not only at the position where the maximum vertical field E_x is required (see Fig. 3.5), but all E_x-values along the channel have to be considered.

To model the effects of the pocket implantation, a new definition of

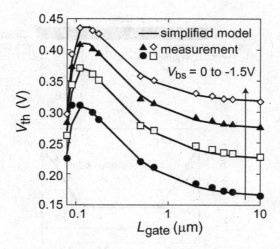

Fig. 3.15 Threshold voltage V_{th} as a function of the gate length L_{gate} for various bulk voltages V_{bs} in the case of a technology with pocket implantations. Measurements are shown by symbols and simulated results are shown by lines with Eq. (3.45)- (3.50) denoted as "simplified model".

V_{th} is derived from an average sheet carrier concentration n_{av} along the channel, which at threshold is required to be equal to the threshold sheet concentration n_{th} [31,32]. Conventionally, with homogeneous impurity profiles, V_{th} is given by the gate voltage V_{gs} when the surface potential ϕ_S is equal to $2\Phi_B$.

The averaged carrier concentration along the channel n_{av} can be written under the charge sheet approximation as

$$n_{av} = \frac{L_{eff}}{\displaystyle\int_0^{L_{eff}} n^{-1}(y)dy} \qquad (3.25)$$

where n_{av} determines the drain current

$$I_{ds} = qW_{eff}n_{av}v \qquad (3.26)$$

with the carrier velocity v. To describe n_{av} analytically, the pocket profile is approximated by a linearly graded function along the channel as depicted in Fig. 3.16. Two model parameters (L_p: length of the pocket extension into the channel; N_{subp}: peak of the pocket impurity concentration) are introduced as also indicated in Fig. 3.16. The basic impurity concentration

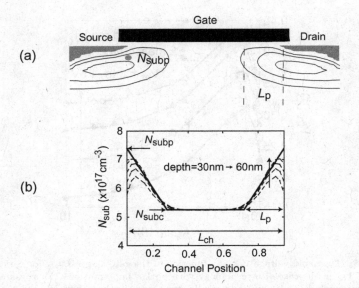

Fig. 3.16 Impurity profiles of a pocket technology extracted from measurements by inverse modeling with a 2D-process simulator: (a) Contours of the pocket impurity concentration; (b) the projection along the channel. The thick solid line is the extracted result with the developed model.

in the substrate is distinguished with the parameter N_{subc} from N_{subp}. The simplified equation

$$n_{av} = \frac{L_{eff}}{\dfrac{L_{eff} - 2L_p}{n_c} + \dfrac{2L_p}{n'_p}} \tag{3.27}$$

is obtained for Eq. (3.25) with this linearly graded approximation, where

$$n'_p = \frac{L_p}{\displaystyle\int_0^{L_p} n(y)^{-1}dy} \tag{3.28}$$

determines the averaged carrier concentration in the pocket regions. The integral in Eq. (3.28) can be solved analytically by approximating n to be a quadratic function of position, resulting in

$$n'_p = \frac{3n_c n_p}{n_c + \sqrt{n_c n_p} + n_p} \tag{3.29}$$

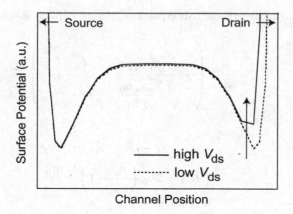

Fig. 3.17   2D-device simulation results of the surface potential along the channel for two different drain voltages V_{ds}, showing pronounced minima at the peak impurity concentration of the pocket implant.

where n_c and n_p are sheet carrier concentrations for homogeneous substrate concentrations of N_{subc} and N_{subp}, respectively. Under the threshold condition the sheet carrier concentration n_c is approximated with the first-order term of a Taylor expansion around the potential giving threshold condition ϕ_{thc}, leading to

$$n_c = \sqrt{\frac{N_{subc}\epsilon_{Si}}{q\beta}} \left(\frac{n_i}{N_{subc}}\right)^2 \frac{\exp(\beta\phi_{Sc})}{\sqrt{2\beta(\phi_{thc} - V_{bs}) - 2}} \tag{3.30}$$

The Gauss law ($\epsilon_{Si}E_s = \epsilon_{ox}E_{ox}$) is applied to derive the relationship between V_{gs} and ϕ_{Sc} resulting in

$$\phi_{Sc} = V_{gs} - V_{fb} + A_c - \sqrt{A_c} \cdot \sqrt{2V_{gs} - 2V_{fb} - 2V_{bs} + A_c} \tag{3.31}$$

$$A_c = \frac{qN_{subc}\epsilon_{Si}}{C_{ox}^2} \tag{3.32}$$

If V_{gs} is equal to V_{thc}, the threshold voltage with N_{subc}, ϕ_{Sc} must be equal to ϕ_{thc}. As shown in Fig. 3.18, to achieve n_{av} equal to n_{th}, the carrier concentration giving the threshold condition for the pocket-implanted case, n_c itself has to be higher than n_{th} at the threshold condition ($V_{gs} = V_{th}$) to compensate the lower concentration in the pocket region. The potential giving $n_{c,th}$ (see Fig. 3.18) is obtained by substituting $n_{c,th}$ for n_c in

Eq. (3.30) and by rewriting this equation into an explicit form for ϕ_{th}

$$\phi_{th} = \frac{1}{\beta} \ln \left(\frac{n_{c,th} \sqrt{2\beta(2\Phi_{Bc} - V_{bs})} - 2}{\sqrt{\frac{N_{subc}\epsilon_{Si}}{q\beta}} \left(\frac{n_i}{N_{subc}} \right)^2} \right) \tag{3.33}$$

where

$$n_{c,th} = \left(L_{eff} + 2L_p \left(\frac{n_c}{n'_p} - 1 \right) \right) \frac{n_{th}}{L_{eff}} \tag{3.34}$$

$$2\Phi_{Bc} = \frac{2}{\beta} \ln \left(\frac{N_{subc}}{n_i} \right) \tag{3.35}$$

After solving Eq. (3.31) for V_{gs} and replacing ϕ_{Sc} to ϕ_{th}, the explicit equation for V_{th}, i. e. V_{gs} at threshold condition, is obtained as

$$V_{th} = \frac{\phi_{th} + V_{fb} - A_c + \sqrt{A_c} \left(\sqrt{V_c} - \frac{V_{thc}}{\sqrt{V_c}} \right)}{1 - \sqrt{\frac{A_c}{V_c}}} \tag{3.36}$$

where

$$V_c = 2V_{thc} - 2V_{fb} - 2V_{bs} + A_c \tag{3.37}$$

and V_{thc} is the threshold voltage of the homogeneous impurity profile with N_{subc} described conventionally as

$$V_{thc} = 2\Phi_{Bc} + V_{fb} + \frac{\sqrt{2qN_{subc}\epsilon_{Si}(2\Phi_{Bc} - V_{bs})}}{C_{ox}} \tag{3.38}$$

To derive a further simplified equation, the averaged impurity concentration N_{subeff} along the channel is considered

$$N_{subeff} = \frac{(L_{eff} - L_p)N_{subc} + L_p N_{subp}}{L_{eff}} \qquad \text{for } L_{eff} \geq L_p \tag{3.39}$$

$$N_{subeff} = N_{subp} + (N_{subp} - N_{subc}) \left(1 - \frac{L_{eff}}{L_p} \right) \qquad \text{for } L_{eff} < L_p \tag{3.40}$$

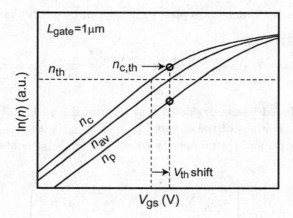

Fig. 3.18 Simulated average sheet carrier concentration in the channel as a function of the gate voltage V_{gs} for two homogeneous impurity concentrations without pocket implantation (n_c and n_p) and with the pocket implantation (n_{av}).

Under the above simplification Eq. (3.33) can be rewritten as

$$\phi_{th} = \frac{1}{\beta} \ln \left(\frac{n_{c,th}\sqrt{2\beta(2\Phi_B - V_{bs}) - 2}}{\sqrt{\dfrac{N_{subeff}\epsilon_{Si}}{q\beta}}} \right) + \frac{2}{\beta} \ln \left(\frac{N_{subeff}}{n_i} \right)$$

$$\simeq \frac{2}{\beta} \ln \left(\frac{N_{subeff}}{n_i} \right) \tag{3.41}$$

By using Eq. (3.41) for the surface potential at threshold in Eq. (3.31) and replacing V_{gs} to V_{th}, Eqs. (3.31) and (3.36), reduce to the well-known form of the V_{th} equation for the homogeneous case

$$V_{th} = 2\Phi_B + V_{fb} + \frac{\sqrt{2qN_{subeff}\epsilon_{Si}(2\Phi_B - V_{bs})}}{C_{ox}} \tag{3.42}$$

$$2\Phi_B = \frac{2}{\beta} \ln \left(\frac{N_{subeff}}{n_i} \right)$$

When the channel length L_{eff} becomes smaller than $2L_p$, the pockets at source and drain start to overlap. This causes effectively an increase of the homogeneous impurity concentration as shown in Fig. 3.19. The elevated bottom value of the substrate plateau N_{subb} can be written as

$$N_{\mathrm{subb}} = 2 \cdot N_{\mathrm{subp}} - \frac{(N_{\mathrm{subp}} - N_{\mathrm{subc}}) \cdot L_{\mathrm{gate}}}{LP} - N_{\mathrm{subc}} \qquad (3.43)$$

which is included in the neglected first term of Eq. (3.41). The different carrier concentrations of the elevated plateau $n_{\mathrm{c,th}}$ (with N_{subb}) and that with the original substrate impurity concentration N_{subc} is modeled as

$$\frac{1}{\beta} \ln \left(\frac{n_{\mathrm{c,th}}(N_{\mathrm{subb}})\sqrt{2\beta(2\Phi_{\mathrm{B}} - V_{\mathrm{bs}}) - 2}}{\sqrt{\dfrac{N_{\mathrm{subeff}}\epsilon_{\mathrm{Si}}}{q\beta}}} \right)$$

$$-\frac{1}{\beta} \ln \left(\frac{n_{\mathrm{c,th}}(N_{\mathrm{subc}})\sqrt{2\beta(2\Phi_{\mathrm{B}} - V_{\mathrm{bs}}) - 2}}{\sqrt{\dfrac{N_{\mathrm{subeff}}\epsilon_{\mathrm{Si}}}{q\beta}}} \right)$$

$$\simeq \frac{1}{\beta} \ln \left(\frac{N_{\mathrm{subb}}}{N_{\mathrm{subc}}} \right) \qquad (3.44)$$

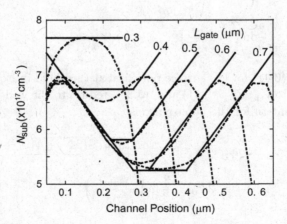

Fig. 3.19   Extracted pocket profile along the channel for different L_{gate}. Dashed curves are 2D-process simulator results at a depth of 50nm.

Thus the final equation is

$$V_{\text{th,P}} = V_{\text{fb}} + 2\Phi_{\text{B}} + \frac{Q_{\text{b}}}{C_{\text{ox}}} + \frac{1}{\beta} \log\left(\frac{N_{\text{subb}}}{N_{\text{subc}}}\right) \qquad (3.45)$$

$$2\Phi_{\text{B}} = \frac{2}{\beta} \ln\left(\frac{N_{\text{subeff}}}{n_{\text{i}}}\right) \qquad (3.46)$$

$$Q_{\text{b}} = \sqrt{2qN_{\text{subeff}}\epsilon_{\text{Si}}\left(2\Phi_{\text{B}} - V_{\text{bs}}\right)} \qquad (3.47)$$

Fig. 3.19 compares the extracted pocket profile, to reproduce the measured V_{th}-L_{gate} characteristics, also with the 2D-process simulator profile. The model profile has a fairly good agreement with this 2D simulated pocket profile at the depletion depth.

The derivation of the average carrier concentration n_{av} assumes a small V_{ds} value (e. g. 50mV). While the presence of a V_{th} dependence on V_{ds} in conventional MOSFETs is an important indication for the existence of short-channel effects, the situation is different for the pocket-implant case, because the V_{th} dependence on V_{ds} occurs even for long-channel transistors. In pocket-implanted MOSFETs a V_{ds} increase induces always a surface potential increase in the drain-side pocket even for long-channel cases (see Fig. 3.17), resulting in the observed V_{th} dependence on V_{ds} for practically all channel length. Fig. 3.20 shows the subthreshold characteristics simulated with a 2D-device simulator for different pocket structures and V_{ds} values at a gate length $L_{\text{gate}} = 1\mu\text{m}$, and Fig. 3.21 summarizes the V_{th}-dependence on V_{ds}. With only a drain-side pocket an enhanced dependence of the subthreshold characteristics on V_{ds} is observed. However, with only a source-side pocket the V_{th} dependence on V_{ds} becomes even smaller than without pocket implants and practically disappears as clearly illustrated in Fig. 3.21. With pockets on both sides an averaged V_{th} dependence on V_{ds} is observed. This is exactly due to our argument that the whole carrier concentration along the channel determines the V_{th} condition.

The effect of a V_{th} dependence on V_{ds}, observed with the pocket implant even for long L_{gate}, is modeled in the same way as the conventional short-channel version of this effect [12,33], but the potential distribution is restricted to be within the region of the pocket-extension length L_p

$$\Delta V_{th,PSC} = (V_{th,P} - V_{th0})\, W_d \frac{\epsilon_{Si}}{C_{ox}} \frac{dE_{y,P}}{dy} \tag{3.48}$$

$$V_{th0} = V_{fb} + 2\Phi_{Bc} + \frac{\sqrt{2qN_{subc}\epsilon_{Si}(2\Phi_{Bc} - V_{bs})}}{C_{ox}} \tag{3.49}$$

$$\frac{dE_{y,P}}{dy} = \frac{2(V_{bi} - 2\Phi_B)}{L_p^2}\left(SCP1 + SCP2 \cdot V_{ds} + SCP3 \cdot \frac{2\Phi_B}{L_P}\right) \tag{3.50}$$

where $V_{th,P}$ and V_{th0}, defined in Eqs. (3.45) and (3.49), are the threshold voltages for the cases with and without pocket-implant, respectively. V_{bi} in Eq. (3.50) is the built-in potential. Eqs. (3.48) to (3.50) show that the modeling of the additional short-channel effects due to the potential minimum caused by the pocket at drain side are done by applying the same concept as for the conventional short-channel case but with the electric field gradient $dE_{y,P}/dy$ in the pocket region. $SCP1$, $SCP2$, and $SCP3$ are model parameters describing the bias dependence of the measured V_{th}. The V_{th} increase with reducing L_{gate} as seen in Fig. 3.2 is automatically included in HiSIM through the impurity concentration N_{subb}. Thus only $\Delta V_{th,PSC}$ has to be incorporated additionally into V_G' as

$$V_G' = V_{gs} - V_{fb} + \Delta V_{th,SC} - \Delta V_{th,R} + \Delta V_{th,PSC} \tag{3.51}$$

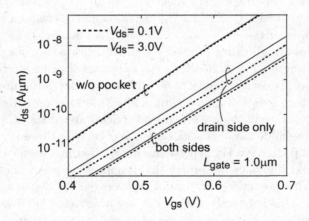

Fig. 3.20 Simulated I_{ds}-V_{gs} subthreshold characteristics with a 2D-device simulator for substrate conditions without pocket implantation, with only drain-side pocket and with pockets at both drain and source.

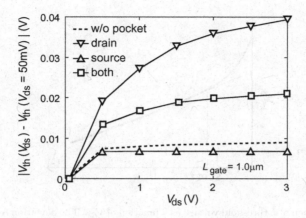

Fig. 3.21 Simulated V_{th} shift from V_{th} at V_{ds}=50mV as a function of drain voltage V_{ds}.

As V_{ds} approaches zero, the V_{th} dependence on V_{ds} deviates from linearity and V_{th} increases drastically as shown schematically in Fig. 3.22. This effect is modeled by a quadratic function of V_{ds} with two model parameters $SCP21$ and $SCP22$ as

$$\Delta V_{\text{th,P}} = \Delta V_{\text{th,P}} - \frac{SCP22}{(SCP21 + V_{\text{ds}})^2} \tag{3.52}$$

where $SCP21$ determines the V_{ds} value at which V_{th} starts to deviate from linearity as a function of V_{ds}. The parameter $SCP22$ determines the gradient of this deviation.

The overlap start of source and drain pockets causes a steep increase of V_{th} as a function of decreasing L_{gate}. This effect enables to extract L_{p} from measurements. Fig. 3.23 compares the V_{th}-L_{gate} characteristics of the developed pocket-implant model with and without inclusion of the short-channel effects. The steep increase at L_{gate}=0.1μm in Fig. 3.23a indicates the starting of the pocket overlap, confirming L_{p}=0.05μm.

In some cases the pocket profile cannot be described by a single linearly decreasing form, but exhibits extensive tails as schematically shown in Fig. 3.24. Therefore, two model parameters $NPEXT$ and $LPEXT$ are introduced to model the case of possible pocket tails as

$$N_{\text{sub}} = N_{\text{sub}} + \frac{NPEXT - N_{\text{subc}}}{\left(\frac{1}{xx} + \frac{1}{LPEXT}\right) L_{\text{gate}}} \tag{3.53}$$

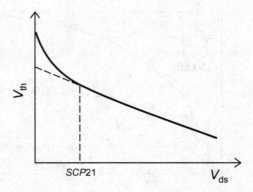

Fig. 3.22   Schematic threshold voltage as a function of V_{ds}. The deviation from linearity for small V_{ds} is modeled with parameters $SCP21$ and $SCP22$.

where the term xx is defined as

$$xx = 0.5 \times L_{gate} - L_P \ . \tag{3.54}$$

$NPEXT$ is the maximum concentration of the pocket tail and $LPEXT$ describes the tail extension characteristics. Usually strong pocket implantation induces not only the modeled lateral impurity distribution but also a vertical distribution at the same time. For fitting the measured results in such cases it is recommended to use the parameter $SCP3$(see Eq. (3.50)) together with parameters $BS1$ and $BS2$(see Eq. (3.24)).

Reliability of the developed model has been tested by varying the pocket concentration N_{subp}. Results are compared with 2D-device simulation results, calculated with explicit 2D impurity profiles in Fig. 3.25. For the comparison only the pocket peak concentration has been varied for both cases. It can be seen that HiSIM predicts the threshold voltage change as a function of N_{subp} qualitatively.

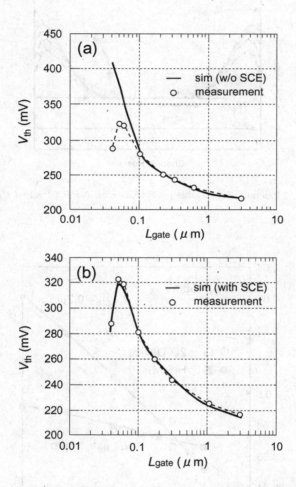

Fig. 3.23 Comparison of measurements and pocket-implant model calculation for V_{th} as a function of L_{gate}. Results (a) without and (b) with short-channel effects (SCE) are shown.

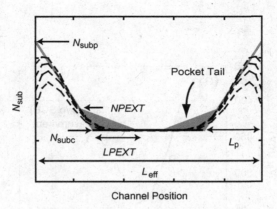

Channel Position

Fig. 3.24   Modeling of a possible pocket tail with $NPEXT$ and $LPEXT$.

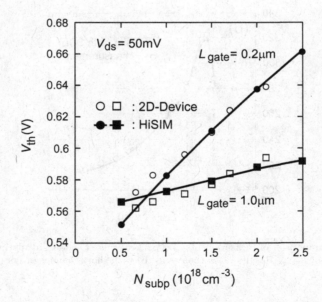

Fig. 3.25   Calculated threshold voltage V_{th} as a function of pockets maximum impurity concentration N_{subp}. For comparison 2D-device simulation results are depicted as well.

3.2 Depletion Effect of the Poly-Si Gate

Polycrystalline Silicon(poly-Si) can have a substantial concentration of impurities with shallow energy levels, such as phosphorus donors or boron acceptors, which are ionized at room temperature and hence can be used to achieve low series-resistance losses. Therefore poly-Si has been used as a gate electrode soon after the invention of the MOSFET to replace the metal gate electrode, mainly the chemically rather reactive aluminum because gold is not only expensive but also difficult to define lithographically to give the fine lines for interconnection. The cross-sectional view of a typical gate structure and the corresponding band diagram are show in Figs. 3.26(a) and (b), respectively. On the other hand, a shortcoming of the developed poly-Si technology has become critical as the gate-oxide thickness became thinner, namely, the depletion of majority carriers in the poly-Si gate at the poly-Si/oxide interface as also indicated in Fig. 3.26(b) for electron depletion at the n^+ poly-Si gate/Oxide interface. Problems

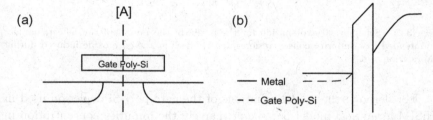

Fig. 3.26 (a) Cross-sectional view of the gate structure and (b) the energy diagram of the cross-section along the line [A].

arise because of this carrier depletion in the gate poly-Si. Carrier depletion by the gate-electric field is compounded by the lowered impurity concentration at the oxide or insulated covered poly-Si due to impurity segregation during high temperature processing, first reported by Grove and Sah in 1965 [34,35] and schematically illustrated in Fig. 3.27. The series resistance of the gate electrode reduces the gate voltage control of the channel and drain-source terminal current, both at a steady-state gate voltage, which could be off-set by additional DC gate voltage, but especially also at high frequencies which cannot be compensated by a higher DC voltage applied between the gate and the source terminals. Presently, this depletion-effect problem of poly-Si leads to a renewed development effort of metals as gate electrode materials for future technologies. An analysis of the depletion effect in the poly-Si gate has to consider that the gate-impurity concentration is usually still much higher than the impurity concentration in the substrate. Therefore, carrier depletion in the poly-Si near the gate-oxide interface starts normally after the formation of the inversion layer in the substrate below the gate as indicated in Fig. 3.28.

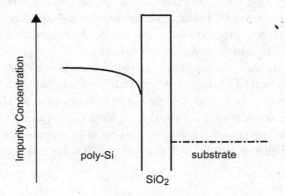

Fig. 3.27 Schematical explanation for the origin of the poly-depletion effect, namely unintended low impurity concentration in the poly-Si at the gate oxide induced during fabrication.

For deriving the model equations of the gate poly-Si depletion used in HiSIM a physical model parameter, namely the impurity concentration in the gate poly-Si (N_{pg}), is introduced. The Poisson equation is then solved in the substrate and in the gate poly-Si simultaneously by iteration [13, 36] using the formulation

$$V_G' - \phi_S - \phi_{Spg} = \frac{\epsilon_{Si} E_{Si}}{C_{ox}} \tag{3.55}$$

where E_{Si} is the electric field perpendicular to the channel surface (see Eq. (2.38)) at the lower gate-oxide interface to the substrate. The electric field E_{pg} in the poly-Si at the upper gate oxide interface to the poly-Si is written as

$$E_{pg} = \sqrt{\frac{2\epsilon_{poly} q N_{pg}}{\beta}} \left[\exp(-\beta\phi_{Spg}) + \beta\phi_{Spg} - 1 \right.$$
$$\left. + \frac{n_{p0,pg}}{p_{p0,pg}} \left\{ \exp(\beta\phi_{Spg}) - \beta\phi_{Spg} - 1 \right\} \right]^{\frac{1}{2}} \tag{3.56}$$

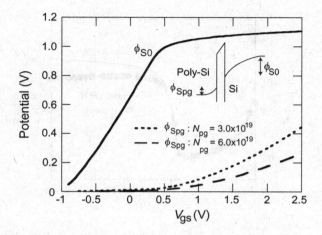

Fig. 3.28   Simulated surface potential at the source end (ϕ_{S0}) as a function of V_{gs}. The poly-depletion potential is also shown for two doping concentrations N_{pg} in the poly-Si.

where $n_{p0,pg}$ and $p_{p0,pg}$ are the carrier concentrations for electrons and holes in the poly-Si at the equilibrium condition, respectively (see [37] for exact derivation of the equation). The permittivity in the poly-Si is approximated to be equal to that of the silicon. A further reasonable approximation is that ϕ_{Spg} never enters the inversion condition under normal operating conditions, thus Eq. (3.56) can be simplified to

$$E_{pg} = \sqrt{\frac{2\epsilon_{pg}qN_{pg}}{\beta}}(\beta\phi_{Spg} - 1)^{\frac{1}{2}} \qquad (3.57)$$

Eqs. (3.55) and (3.57) are solved iteratively under the boundary condition of $\epsilon_{Si}E_{Si} = \epsilon_{poly}E_{pg}$ as schematically shown in Fig. 3.30. The permittivity in the poly-Si is approximated by that of Silicon, thus $\epsilon_{poly}E_{pg} \simeq \epsilon_{Si}E_{pg}$. Fig. 3.28 shows calculation results of ϕ_{Spg} together with ϕ_{S0} as a function of V_{gs}. A lower N_{pg} results in a stronger potential drop in the gate poly-Si. This potential drop causes a smaller increase of the surface potential ϕ_{Si} in the substrate. The poly-Si gate depletion also results in a reduction of the gate capacitance, C_{gate} [38–40]. Thus the impurity concentration N_{pg} at the upper gate-oxide interface is fitted to measured gate capacitance C_{gate} characteristics. Calculated C_{gate} is compared with measurements in Fig. 3.29 and verifies a sufficient capability of the described compact model for the poly-depletion effect to reproduce real data.

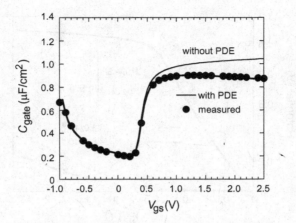

Fig. 3.29 Comparison between measured C-V characteristics and simulation results with and without gate-poly depletion effect (PDE).

As shown in Fig. 3.30 the potential drop in the gate poly-Si is calculated by a second iteration in addition to the iteration for calculating the surface potential in the substrate. This increases the MOS transistor device and MOS circuit simulation costs. To eliminate the otherwise necessary iteration procedure for the circuit-simulation application, ϕ_{Spg} is further approximated as a function of V_{gs} and V_{ds} by the simple formula of Eq. (3.58), and is included in the ΔV_{th} determination as a potential drop of V_{gs}.

$$\phi_{\mathrm{Spg}} = PGD1 \left(1 + \frac{1}{L_{\mathrm{gate}} \cdot 10^4}\right)^{PGD4} \exp\left(\frac{V_{\mathrm{gs}} - PGD2 - PGD3 \cdot V_{\mathrm{ds}}}{V}\right)$$

(3.58)

In Eq. (3.58), $PGD1$ describes the strength of the poly-depletion effect, $PGD2$ represents the threshold voltage of the onset of the poly-depletion, and $PGD3$ is introduced to take into account of the weakened poly-Si gate depletion at large V_{ds}. The reason for adopting an exponential function for the V_{gs} dependence is the exponential ϕ_{Spg}-V_{gs} characteristic obtained when solving Eqs. (3.55) and (3.57) iteratively. With large increased V_{gs} values the depletion in the poly-Si enters the inversion condition, which prevents a further potential drop. Thus the function of Eq. (3.58) has to smoothly enter the saturation value so that ϕ_{Spg} does not exceed ϕ_{S0} even for very large V_{gs}. A further parameter $PGD4$ has been introduced to represent the channel-length dependence of the gate poly-depletion effect,

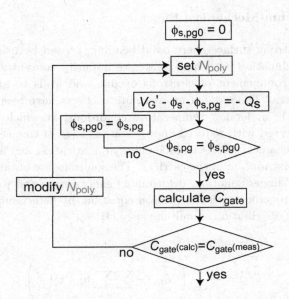

Fig. 3.30 Extraction flow chart of poly-depletion effect, where $V_G' = V_{gs} - V_{fb} - \Delta V_{th}$.

if it is necessary. V in the exponential term is introduced just to eliminate its dimension.

The model parameters of $PGD1$ and $PGD2$ are extracted by C_{gate} measurements, while the rest of the parameters are extracted by measured derivatives of the I-V characteristics, where not-negligible effects are observed not only for short-channel transistors but also for long-channel transistors.

3.3   Quantum-Mechanical Effects

Due to the strong surface enery-band bending, caused by using thin gate oxide in combination with high substrate impurity concentration, a two-dimensional confinement of electrons occurs, and leads to quantum mechanical effects. These quantum mechanical effects have been intensively studied mostly under low temperature conditions, at which the confinement is preserved with reduced energy-state change of the electrons. The quantum-mechanical effects include energy quantization and an existence-probability distribution of the carriers. The solutions are obtained by solving the Schrödinger equation, determining energy state and wave function ψ, and then together with the Poisson equation, by determining potential ϕ and carrier distribution, simultaneously [41–44]

$$H_j \psi_{ij} = E_{ij} \psi_{ij} \tag{3.59}$$

$$\nabla^2 \phi = -\frac{q}{\epsilon_{Si}} \left(N_{dep} - \sum_i \sum_j n_{ij} \mid \psi_{ij} \mid^2 \right) \tag{3.60}$$

where E_{ij} is the quantized energy state, N_{dep} is the depletion charge. The coefficients n_{ij} in the double sum of Eq. (3.60) are written as

$$n_{ij} = \frac{N_v m_j kT}{\pi \hbar^2} \ln \left\{ 1 + \exp \left(\frac{E_f - E_{ij}}{kT} \right) \right\} \tag{3.61}$$

where m_j and $\hbar$ are the electron mass and the reduced Planck constant, respectively. The Fermi level is described by E_f. The subscript i and j refer to number of energy sub-band state and crystal direction, respectively. N_v is the degeneracy of the sub-bands, which is equal to 2 for the longitudinal direction (j=1) and equal to 4 for the transverse direction (j=2). The Hamiltonian H consists of the kinetic and potential energies, written as

$$H_j = -\frac{\hbar^2}{2m_j} \frac{d^2}{dx^2} + \phi(x) \tag{3.62}$$

where x determines the position vertical to the channel surface. A solution of Eqs. (3.59)– (3.62) is depicted in Fig. 3.31.

It has been expected that the carrier confinement due to the energy quantization, referred to as the 2-dimensional electron gas, enhances the carrier mobility and suppresses carrier scattering during movement. However, the energy quantization is hardly observed in real device operations at room temperature, because the energy-level split into different states is

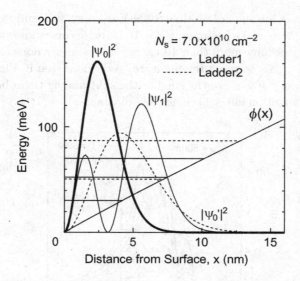

Fig. 3.31 Calculated energy states, wavefunctions and charge density in the direction vertical to the MOSFET channel by solving the Schrödinger equation together with the Poisson equation. Ladder 1 and 2 show results with different effective masses corresponding two different degenerate directions, referring to the longitudinal direction and the transverse direction, respectively [45].

not so large in comparison with the thermal voltage of 26mV as can be seen in Fig. 3.31.

Reduction of the thermal voltage kT/q accompanied with temperature reduction enhances the quantization effect [46, 47]. Fig. 3.32 shows measured transconductance oscillations at 39K together with those at 50K. The fabricated device includes strong quantum confinement of carriers due to a very thin substrate thickness of about 5nm as shown in Fig. 3.33. The device structure is called Silicon-on-Insulator (SOI) MOSFET structure and is different from the conventional bulk MOSFET due to an oxide layer beneath the active device structure. For such a case the energy-level split becomes larger than that shown in Fig. 3.31. The fabricated device includes thickness variations of the Si-layer along the channel, which results in an energy-level variation along the channel. This is observed as a current oscillation due to reduced transmission probabilities of carriers at the different energy levels in the channel. Thus the origin of the measured transconductance oscillations in Fig. 3.32 is explained by the existnece of the different energy levels acting as barriers. However, the energy differences of the quantized levels are much less than 100mV even at 39K and immediately disappear with increased temperature. It is furthermore known that carriers normally occupy more than 10 energy states under normal operation voltages for V_{gs} [48] at room temperature. As can be seen in Fig. 3.31, the higher the energy states are the smaller the split among them becomes, so that the quantization effect becomes less obvious.

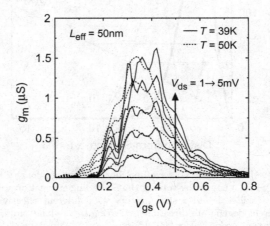

Fig. 3.32   Measured transconductance g_m oscillations as a function of V_{gs} at 39K (solid lines) and 50K (dotted lines) for different small V_{ds} values. The channel length and width are 50nm and 2μm, respectively.

Consequently, the main quantum-mechanical phenomenon, which has to be included into a MOSFET model for circuit simulation, is the repulsion of the channel carrier-density peak into the substrate away from the surface of the substrate as schematically shown in Fig. 3.34. The repulsion of the channel carrier density away from the surface can be described phenomenologically by an increased effective oxide thickness $T_{ox,eff}$ $(=T_{ox} + \Delta T_{ox})$, where ΔT_{ox} is the averaged distance from the surface

$$\Delta T_{ox,i} = \int_0^\infty x|\psi_i(x)|^2 dx \qquad (3.63)$$

Figure 3.35 shows the exactly calculated ΔT_{ox}-V_{gs} characteristics as a function of V_{gs} by solid circles for the lowest energy level. The calculation is done by solving the Schröding equation together with the Poisson equation iteratively. Extension of this treatment to practical applications with normal operation conditions, where the Schrödinger equation and the Poisson

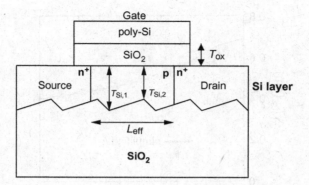

Fig. 3.33 Schematic of the SOI-MOSFET structure investigated with the silicon-layer thickness of 5nm. The Si-layer thickness varies between $T_{Si,1}$ of 6nm and $T_{Si,2}$ of 4nm along the channel. This terrace-like thickness variation is caused during the fabrication process of the SOI layer in older technologies, and is reduced below 1nm in advanced technologies by better technology control.

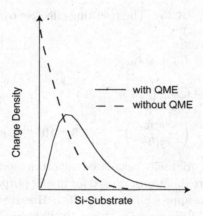

Fig. 3.34 Schematical comparison of carrier distributions perpendicular to the channel surface with the quantum-mechanical effect and without.

equation have to be solved simultaneously, is not possible mainly due to the simulation time problem [49, 50]. Therefore, two major approximations are introduced to derive a simple set of equations for $T_{ox,eff}$: First, a triangular potential perpendicular to the channel (shown by a dotted line in Fig. 3.36) is approximated, and that carriers occupy only the lowest quantized energy level (shown by a horizontal thick solid line in Fig. 3.31) is the second ap-

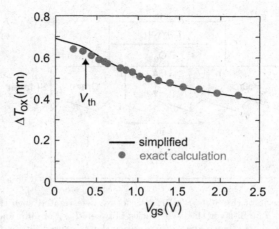

Fig. 3.35   Calculated effective T_{ox} increase due to the quantum mechanical effect. The solid line shows model results with Eqs. (3.68) and (3.69). Symbols are exact calculation results by solving the Poisson equation and the Schrödinger equation simultaneously.

proximation (see Figure 3.36). The resulting effective oxide thickness $T_{ox,eff}$ can be written as [51–53]

$$T_{ox,eff} = T_{ox} + \Delta T_{ox}$$

$$\Delta T_{ox} = Q_{ealp} \left(Q_b + \frac{11}{32} Q_i \right)^{-\frac{1}{3}}$$

$$Q_{ealp} = \left(\frac{48\pi m_e q}{\epsilon_{Si} \hbar^2} \right)^{-\frac{1}{3}} = 3.5 \times 10^{-10} (\mathrm{C\,cm})^{\frac{1}{3}}$$

(3.64)

The coefficient Q_{ealp}, originally calculated quantum mechanically under the above mentioned approximations, is used for fitting purposes to compensate inaccuracies due to the approximations applied. However, the value of Q_{ealp} cannot not be much different from $3.5 \times 10^{-10} (\mathrm{C\,cm})^{\frac{1}{3}}$. With measured C_{gate}-V_{gs} characteristics Q_{ealp} and $T_{ox,eff}$ can be extracted (see Fig. 3.37) in practice. This extraction is performed at $V_{ds} = 0$ and results in position independent values for Q_b and Q_i. However, as can be seen from the above $T_{ox,eff}$ definition in Eq. (3.64), Q_b and Q_i are required to calculate $T_{ox,eff}$, while on the other hand, the Q_b and Q_i calculations require the knowledge of $T_{ox,eff}$. Therefore, the $T_{ox,eff}$ extraction procedure has to be carried out iteratively as depicted in Fig. 3.37.

The described iteration would required also during circuit simulations to estimate the correct quantum-mechanical effect, and is therefore not

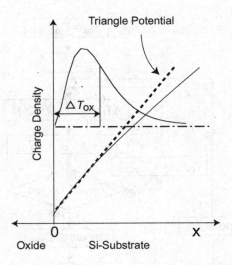

Fig. 3.36 Increase of the gate-oxide thickness ΔT_{ox} is obtained by averaging the wave function of the lowest energy level. This lowest energy level is determined on the basis of an approximate triangular potential distribution.

practical for a compact model. However, the characteristics observed in Fig. 3.35 suggests that the ΔT_{ox} dependence on V_{gs} can be approximated by the simple equation

$$\Delta T_{ox} = a(V_{gs} - V_{th} - b)^2 + c \tag{3.65}$$

where a, b, and c are fitting parameters and

$$V_{th} = 2\Phi_B + V_{fb} + \frac{T_{ox} + \Delta T_{ox}}{\epsilon_{ox}} q N_{sub} W_d \tag{3.66}$$

Here the V_{th} calculation requires again the knowledge of ΔT_{ox}. By substituting Eq. (3.66) into Eq. (3.65), ΔT_{ox} is obtained analytically after some simplifications as

$$\Delta T_{ox} = a \left(V_{gs} - V_{th}(T_{ox,eff} = T_{ox}) - b' \right)^2 + c' \tag{3.67}$$

Consequently, the final equations implemented into HiSIM for describing the quantum-mechanical effects are

$$T_{ox,eff} = T_{ox} + \Delta T_{ox} \tag{3.68}$$

$$\Delta T_{ox} = QME1 \left[V_{gs} - V_{th}(T_{ox,eff} = T_{ox}) - QME2 \right]^2 + QME3 \tag{3.69}$$

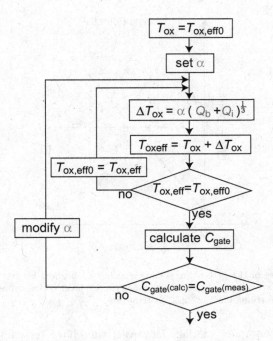

Fig. 3.37   Procedure required for exact extraction of the quantum-mechanical effect from the measured gate capacitance C_{gg}.

where $QME1$, $QME2$, and $QME3$ are the quantum-effect model parameters introduced in HiSIM. $QME1$ describes the magnitude of the quantum-mechanical effect. $QME2$ and $QME3$ describe the onset of the quantum-mechanical effect, which is near the threshold voltage, and the minimum ΔT_{ox}, respectively. Calculated C_{gate} with the quantum-mechanical effect is shown in Fig. 3.38 as a function of the gate voltage V_{gs}.

In our derivation of the above model equations all discussion has ignored the V_{ds} contribution. For real applications of the quantum-mechanical description under normal device operation, the quantization along the channel would have to be solved for the potential distribution along the channel with non-zero V_{ds}. Extension to such a more exact treatment, requires to solve the Schrödinger equation together with the Poisson equation in 2 dimensions and is not possible until now mainly due to the excessive simulation time problems [49, 50] A phenomenological method has been developed by introducing an energy-band bending at the oxide/silicon interface to take the repulsion of the channel charge distribution deeper into the silicon base

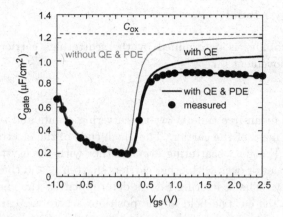

Fig. 3.38 Comparison of measured C-V characteristics with simulation results for different model complexities with and without quantum effect (QE) or poly-Si depletion effect (PDE).

layer into account [54]. This approximation shows that the longitudinal variation of the channel width from V_{ds} can be ignored [55], because the carrier density or current density is concentrated at the oxide/Si interface, most at the source, and spread deep into the base layer or bulk near the drain due to the smaller electric field or oxide voltage.

3.4   Mobility Model

The carrier mobility is determined by the scatterings carrier experience during their movement [37]

$$\mu = \frac{q\tau}{m} \tag{3.70}$$

where τ is the mean-free time between scattering events and m is the mobility effective mass of the carrier. The magnitude of the electron charge is denoted by q. Frequent scattering events either with the crystal lattice or ionized impurities or even with other carriers shorten τ, and thus reduce the carrier mobility. The scattering rate is determined by the carrier energy, obtained from the electric field at the positions where the carriers are located. Therefore μ is modeled usually as a function of the electric field, and the maximum carrier mobilities in Silicon are about $1450 \text{cm}^2/(\text{Vs})$ for electrons and $450 \text{cm}^2/(\text{Vs})$ for holes, respectively [57]. These mobility values are often referred as drift mobilities, determing the device characteristics mostly under strong inversion condition.

The MOSFET is mainly controlled by the vertical electrical field E_x in the channel region as shown in Fig. 3.39, which is determined by the gate voltage V_{gs} [56]. The drain voltage V_{ds} induces a field E_y along the channel from which the carriers get energy to flow between source and drain. Consequently, the field which the carriers experience has to be separated into these two components, namely the vertical field E_x and the longitudinal (lateral) field E_y [58–60]. Under the low V_{ds} condition, the effect of the longitudinal field on the carrier mobility can be ignored, and the carrier mobility is determined only by the vertical electric field. This case is referred as the low field mobility. The carrier mobility under a high longitudinal electric field is conventionally called the high field mobility.

3.4.1   *Low Field Mobility*

The existence of a universal relationship between the carrier mobility in the inversion layer of a MOSFET channel and the effective electric field, $E_{x,\text{eff}}$, or in short E_{eff}, perpendicular to the channel has been well confirmed experimentally for long-channel MOSFETs at small V_{ds} for the first time by Savinis and, Clemens for the first time [61–64]. This relationship is, as mentioned generally known as the universal mobility and provides an important link between the macroscopic carrier transport in MOSFETs and theoretical descriptions based on microscopic quantum physics. It thus as-

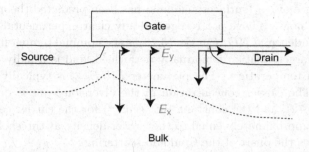

Fig. 3.39 Schematic diagram of the electric field components in the MOSFET channel. Under most operating conditions the vertical field E_x is large in comparison to the longitudinal field E_y.

sures the validity of the macroscopic carrier transport model, and the fact that the carrier mobility in the channel is indeed governed only by the electric field applied. The validity of the mobility universality for scaled MOSFETs down to below the 100nm L_{gate} node can be preserved in the compact model, if the additional effects that are observed in scaled MOS-FETs, like quantum or poly-depletion effects, are included in the model equations in a physically correct way [36]. This is a key to the successful application of the compact model and requires the precise calculation of the depletion-charge density Q_b and the inversion-charge density Q_i, which should be very much dependent on the fabrication technology.

The mobility in MOSFET channels is controlled by three dominating mechanisms: the Coulomb scattering by impurity ions, the phonon scattering, and the surface-roughness scattering. With the Matthiessen rule the low field mobility μ_0 is written as [43]:

$$\frac{1}{\mu_0} = \frac{1}{\mu_{CB}} + \frac{1}{\mu_{PH}} + \frac{1}{\mu_{SR}} \tag{3.71}$$

$$\mu_{CB}(\text{Coulomb}) = MUE_{cb0} + MUE_{cb1}\frac{Q_i}{q \times 10^{11}} \tag{3.72}$$

$$\mu_{PH}(\text{phonon}) = \frac{MUE_{ph1}}{(T/TNOM)^{MUE_{tmp}}E_{eff}^{MUE_{ph0}}} \tag{3.73}$$

$$\mu_{SR}(\text{surface roughness}) = \frac{MUE_{sr1}}{E_{eff}^{MUE_{sr0}}} \tag{3.74}$$

where MUE_{cb0}, MUE_{cb1}, MUE_{ph0}, MUE_{ph1}, MUE_{tmp}, MUE_{sr0}, and MUE_{sr1} are model parameters. $\mu_{PH}(\text{phonon})$ is dependent on the temperature T as can be seen in the corresponding Eq. (3.73). The temperature dependence of μ_{SR} (surfaceroughness) has been predicted theoretically [65], but until now we have not recognized any clear experimental evidence for this dependence. $TNOM$ in Eq. (3.73) is the nominal temperature at which device characterization is normally performed, and is usually chosen to be the room temperature. The parameter MUE_{tmp} is typically observed to be 1.6. The carrier concentration in the inversion layer is denoted as Q_i in Eq. (3.72), and the constant 10^{11} (cm^{-2}) for the carrier concentration, which is approximately equal to the carrier density at threshold condition, determines the onset of the Coulomb scattering.

Validity of the Matthiessen rule has been investigated by including the valley repopulation effect [63, 66]. It has been verified that the exponents of E_{eff} in the phonon and the surface-roughness terms are empirical constants in a narrow practical range of E_{eff} with the values of $MUE_{ph0} \simeq 0.3$, which comes from 2-D scattering in the electric field narrowed channel, and $MUE_{SR0} \simeq 2$ which comes from the proximity of the channel electrons to the localized scatterers. Furthermore, the effective field $E_{eff}(y)$ as a function of the position along the channel is related to depletion charge Q_b and inversion charge Q_i by the closed-form equation

$$E_{eff}(y) = \frac{1}{\epsilon_{Si}} [\gamma \cdot Q_b(y) + \eta \cdot Q_i(y)] \tag{3.75}$$

which is derived quantum mechanically with $\gamma=1$ and $\eta=11/32$ [41]. For n-MOSFETs $\eta=1/3$ is obtained, if the channel electrons occupy only the lowest subband (low temperature case). For the (100)-Si surface at room temperature, where channel electrons also populate higher subbands, η is predicted theoretically to be 1/2 by Matsumoto and Uemura [67], which has been verified experimentally by Lee et al [63]. For holes, $\eta=1/3$ has been observed [68].

The constant values of MUE_{ph0}, MUE_{sr0}, γ, and η, which are independent of N_{sub}, V_{bs} and other technology variations, constitute the mobility universality [61, 62]

$$MUE_{ph0} \simeq 0.3 \tag{3.76}$$
$$MUE_{sr0} = 2.0 \tag{3.77}$$
$$\gamma = 1.0 \tag{3.78}$$
$$\eta = 0.5 \tag{3.79}$$

as illustrated in Fig. 3.40. Increase of the mobility at smaller E_{eff} is due to a larger separation of the electrons from the scatterers. Increase of the mobility at larger E_{eff} comes from an increased carrier screening of the Coulomb scattering by the ions. Once the phonon scattering and the surface-roughness scattering dominate the mobility, a universal relationship is therefore observed, independent of the substrate impurity concentration N_{sub}. However, in actual applications of HiSIM, the parameters of Eqs.(8.62)-(8.65) are used for fitting purposes [69], if it is necessary. Furthermore, γ and η are renamed as model parameters $NDEP$ and $NINV$, respectively. Because of the aggressive reduction of T_{ox} in advanced technologies, quantum modification of the electric field in the oxide, E_{ox}, and roughness scattering from the top interface, the Gate-metal/oxide interface, become important. These effects can be modeled empirically with γ [70,71].

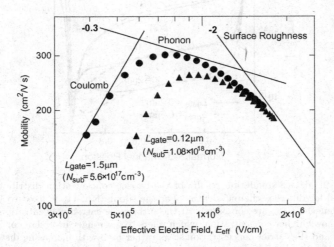

Fig. 3.40   Calculated mobility as a function of effective field E_{eff} for different MOSFET devices.

In real devices with very short L_{gate}, the conditions applied to derive the universal mobility description are not completely valid any more. Therefore, $NDEP$ and $NINV$ are treated as L_{gate} dependent in HiSIM

$$NDEP = NDEP \left(1 - \frac{NDEPL}{NDEPL + (L_{\text{gate}} \cdot 10^4)^{NDEPLP}} \right) \quad (3.80)$$

$$NINV = NINV \left(1 - \frac{NINVL}{NINVL + (L_{\text{gate}} \cdot 10^4)^{NINVLP}} \right) \quad (3.81)$$

The multiplication factor 10^4 of L_{gate} has the purpose to normalize the length units. The parameters $NDEPL$ and $NDEPLP$ in Eq. (3.80) can additionally be used to reproduce the V_{bs} dependence for short-channel MOSFETs. Carriers spread deeper into the bulk near the drain end of shorter-channel MOSFETs as clearly observable in the inversion charge density of Fig. 3.41, even at small V_{ds}. This has been known as "carrier sinking" and it further modifies E_{eff}.

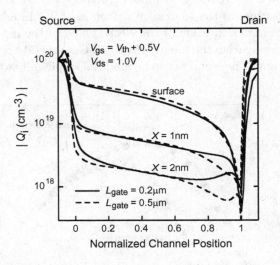

Fig. 3.41 2D-device simulation results of the carrier concentration distribution along the channel for different channel depths. For the simulation, V_{gs} is fixed to $V_{th}+0.5$V. Two gate lengths L_{gate} are compared. At the surface the carrier concentration for L_{gate} = 0.5μm shows higher concentration, which refers to a stronger inversion condition. On the other hand the carrier concentration stays rather high with increasing distance from the surface for the L_{gate} = 0.2μm case, especially near the drain region.

Due to the carrier flow at increasing distance from the surface with reducing L_{gate}, not only the electric field experienced by the carriers is different from the field in the long L_{gate} case, but also the effective impurity concentration can be different. This results in a modification of MUE_{PH1}, which is modeled as

$$MUE_{ph1} = MUE_{ph1} \times \left(1 + \frac{MUEPHL}{(L_{gate} \cdot 10^4)^{MUEPLP}}\right) \qquad (3.82)$$

The surface-roughness coefficient MUE_{SR0} includes also the gate length dependence

$$MUE_{sr0} = MUE_{sr0} \times \left(1 + \frac{MUESRL}{(L_{gate} \cdot 10^4)^{MUESLP}} \right) \qquad (3.83)$$

because the carrier flow deeper into the substrate causes a reduced surface-roughness scattering.

3.4.2 *High Field Mobility*

The high-field mobility is modeled phenomenologically as [72]

$$\mu = \frac{\mu_0}{\left(1 + \left(\frac{\mu_0 E_y}{V_{max}} \right)^{BB} \right)^{\frac{1}{BB}}} \qquad (3.84)$$

where the maximum drift velocity V_{max} is temperature dependent due to carrier scatterings by optical and intervalley phonon emissions as described and modeled in Section 3.8. BB is usually fixed to 2 for electrons and to 1 for holes. The longitudinal electric field E_y is derived from the calculated ϕ_S values at the source end and the drain end. The maximum velocity V_{max} should be the maximum electron-saturation velocity [73] ($\simeq 1 \times 10^7$cm/s [74] or $\simeq 6.5 \times 10^6$cm/s [75]), which is however exceeded at reduced L_{gate}. This phenomenon, called the velocity overshoot, is included in the mobility model in the following manner

$$V_{max} = V_{max} \cdot \left(1 + \frac{VOVER}{(L_{gate} \cdot 10^4)^{VOVERP}} \right) \qquad (3.85)$$

3.5   Channel-Length Modulation

When the drain voltage V_{ds} becomes larger, the gate-voltage control becomes weaker at the drain end of the channel. This region, where gate control of the inversion carrier density is lost and where the control of the carrier flow is by the voltage applied to the drain relative to the gate, is known as the pinch-off region in the junction-gate FET [73], in which the volume channel is actually pinched off to zero electrical width. There is no such reduction of the electrical width in the MOSFET surface channel, which is in fact widened to the back contact or the back gate junction. So, we should call this the depletion region, following the terminology for the reverse-biased p/n collector junction of the Bipolar Junction transistor. In this depletion region, the carrier drift velocity reaches the phonon-emission-scattering-limited saturation value, while the carrier concentration reduces to the value necessary to maintain the continuity of the drift current. However, due to the inefficient energy dissipation of intervalley phonon scattering at higher electric field, an actual increase of the velocity above the 'saturation value', known as velocity overshoot, is caused as the electric field increases further ($\simeq 1 \times 10^7 \mathrm{cm/s}$ [74] or $\simeq 6.5 \times 10^6 \mathrm{cm/s}$ [75]). The main effect on the MOSFET operation under this condition is a saturated drain current as shown in Fig. 3.42. In the surface-potential modeling a smooth transition from the linear range to the saturation range is achieved automatically due to the smooth transition of the surface potential ϕ_{SL} into the saturation as can be seen in Fig. 2.11. However, additional phenomena occurring in this depletion region after the end of the inverted channel region become obvious as the channel shortens.

A typical surface potential distribution along the channel under the saturation condition is shown in Fig. 3.43 [3, 37, 76]. In the following we will call the depletion region after the end of the inverted channel region by the name pinch-off region because this term has become widely used in the literature although it is not quite correct. Accurate treatment has been proposed by Sah's group [77].

For MOSFET modeling based on the surface potential, the gradual-channel approximation is applied to derive analytical equations for describing the MOSFET characteristics. Here, ϕ_{SL} determined as the surface-potential value at the drain end gives the value at the end of the gradual-channel approximation, namely at the starting point of the depletion region. A steep potential increase is clearly seen in Fig. 3.43 beyond this depletion point. Thus the gradual-channel approximation is not valid anymore for

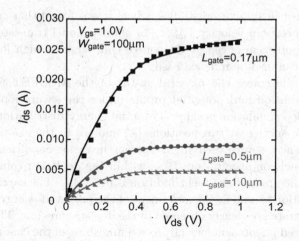

Fig. 3.42 I_{ds}-V_{ds} characteristics exhibiting I_{ds} saturation due to inversion-layer pinch-off and channel-length modulation.

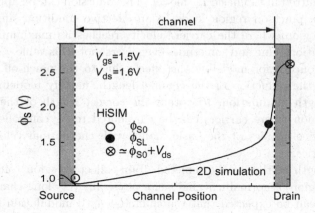

Fig. 3.43 Calculated surface potential by HiSIM at source end ϕ_{S0} and at drain ϕ_{SL} under the saturation condition. Simulated surface-potential distribution along the channel with a 2D-device simulator is also depicted for comparison.

large V_{ds}, inducing the depletion condition in the channel. The carrier velocity v ($= \mu E_y$, where μ is the carrier mobility and E_y is the electric field along channel.) also increases steeply in the depletion region due to the steep potential increase. It has been observed as the so called the velocity overshoot effect in simulation studies, where it was found that carriers can exceed the maximum velocity V_{max}. For short-channel transistors the velocity overshoot occurs even at the source end, where a high longitudinal electric field can also be induced easily.

Fig. 3.44 illustrates the physical status of the MOSFET for carrier-density distribution and potential profile under the saturation condition with 2D-device simulation results. Potential-energy distributions perpendicular to the surface at two positions [A] and [B] in the channel are depicted: [A] is near source verifying the strong inversion condition and valid gradual-channel conditions, and [B] is within the depletion (pinch-off) region, where the gradual-channel condition is not valid. The carrier concentration reduction as well as the movement of the potential-energy minimum into the substrate are·clearly observed in the depletion region. These effects are often viewed phenomenologically as a diminishing of the channel [78,79].

For compact modeling the drain current as a function of V_{ds} is described analytically under the gradual-channel approximation, and the pinch-off region is omitted in the basic I_{ds} model. This omission can be appropriate, because the pinch-off region leads to an effective condition similar to a depletion region, where the carrier velocity reaches its maximum steady-state saturation value and sometimes even overshoots this value. Therefore, the main concern for modeling the effects due to the pinch-off region is to capture the reduction of the channel length, namely modeling of the channel-length modulation, to obtain the correct length of the inverted channel region where carriers can be considered to be controlled by the normal device physics of the basic model with the normal velocity–field characteristics.

The length of the nearly depleted drain-collector region (often called pinch-off region) is normally negligible for physically very long-channel transistors. It can be expected that the channel-length modulation (CLM) is dependent on bias conditions as well as on the channel/drain junction profile. Indeed, the gradual increase of the drain current I_{ds} under the saturation condition (see Fig. 3.42) is attributed to the V_{ds} dependence of the CLM effect. Therefore, the CLM is expected to be observed more clearly in the channel conductance g_{ds} ($= \partial I_{ds}/\partial V_{ds}$). Fig. 3.45 compares 2D-device simulation results of g_{ds} for two different gate lengths L_{gate}. The saturation behavior of g_{ds} in Fig. 3.45a for a technology without pocket implantation is quite different for these two L_{gate} values. It reduces for increased V_{ds} values by 3 orders of magnitude from the $V_{ds}=0$ value in the long-channel

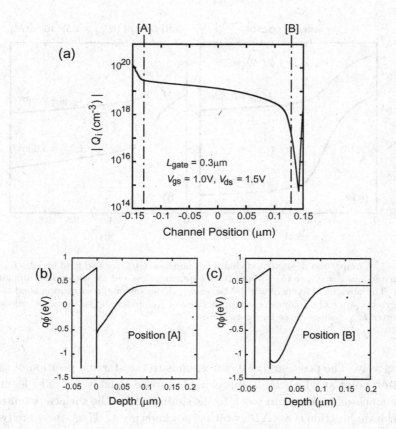

Fig. 3.44 (a) shows the carrier density distribution along the channel. (b) and (c) are giving the potential distributions along the depth direction at the positions [A] and [B], as depicted in (a), respectively.

case with $L_{\text{gate}} = 3\mu m$, whereas it stays at a relatively large value in the short-channel case with $L_{\text{gate}} = 0.2\mu m$. This difference has to be captured in the compact model accurately. Fig. 3.45b shows the simulation results for an advanced technology with a pocket implantation. Since the pocket implantation induces potential minima at source and drain (see Fig. 3.17) and because the drain end minimum increases with larger V_{ds}, this leads to relative large g_{ds} even in the long-channel case.

The position of the pinch-off point in the channel is modeled using the Gauss law in the pinch-off region as schematically shown in Fig. 3.46 [5, 80], where E_y and E_x denote the longitudinal and the transverse electric field,

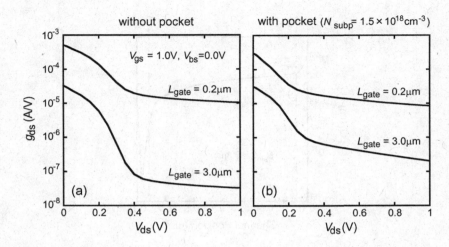

Fig. 3.45   Comparison of simulated channel conductance g_{ds} for two gate lengths L_{gate} = $3\mu m$ and $L_{gate} = 0.2\mu m$ (a) without pocket implantation and (b) with pocket implantation. The obvious differences are the larger g_{ds} values under the saturation condition for shorter L_{gate}. The stronger gradual increase of I_{ds} under the saturation condition for shorter L_{gate} causes these larger g_{ds} values.

respectively. The position $y = 0$ corresponds to the starting position of the pinch-off point at which E_y is approximated to be equal to E_x. The length of the pinch-off region from $y = 0$ to the drain is ΔL. The surface potential at the drain junction is $\phi_S(\Delta L)$, which is not known yet. Thus, there are two unknown values, namely as ΔL and $\phi_S(\Delta L)$, which have to be determined. Following the conventional modeling approach, we derive an equation by applying the Gauss law in the ΔL region under the assumption that the longitudinal electric field E_y and the carrier density are homogeneously distributed along the depth direction as well as along ΔL, respectively [5,81]

$$W_d \frac{dE_y(y)}{dy} + \frac{\epsilon_{ox}}{\epsilon_{Si}} E_x(y) = \frac{1}{\epsilon_{Si}}(qN_{sub}W_d + Q_i') \qquad (3.86)$$

$$E_x(y) = \frac{V_{gs} - V_{fb} - \phi_S(y)}{t_{ox}}$$

Here W_d is the depletion width and Q_i' is the integrated carrier density along the inversion-layer thickness direction, which corresponds to the charge density under the charge-sheet approximation.

Usually a closed-form solution for ΔL can be derived with the boundary

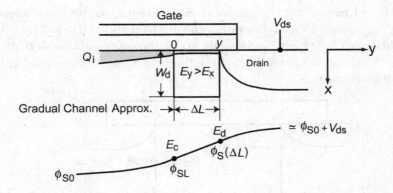

Fig. 3.46 Schematic showing the correlation among physical quantities in the pinch-off region.

conditions of $\phi_S(0)=V_{\mathrm{dsat}}$ and $\phi_S(\Delta L)=V_{\mathrm{ds}}$ as

$$\Delta L = l \ln\left[\frac{V_{\mathrm{ds}} - V_{\mathrm{dsat}}}{lE_{\mathrm{c}}} + \frac{E_{\mathrm{d}}}{E_{\mathrm{c}}}\right] \tag{3.87}$$

$$l^2 = \frac{\epsilon_{\mathrm{ox}}}{\epsilon_{\mathrm{Si}}} t_{\mathrm{ox}} W_{\mathrm{d}} \tag{3.88}$$

$$E_{\mathrm{d}}^2 = E_{\mathrm{c}}^2 + \left(\frac{V_{\mathrm{ds}} - V_{\mathrm{dsat}}}{l}\right)^2 \tag{3.89}$$

where V_{dsat} and E_{c} are the potential value and the longitudinal (lateral) electric field at $y=0$, respectively. However, $\phi(\Delta L)=V_{\mathrm{ds}}$ is not true under the saturation condition, because the potential usually reaches V_{ds} in the overlap region beneath the gate as schematically shown in Fig. 3.46. We therefore consider the potential reduction at $y=\Delta L$ explicitly, namely as $\phi_S(\Delta L)$ which is smaller than V_{ds}.

The value of $\phi_S(\Delta L)$ is known to lie between ϕ_{SL} and $\phi_{\mathrm{S0}} + V_{\mathrm{ds}}$ as schematically shown in Fig. 3.47. Its specific magnitude is expected to depend on the junction profile between channel and drain contact. The described features of $\phi_S(\Delta L)$ are modeled with a channel-length-modulation (CLM) parameter $CLM1$ as

$$\phi_S(\Delta L) = CLM1(\phi_{\mathrm{S0}} + V_{\mathrm{ds}}) + (1 - CLM1)\phi_{\mathrm{SL}} \tag{3.90}$$

$CLM1$ can be interpreted to represent the hardness of the channel/drain junction and must fulfill the condition $0 \leq CLM1 \leq 1$. The condition of

$CLM1 = 1$ means that the contact profile is abrupt, and potential increase from ϕ_{SL} to $\phi_{S0} + V_{ds}$ occurs within the pinch-off region $0 \leq y \leq \Delta L$. The condition of $CLM1 = 0$ corresponds to the opposite condition so that $\Delta L = 0$ is valid and the potential increase from ϕ_{SL} to $\phi_{S0} + V_{ds}$ occurs within the drain contact.

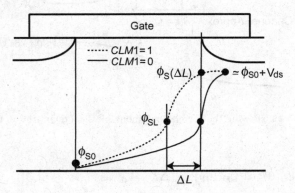

Fig. 3.47   Schematic of the surface-potential distribution along the channel.

To derive a closed form for ΔL in the case of the more general boundary condition of Eq. (3.90), the approximation that $E_x(y)$ is smaller than $E_y(y)$ in the ΔL region is introduced. As a result a simplified description is obtained as

$$\oint \boldsymbol{E} dS = -\frac{1}{\epsilon_{Si}}(qN_{sub}W_d + Q_i')\Delta L \tag{3.91}$$

$$\simeq W_d(E_c - E_d)$$

where E_d is the longitudinal electric field at $y = \Delta L$. An explicit analytical expression for E_d can be also derived under the assumption used for Eq. (3.91). Starting point is the Poisson equation

$$\frac{d^2\phi_S}{dy^2} = \frac{1}{\epsilon_{Si}}(qN_{sub} + Q_i) \tag{3.92}$$

where Q_i is the carrier density in the pinch-off region. By integrating Eq. (3.92) from $y=0$ to $y = \Delta L$, we obtain

$$E_d^2 - E_c^2 = \frac{2}{\epsilon_{Si}}(qN_{sub} + \frac{Q_i'}{W_d})(\phi_S(\Delta L) - \phi_S(0)) \tag{3.93}$$

where Q_i' is the integrated carrier density in the pinch-off region. This can be further simplified

$$E_d^2 \simeq E_c^2 + \frac{2qN_{\text{sub}}}{\epsilon_{\text{Si}}}(\phi_S(\Delta L) - \phi_S(0)) \tag{3.94}$$

where $\phi_S(0) = \phi_{\text{SL}}$. The neglection of $E_x(y)$ in Eq. (3.86) results in a linear dependence of E_d^2 on the potential difference in the pinch-off region, whereas the inclusion of $E_x(y)$ results in a quadratic dependence as can be seen in Eq. (3.89). The steep potential increase occurs only at drain in a small region, and if the longitudinal electric field exceeds the value sustained at drain, the potential increase is distributed into the channel. Therefore, the influence of the different E_d dependences on the potential difference is expected to be small.

The validity of the gradual-channel approximation at $y = 0$ is exploited to obtain an analytical description of E_c in the framework of HiSIM as

$$E_c = \frac{I_{\text{dd}}}{\beta(L_{\text{eff}} - \Delta L)Q_i'} \tag{3.95}$$

$$\begin{aligned} I_{\text{dd}} &= C_{\text{ox}}(\beta V_G' + 1)(\phi_{\text{SL}} - \phi_{\text{S0}}) - \frac{\beta}{2}C_{\text{ox}}(\phi_{\text{SL}}^2 - \phi_{\text{S0}}^2) \\ &\quad - \frac{2}{3}const0\left[\{\beta(\phi_{\text{SL}} - V_{\text{bs}}) - 1\}^{\frac{3}{2}} - \{\beta(\phi_{\text{S0}} - V_{\text{bs}}) - 1\}^{\frac{3}{2}}\right] \\ &\quad + const0\left[\{\beta(\phi_{\text{SL}} - V_{\text{bs}}) - 1\}^{\frac{1}{2}} - \{\beta(\phi_{\text{S0}} - V_{\text{bs}}) - 1\}^{\frac{1}{2}}\right] \end{aligned}$$

$$V_G' = V_{\text{gs}} - V_{\text{fb}} - \Delta V_{\text{th}} \tag{3.96}$$

$$const0 = \sqrt{\frac{2\epsilon_{\text{Si}}qN_{\text{sub}}}{\beta}} \tag{3.97}$$

where I_{dd} is the integrated product of the carrier concentration and the electric field, determining the drain current as (see Eq. (2.58))

$$I_{\text{ds}} = \frac{W_{\text{eff}}}{L_{\text{eff}} - \Delta L}\frac{\mu I_{\text{dd}}}{\beta} \tag{3.98}$$

After integrating the Poisson equation of Eq. (3.92) in the ΔL region under neglection of the vertical electric field E_x, we obtain [69]

$$\Delta L = \epsilon_{\text{Si}}\frac{E_d - E_c}{qN_{\text{sub}} + Q_i'/W_d} \tag{3.99}$$

where E_d is determined in the simple form of Eq. (3.93)

The final equation for ΔL is obtained by solving Eqs. (3.99), (3.94) and (3.95), simultaneously. The dependence of E_c on ΔL, which becomes more significant for sub-100nm devices, is consistently included in the calculation. Thus the equation for ΔL is reduced to

$$(\Delta L)^2 + \frac{1}{L_{eff}}\left(2\frac{I_{dd}}{\beta Q_i'}z + 2\frac{N_{sub}}{\epsilon_{Si}}(\phi_S(\Delta L) - \phi_{SL})z^2 + E_0 z^2\right)\Delta L$$
$$-\left(2\frac{N_{sub}}{\epsilon_{Si}}(\phi_S(\Delta L) - \phi_{SL})z^2 + E_0 z^2\right) = 0 \qquad (3.100)$$

by taking into account only up to quadratic terms of ΔL, and the final ΔL is written as

$$\Delta L = \frac{1}{2}\left[-\frac{1}{L_{eff}}\left(2\frac{I_{dd}}{\beta Q_i}z + 2\frac{N_{sub}}{\epsilon_{Si}}(\phi_S(\Delta L) - \phi_{SL})z^2 + E_0 z^2\right)\right.$$
$$+ \left\{\frac{1}{L_{eff}^2}\left(2\frac{I_{dd}}{\beta Q_i}z + 2\frac{N_{sub}}{\epsilon_{Si}}(\phi_S(\Delta L) - \phi_{SL})z^2 + E_0 z^2\right)^2\right.$$
$$\left.\left.- 4\left(2\frac{N_{sub}}{\epsilon_{Si}}(\phi_S(\Delta L) - \phi_{SL})z^2 + E_0 z^2\right)\right\}^{\frac{1}{2}}\right] \qquad (3.101)$$

where z is given by

$$z = \frac{\epsilon_{Si}}{CLM2 \cdot Q_b + CLM3 \cdot Q_i} \qquad (3.102)$$

and E_0, fixed to 10^5, refers to the longitudinal electric field at the pinch-off point, E_c. The simplification of a constant depletion width W_d in the ΔL region, which was applied to derive Eq. (3.89), might induce an inaccuracy. Therefore, the two model parameters $CLM2$ and $CLM3$ are introduced to consider the uncertainty of Q_i in the pinch-off region and to counterbalance the two contributions from Q_b ($= qN_{sub}W_d$) and Q_i. Though ΔL is determined mostly by $\phi_S(\Delta L)$, namely $CLM1$, the combination between $CLM2$ and $CLM3$ slightly influences device characteristics such as g_{ds} as well as the overlap capacitance and the longitudinal-field-induced capacitance C_{Q_y} described in the Sections 4.2 and 4.3, respectively.

Simulated g_{ds} results are compared in Fig. 3.48 with measurements for two different gate lengths, (a) L_{gate} of $10\mu m$ and (b) L_{gate} of $0.1\mu m$, fabricated by a $0.1\mu m$ pocket-implant technology. The influence of the channel-length modulation on the g_{ds} value in the saturation region is visible for the long-channel case, but is as expected very small. For the short-channel case, on the other hand, the channel-length modulation (CLM) effect is strongly enhanced. However, not only the enhanced channel-length-modulation effect, but also the short-channel effect of a V_{ds} dependent threshold voltage influences the g_{ds} characteristics [82]. Fig. 3.48b shows these two contributions (for g_{ds}) separately. It is commonly believed that the channel-length modulation is the main cause for the large g_{ds} in the saturation range of short-channel transistors [83]. However, the calculation results of Fig. 3.48b with HiSIM show that the short-channel effect is the dominant origin for the large g_{ds} even in the saturation region. The channel-length modulation occurs always under the saturation condition for any channel lengths. However, the effect is not obvious for long-channel transistors, because the modulation length is relatively short in comparison to the channel length. Whereas the short-channel effect occurs only when the drain voltage influences not only on the channel potential at the drain end but also on that at the source end. This happens when the channel length becomes short and the potential increase at the drain extends to the souce side. The feature of the saturation behavior is more visible in the output resistance R_{out}, which is determined as

$$R_{out} = \frac{1}{g_{ds}} \tag{3.103}$$

The large g_{ds} value degrades the desired high output resistivity of MOSFETs under the saturation condition as shown in Fig. 3.49.

Different from conventional models, HiSIM solves the surface potentials with the inclusion of the longitudinal electric field in ΔV_{th} according to Eq. (2.17). Therefore, the high field effect is automatically included in the surface potential, which results together with the channel-length-modulation, in the large g_{ds}. Fig. 3.50 compares calculated ΔL by the described model and by a 2D-device simulator for different gate lengths and applied biases. The parameters used in the HiSIM calculation are extracted from I-V measurements. On the other hand, the device profile used in the 2D-device simulation is constructed to reproduce measured I-V characteristics. In the 2D-device simulation, ΔL is extracted directly as the length between the pinch-off point and the channel/drain contact. The pinch-off point is determined as the position where the sign of E_x at the channel surface changes. It is seen that the magnitude ΔL of the channel-length modulation is at maximum about 20nm for the 0.1μm pocket-implant technology.

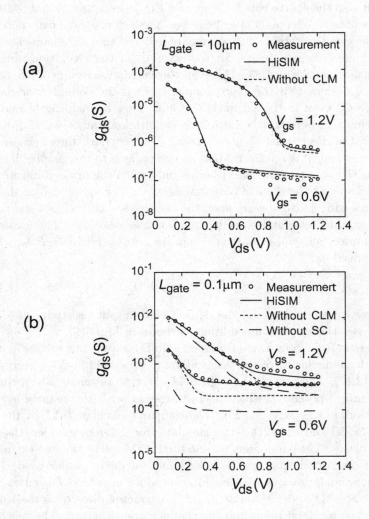

Fig. 3.48 g_{ds} modeling results as compared to measurement for (a) $L_{gate} = 10\mu m$ and (b) $L_{gate} = 0.10\mu m$. SC denotes the previously described short-channel effects. SC relates the longitudinal-field gradient to the reduction of the threshold-voltage at reduced L_{gate}. The effect of the channel-length modulation (CLM) is small for large L_{gate}, but vital at reduced L_{gate}.

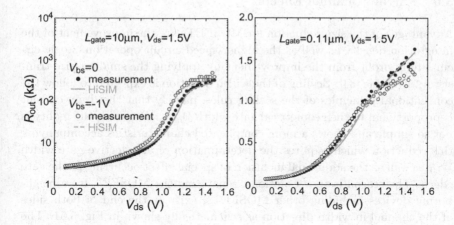

Fig. 3.49 Comparison of calculated channel resistance $R_{\text{out}} = g_{\text{ds}}^{-1}$ with measurements. A long and a short gate lengths are compared.

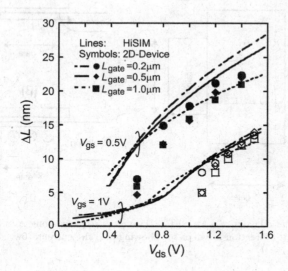

Fig. 3.50 Calculated ΔL with HiSIM for different gate lengths and bias conditions as compared with extracted values from a 2D-device simulations. The pinch-off point is determined as the position where the sign of E_{x} at the channel surface changes in the 2D-device simulation.

3.6   Narrow-Channel Effects

Advantages of the size reduction for MOSFETs are the improvement of the integration density as well as the higher-speed circuit operation. Logic circuits fully profit from the improvement by applying the smallest-size transistors available [84]. Scaling of the width direction is expected to follow the conventional principles of the scaling rules, namely that the drain current is proportional to the geometrical gate width W_{gate}. However, the reality is not so simple and shows a more complicated characteristics accompanying this reduction which requires the determination of an effective gate width W_{eff} as well as the additional inclusion of special effect occurring at the gate edges. The reason is that isolations, separating the MOSFET from neighboring devices including other MOSFETs, exist at the end of both sides of the channel in width direction as schematically shown in Fig. 3.51. The

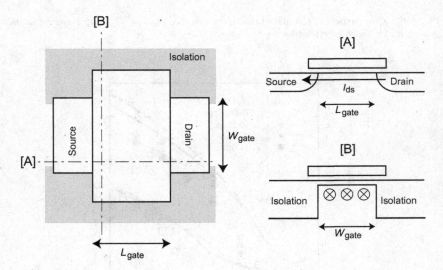

Fig. 3.51   Schematic layout of a MOSFET. Two cross-sections along A in length direction and B in width direction are depicted showing also the current flow.

electrical properties of the MOSFET edge at the interface to the type of the isolation can be expected to be different from the inner MOSFET part and may show additional dependencies on the isolation technology. Two major isolation technologies have been developed and applied in practice, the local oxidation of silicon (LOCOS) technology [85] and the shallow-trench-isolation technology (STI) technology [86]. Since the STI technology is developed for aggressive scaling purposes, the main focus is given here on this technology. Nevertheless, the developed narrow-channel-effect model is also applicable for the LOCOS technology.

From the modeling point of view there are four main aspects to be considered:

- Threshold Voltage Shift
- Mobility Change
- Leakage Current due to Shallow-Trench Isolation (STI)
- Small-Geometry Effects

3.6.1 *Threshold Voltage Shift*

The threshold voltage shift can be understood by considering additional capacitance contributions crowded at edges in width direction as schematically shown in Fig. 3.52.

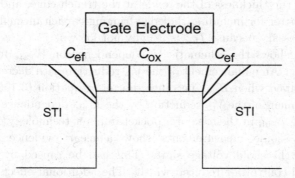

Fig. 3.52 Schematic of the gate capacitance distribution along the width direction for a shallow-trench isolation technology.

The threshold voltage V_{th} is conventionally written as [37, 87–90]

$$V_{\text{th}} = V_{\text{fb}} + 2\Phi_{\text{B}} + \frac{\sqrt{2qN_{\text{sub}}\epsilon_{\text{Si}}(2\Phi_{\text{B}} - V_{\text{bs}})}}{C_{\text{ox}}} \tag{3.104}$$

where $2\Phi_{\text{B}}$ is the surface potential value giving the threshold condition. For the STI case, C_{ox} is consequently replaced by

$$C_{\text{ox,STI}} = C_{\text{ox}} + \frac{2 \cdot C_{\text{ef}}}{L_{\text{eff}} W_{\text{eff}}} \tag{3.105}$$

to include the edge-fringing capacitances C_{ef}, where C_{ef} can be determined only as an integrated value. As a result, the shallow-trench-isolation technology induces a corresponding V_{th} reduction for reduced gate width (W_{gate}). In a surface-potential-based model, V_{th} does not appear explicitly in the model equations, so that the threshold voltage shift from the threshold-voltage value of a relatively wide transistor $\Delta V_{\text{th,W}}$ is modeled as

$$\Delta V_{\text{th,W}} = \left(\frac{1}{C_{\text{ox}}} - \frac{1}{C_{\text{ox}} + 2C_{\text{ef}}/(L_{\text{eff}}W_{\text{eff}})} \right) qN_{\text{sub}}W_{\text{d}} \qquad (3.106)$$

where

$$C_{\text{ef}} = \frac{2\epsilon_{\text{ox}}}{\pi} L_{\text{eff}} \ln \left(\frac{2T_{\text{fox}}}{T_{\text{ox}}} \right) = \frac{WFC}{2} L_{\text{eff}} \qquad (3.107)$$

$$W_{\text{d}} = \sqrt{\frac{2\epsilon_{\text{Si}}(2\Phi_{\text{B}} - V_{\text{bs}})}{qN_{\text{sub}}}} \qquad (3.108)$$

Here, T_{fox} is the thickness of the oxide at the trench edge, and WFC is a model parameter for including the edge-fringing-capacitance effects, which becomes necessary because T_{fox} is usually not known.

Fig. 3.53 shows the schematic V_{th} dependence on W_{gate} for two gate lengths, L_{gate}. An enhancement of the V_{th} reduction with decreasing W_{gate} is often observed when L_{gate} becomes shorter. Equation (3.106) describes this enhancement of the V_{th} reduction by the L_{gate} dependence of the substrate doping N_{sub} in the case of a pocket-implant technology.

However, some measurements show a clear evidence of a L_{eff}-independent threshold voltage shift. This can be caused by insufficient technological control for reduced width. The additional effect is modeled by introducing another model parameter $WVTH0$ in the form

$$\Delta V_{\text{th,W}} = \left(\frac{1}{C_{\text{ox}}} - \frac{1}{C_{\text{ox}} + 2C_{\text{ef}}/(L_{\text{eff}}W_{\text{eff}})} \right) qN_{\text{sub}}W_{\text{d}} + \frac{WVTH0}{W_{\text{gate}} \cdot 10^4}$$
$$(3.109)$$

The enhancement of the threshold-voltage shift due to the narrow-channel effect for reduced channel length becomes more pronounced for the pocket technology. The additional pocket related effect, is included in the model as a further extension, which modifies the pocket impurity concentration to become a function of the width

$$N_{\text{subp}} = NSUBP \cdot \left(1 + \frac{NSUBP0}{(W_{\text{gate}})^{NSUBWP}} \right) \qquad (3.110)$$

The final ΔV_{th} (see Eq. (2.17)), under inclusion of the narrow-channel effects, becomes then

$$\Delta V_{\text{th}} = \Delta V_{\text{th,SC}} + \Delta V_{\text{th,R}} + \Delta V_{\text{th,P}} + \Delta V_{\text{th,W}} - \phi_{\text{Spg}} \qquad (3.111)$$

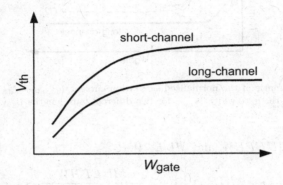

Fig. 3.53 Schematic of the V_{th} characteristic as a function of the gate width W_{gate} for two different gate lengths, L_{gate}.

3.6.2 *Mobility Modification due to a Narrow Gate*

Fig. 3.54 shows a schematic of the measured normalized-saturation current as a function of W_{gate} often observed in 100nm and shorter gate-length technologies. Reduction of the saturation current $I_{\text{ds,sat}}$ is generally not obvious for the long-channel case, whereas it is quite pronounced for the short-channel case. It is also known that the shallow-trench isolation induces mainly compressive mechanical stress in the channel as depicted in Fig. 3.55 due to the larger lattice constant of SiO_2 in comparison to that of Si. This stress results in a degradation of the carrier mobility [91], and causes a reduction of $I_{\text{ds,sat}}$ with reducing W_{gate} as indicated by the curve C1 for this mobility degradation case due to compressive stress. The magnitude of the degradation or the degradation-causing stress is very much dependent on the technology used. The stress is easily relaxed for a larger gate area because the stress is highly localized at the edges or the two width sides of the gate area, so that a diminished stress effect is observed for the narrow gate but also long gate or long channel transistors. The described effect is modeled as a decreasing phonon mobility with two model

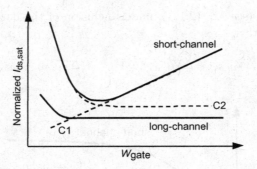

Fig. 3.54   Schematic of the normalized saturation current $I_{ds,sat}$ $(=I_{ds,sat} \cdot L_{eff}/W_{eff})$ as a function of the gate width W_{gate} for two different gate lengths L_{gate}.

parameters $MUEPHW$ and $MUEPWP$

$$MUE_{ph1} = MUE_{ph1} \left(1 - \frac{MUEPHW}{(W_{gate} \cdot 10^4)^{MUEPWP}} \right) \qquad (3.112)$$

However, the I_{ds}-W_{gate} characteristic does not always show monotonous decrease but sometimes starts to increase for narrower W_{gate} as indicated by the curve C2. This can be understood with the assumption that high applied stress induces a non-flat surface causing a carrier flow deeper in the substrate. As a result a reduction of the surface-roughness scattering can be expected. Consequently, this additional effect of an $I_{ds,sat}$ increase for small W_{gate} is modeled as a change of the surface-roughness contribution to the mobility with model parameters $MUESRW$ and $MUESWP$ as

$$MUE_{sr0} = MUE_{sr0} \left(1 + \frac{MUESRW}{(W_{gate} \cdot 10^4)^{MUESWP}} \right) \qquad (3.113)$$

Advanced technologies exploit stress effects actively and in a positive way. It has been discovered that the uniaxial stress induced with a strained-Si technology can enhance the p-MOSFET performance [92] by more than 50%. The uniaxial mechanical strain removes the valence band degeneracy, decreasing the hole effective masses of the originally degenerate valence bands. Lighter holes give a lower final density of state, hence lower the scattering rate of holes to the final states, which results in the observed higher hole mobility [93]. This effect of a stress-enhanced carrier mobility can be included by adjusting $MUEPHW$ to the actually measured drain current.

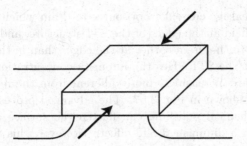

Fig. 3.55 Schematic of the shallow-trench-isolation stress, which is mainly compressive in the width direction.

3.6.3 *Leakage Current due to STI Technology*

The shallow-trench isolation (STI) can induce also an undesired hump in the subthreshold region of the I_{ds}-V_{gs} characteristics as shown in Fig. 3.56. The observed behavior is that such a hump becomes increasingly obvious for large $|V_{bs}|$ (with negative sign for n-MOSFET and positive sign for p-MOSFET, so that the depletion width becomes thicker) in the case of an STI technology. By increasing $|V_{bs}|$ the threshold voltage of the regular (non-parasitic) part of the MOSFET increases faster than the threshold voltage of the parasitic edge part, determining the position of the hump, which makes the hump more obvious. The observed hump structure is

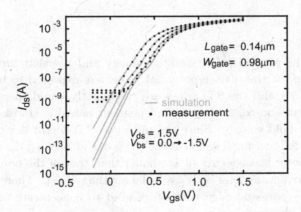

Fig. 3.56 Measured *I-V* characteristics of a MOSFET fabricated with STI technology as a function of V_{gs}. A current hump is observed in sub-threshold characteristics, which becomes more pronounced with larger V_{bs}.

explained by a leakage current from source to drain which is due to an increased electric field at the edges of the STI structure, and which turns on the current flow to the drain earlier at the edges than in the regular MOSFET [94, 95]. At the STI edges the impurity concentration as well as the oxide thickness are increasingly more different from the regular MOSFET as schematically shown in Fig. 3.57. The advanced processing technology of chemical-mechanical polishing can flatten the surface after the oxidation, resulting in a diminished STI effect. However, this STI effect of an additional subthreshold leakage-current generation is not completely under control yet.

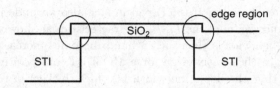

Fig. 3.57   Schematic of the cross-section of a MOSFET, fabricated with STI technology, along the width direction. The edge-region structure with uneven geometry of the layer interfaces is responsible for the observed narrow channel effects.

Due to the applied processing technology and the structural changes at the STI edges, the surface potential values are expected to be different in their vicinity and the STI edges are cosequently found to constitute a leakage-current source, namely an independent parasitic transistor with a reduced threshold voltage. Since the influence of different device parameters is restricted to a small region at the edges of width W_{STI}, the magnitude of this perimeter leakage current is smaller than that of the non-perimeter MOSFET current, as long as W_{gate} is larger than W_{STI}. Therefore, modeling of these parasitic edge transistors need to concentrate only on the subthreshold characteristics.

Instead of solving the Poisson equation iteratively, analytical equations for calculating the surface potential are introduced for these parasitic STI-edge transistors. Solving the Poisson equation using the Gauss law along the x-axis gives the well-known voltage equation [82]

$$C_{\text{ox}}(V'_{\text{G}} - \phi_{\text{S}}(y)) = \sqrt{\frac{2\epsilon_{\text{Si}}qN_{\text{sub}}}{\beta}} \cdot \left[\beta(\phi_{\text{S}}(y) - V_{\text{bs}}) - 1\right.$$

$$\left. + \frac{n_{\text{p0}}}{p_{\text{p0}}}\left\{\exp\left(\beta(\phi_{\text{S}}(y) - \phi_{\text{f}}(y))\right) - \exp\left(\beta(V_{\text{bs}} - \phi_{\text{f}}(y))\right)\right\}\right]^{\frac{1}{2}}$$

(3.114)

The equation (3.114) can be simplified under two limiting conditions [82], namely the depletion condition and the strong-inversion condition, which derive analytical descriptions for these conditions:

- Depletion Condition ignoring the inversion term

$$\phi_{\text{S,STI}} = V'_{\text{gs,STI}} + \frac{\epsilon_{\text{Si}}Q_{\text{STI}}}{C'^{2}_{\text{ox}}}\left[1 - \sqrt{1 + \frac{2C'^{2}_{\text{ox}}}{\epsilon_{\text{Si}}Q_{\text{STI}}}\left(V'_{\text{gs,STI}} - V_{\text{bs}} - \frac{1}{\beta}\right)}\right]$$

(3.115)

- Strong-Inversion Condition ignoring the depletion term

$$\phi_{\text{S,STI}} = \frac{\ln\left[\left(\frac{C'_{\text{ox}}}{const0'}\right)^{2}\frac{n_{\text{p0}}}{n_{\text{p0}}}V'^{2}_{\text{gs,STI}}\right]}{\beta + \frac{2}{V'_{\text{gs,STI}}}}$$

(3.116)

The used terms Q_{STI}, $V_{\text{gs},STI}$ and $const0'$ in Eqs. (3.115) and (3.116) are defined as

$$Q_{\text{STI}} = q \cdot N_{\text{STI}} \cdot W_{\text{d,STI}}$$ (3.117)

$$V'_{\text{gs,STI}} = V_{\text{gs}} - V_{\text{fb}} + VTHSTI + \Delta V_{\text{th,SCSTI}}$$ (3.118)

$$const0' = \sqrt{\frac{2\epsilon_{\text{Si}}qN_{\text{STI}}}{\beta}}$$ (3.119)

and

$$\Delta V_{\text{th,SCSTI}} = \frac{\epsilon_{\text{Si}}}{C'_{\text{ox}}}W_{\text{d,STI}}\frac{dE_{y}}{dy}$$ (3.120)

describes the short-channel effect in the STI parasitic part at the width edges. Here, $W_{\text{d,STI}}$ is the depletion-layer thickness written as

$$W_{\text{d,STI}} = \sqrt{\frac{2\epsilon_{\text{Si}}(2\Phi_{\text{B,STI}} - V_{\text{bs}})}{qN_{\text{STI}}}}$$ (3.121)

To reduce the number of model parameters, the gate capacitance C'_{ox} is considered to be the same as C_{ox} and only N_{STI} is distinguished from N_{sub}. The parameter $VTHSTI$ describes the threshold voltage shift of the STI transistor, originating from all other technology related parameters, in a summarized form.

$\dfrac{dE_y}{dy}$ is described with model parameters in the same form as in section 3.1.1 on the short-channel effects with model parameters $SCSTI1$, $SCSTI2$, and $SCSTI3$ as

$$\frac{dE_y}{dy} = \frac{2(V_{bi} - 2\Phi_{B,STI})}{(L_{gate} - PARL2)^2}$$
$$\cdot \left(SCSTI1 + SCSTI2 \cdot V_{ds} + SCSTI3 \cdot \frac{2\Phi_{B,STI} - V_{bs}}{L_{gate}} \right)$$

$$\tag{3.122}$$

$$2\Phi_{B,STI} = \frac{2}{\beta} \ln \left(\frac{N_{STI}}{n_i} \right) \tag{3.123}$$

The two limiting solutions of $\phi_{S,STI}$ are combined with a mathematical smoothing function

$$y = y_2 - \frac{1}{2} \left(y_{12} + \sqrt{y_{12}^2 + 4\delta y_2} \right) \tag{3.124}$$

$$y_{12} = y_2 - y_1 - \delta \tag{3.125}$$

where y_1 and y_2 are the limiting solutions given in Eqs. (3.115) and (3.116), and δ is fixed to 2×10^{-3}.

The modeling of the parasitic-transistor leakage for STI technologies is done by considering only on the subthreshold region, where the current is governed mostly by the diffusion term. In the subthreshold region the potential value ϕ_S is small and its variation along the length of the channel is even much smaller and therefore neglected, namely $\phi_{SL} - \phi_{S0} \simeq 0$ is approximated. Under this condition the current equation can be simplified as

$$I_{ds,STI} = 2\frac{W_{STI}}{L_{eff} - \Delta L}\mu\frac{Q_{i,STI}}{\beta}\left[1 - \exp(-\beta V_{ds})\right] \tag{3.126}$$

where W_{STI} determines the width of the high-field edge region, and is expected to be much smaller than W_{gate}. The carrier concentration $Q_{i,STI}$ is also calculated in a simplified form as

$$Q_{i,\text{STI}} = \sqrt{\frac{2\epsilon_{\text{Si}} q N_{\text{STI}}}{\beta}} \left\{ \left[\beta(\phi_{\text{S,STI}} - V_{\text{bs}}) - 1 + \frac{n_{\text{p0}}}{p_{\text{p0}}} \exp(\beta \phi_{\text{S,STI}}) \right]^{\frac{1}{2}} - \sqrt{\beta(\phi_{\text{S,STI}} - V_{\text{bs}}) - 1} \right\} \tag{3.127}$$

The gate length dependence of W_{STI} is furthermore included for the technology cases, where it is needed as

$$W_{\text{STI}} = WSTI \left(1 + \frac{WSTIL}{(L_{\text{gate}} \cdot 10^4)^{WSTILP}} \right) \tag{3.128}$$

Here $WSTI$, $WSTIL$, and $WSTLP$ are model parameters. Calculated $I_{\text{ds,STI}}$ simulation results with the developed edge-leakage model are compared in Fig. 3.58 to measurements.

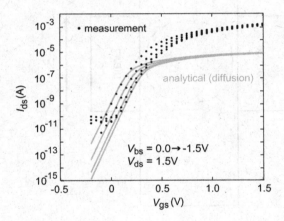

Fig. 3.58 Comparison of measured I_{ds}-V_{gs} (solid circles) and simulated results (lines) with the developed model. The simulation accounts only for the subthreshold leakage at the transistor edges.

3.6.4 *Small-Geometry Effects*

The size reduction of the MOSFET gives increased speed and higher density of transistors and circuit functions, but its small dimensions and volumes also cause unpredictable statistical fluctuations due to a degrading reproducibility of manufacturing processing steps. The characteristics affected by these statistical fluctuations do not scale the same way as the channel length or gate width in larger area and volume transistors. The main cause is the lithography inaccuracy at smaller dimensions shown in Fig. 3.59 [96]. It is impossible to accurately model the device electrical characteristics based on device physics due to such statistical fluctuation. In addition, there is the direct consequence from statistical fluctuations that give rise to electrical noise, which will be explained in Section 6.1. The random nature the small-geometry effects can therefore only be modeled by empirical parameters which are not firmly based on device physics, not only increasing the number of model parameters but also further empiricalizing the compact transistor model as well as the circuit simulation.

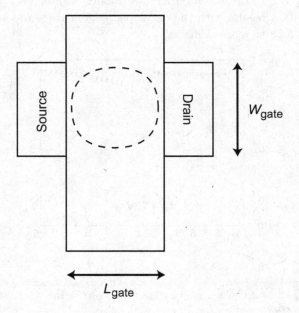

Fig. 3.59   Schematic illustration of the effect of the resolution inaccuracy of the lithography on the MOSFET channel of small-size devices. The rectangular structure disappears due to a rounding effect at the corners.

Measurements often show that the drain current does not decrease linearly with the reciprocal gate width for a short gate as shown in Fig. 3.60. Thus, in HiSM, the small geometry effects are modeled in the threshold voltage by two phenomenological parameters, $WL2$ and $WL2P$ as

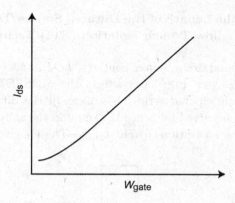

Fig. 3.60   Schematic graph of the flattening of the linear drain current I_{ds} decrease with decreasing gate width W_{gate}, when W_{gate} becomes very small.

$$L_{gate} = L_{gate} + \frac{WL2}{WL^{WL2P}} \tag{3.129}$$

where

$$WL = (W_{gate} \cdot 10^4)(L_{gate} \cdot 10^4) \tag{3.130}$$

The modified L_{gate} is used for the threshold voltage shift calculation given in Eqs. (3.120) and (3.123). The mobility modification due to the small device geometry is additionally modeled in the phonon scattering as

$$MUE_{PH1} = MUE_{ph1} \left(1 + \frac{MUEPHS}{WL^{MUEPSP}} \right) \tag{3.131}$$

$$V_{max} = V_{max} \cdot \left(1 + \frac{VOVERS}{(L_{gate} \cdot 10^4)^{VOVERSP}} \right) \tag{3.132}$$

3.7   Effects of the Length of the Diffused Source/Drain Contacts in Shallow-Trench Isolation (STI) Technologies

Length of the diffused drain/source contacts, LOD, between the MOSFET gate and STI edge (see Fig. 3.61) affects the MOSFET characteristics. Asides from the anticipated series resistances in the diffused source and drain regions, the observed influence is mainly in the shift of the threshold voltage V_{th} and the saturation current $I_{\text{ds,sat}}$. The V_{th} change is attributed

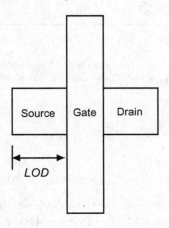

Fig. 3.61   Schematic explanation of meaning of the length of diffusion (LOD) in case of a MOS transistor.

to a change of the pocket impurity concentration and is modeled by

$$N_{\text{substi}} = 1 + \frac{1}{1 + NSUBPSTI2} \cdot \left(\frac{NSUBPSTI1}{LOD}\right)^{NSUBPSTI3} \tag{3.133}$$

Eq. (3.133) is used to modify the pocket concentration N_{subp} as

$$N_{\text{subp}} = N_{\text{subp}} \times N_{\text{substi}} \tag{3.134}$$

The saturation-current change is attributed to a change of the carrier mobility and is modeled by

$$M_{\text{uesti}} = 1 + \frac{1}{1 + MUESTI2} \cdot \left(\frac{MUESTI1}{LOD}\right)^{MUESTI3} \tag{3.135}$$

Eq. (3.135) is used to modify the phonon mobility parameter M_{uephonon} as

$$M_{\text{uephonon}} = M_{\text{uephonon}} \times M_{\text{uesti}} \tag{3.136}$$

3.8   Temperature Dependences

MOSFET characteristics are dependent on three temperatures, which represents the average kinetic energy of the three quasi-particle species: electrons, holes and phonons [97]. The carriers, either electron or hole, with a carrier temperature several times and more higher than the lattice temperature are called hot carriers [98]. This section describes the dependence of the transistor characteristics on the lattice temperature without considering the carrier temperature as being different from the lattice temperature. Such a treatment proves sufficiently accurate for the purpose of constructing a compact model for circuit simulation.

Different from a metal, the number of carriers in a semiconductor increases as the temperature increases. However, the temperature dependences of the output characteristics are not that simple in real devices as shown in Fig. 3.62. The *I-V* characteristics of a MOSFET at two different temperatures cross at a gate voltage a little bit above the threshold condition. The higher drain current in the subthreshold region at higher temperature is mostly due to an higher intrinsic carrier concentration n_i through bandgap narrowing. However, the phonon scattering rate increases with temperature due to larger phonon density and higher lattice vibration amplitude. More frequent phonon scattering reduces the electron mobility, hence also the drain current, at a higher gate-source voltage V_{gs}. Fig. 3.63 gives the measured I_{ds}-V_{gs} characteristics of a MOSFET in linear-linear scales. The experimental data demonstrate the crossing at one single $I-V$ point of curves at several temperatures depicted in Fig. 3.62, first described by us in 2003 [99].

The crossing point, known as the Temperature-Independence Point (TIP), is quite universal and is located about 0.3V above the threshold voltage, V_{th}, as seen in Fig. 3.64. This bias point of temperature independence is exactly the preferred operating point for many analog circuits which require accurate prediction and small magnitude of the temperature dependence.

The main temperature dependence above threshold comes from the normalized surface potential, ϕ_S/kT. The transistor's temperature dependences also come from the temperature dependence of the energy gap, the intrinsic carrier concentration, the carrier mobility and the carrier saturation velocity. The following relations for these temperature dependencies are used in HiSIM:

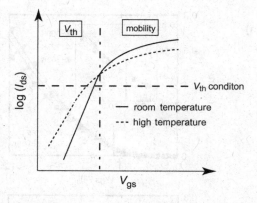

Fig. 3.62 *I-V* characteristics for two different temperatures showing a crossing point which separates operating regions with opposite temperature behavior. The region where the temperature dependence of "V_{th}" is responsible is notified by V_{th} and the region where the mobility degradation is responsible is notified by "mobility".

- Thermal Voltage

$$\frac{kT}{q} = \beta^{-1} \tag{3.137}$$

- Bandgap [100]

$$E_{\text{g}} = EG0 - BGTMP1 \cdot T - BGTMP2 \cdot T^2 \tag{3.138}$$

- Intrinsic Carrier Density

$$n_{\text{i}} = n_{\text{i}0} \cdot T^{\frac{3}{2}} \cdot \exp\left(-\frac{E_{\text{g}}}{2q}\beta\right) \tag{3.139}$$

- Phonon Scattering

$$\mu_{\text{PH}}(\text{phonon}) = \frac{MUE_{\text{PH1}}}{\left(\dfrac{T}{TNOM}\right)^{MUETMP} \times E_{\text{eff}}^{MUE_{\text{PH0}}}} \tag{3.140}$$

- Saturation Velocity [72]

$$V_{\max} = \frac{V_{\max}}{1.8 + 0.4\left(\dfrac{T}{TNOM}\right) + 0.1\left(\dfrac{T}{TNOM}\right)^2 - VTMP \cdot \left(1 - \dfrac{T}{TNOM}\right)} \tag{3.141}$$

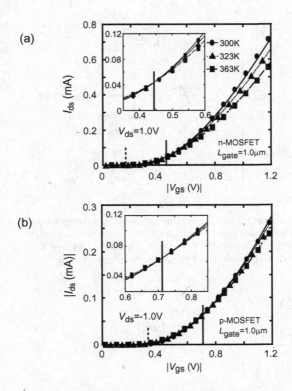

Fig. 3.63 Measured *I-V* characteristics of (a) n-MOSFET and (b) p-MOSFET for three different temperatures. The Temperature Independence Point (TIP) is shown by solid perpendicular lines. Dashed perpendicular lines show the threshold voltages.

Here T is the given temperature in Kelvin and $TNOM$ gives the temperature taken as the nominal value, which is usually the room temperature. The reason for determining $TNOM$ in compact modeling is due to a convenience for parameter extraction as discussion in Section 7.5. $EG0$, $BGTMP1$, and $BGTMP2$ are model parameters describing the temperature dependence of the bandgap.

The temperature dependence of the intrinsic carrier concentration mainly determines the temperature dependence of the threshold voltage, and the simulated dependence with a 2D-device simulator is depicted for 2 temperatures in Fig. 3.65. The threshold voltage V_{th} is written

$$V_{\text{th}} = V_{\text{fb}} + 2\Phi_{\text{B}} + \frac{\sqrt{2qN_{\text{sub}}\epsilon_{\text{Si}}(2\Phi_{\text{B}} - V_{\text{bs}})}}{C_{\text{ox}}} \qquad (3.142)$$

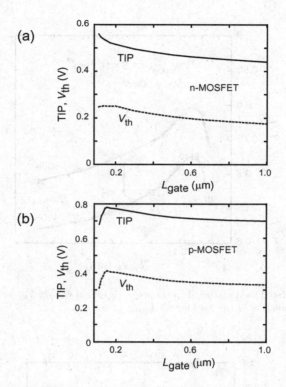

Fig. 3.64 Measured crossing point of the *I-V* characteristics for different temperatures as a function of gate length L_{gate}. The Temperature Independence Point (TIP) is always located about 0.3V above the threshold voltage V_{th}.

where

$$2\Phi_{\text{B}} = \frac{2}{\beta} \ln \left(\frac{N_{\text{sub}}}{n_{\text{i}}} \right) \tag{3.143}$$

Equations (3.142) and (3.143) confirm that the temperature dependence of V_{th} is mainly due to the temperature dependence of n_{i} caused by the temperature dependence of the bandgap. The temperature dependence of the mobility and the temperature dependence of the saturation velocity have a major influence on the temperature dependence of the I_{ds}-V_{ds} characteristics under the on-current condition. Fig. 3.66 shows again the magnitude of this temperature dependence.

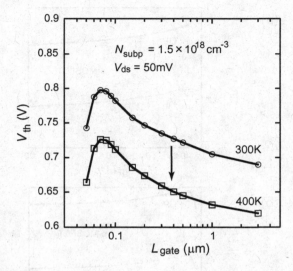

Fig. 3.65   Simulated temperature dependence of threshold voltage V_{th} with a 2D-device simulator as a function of the gate length L_{gate}.

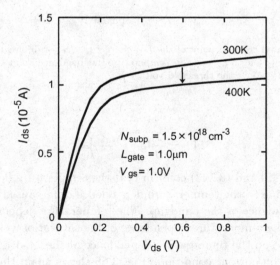

Fig. 3.66   Simulated temperature dependence of the drain current I_{ds} as a function of the drain voltage V_{ds} for fixed gate voltage V_{gs}.

3.9   Conservation of Symmetry at $V_{ds} = 0$

A MOSFET can be driven symmetrically with a supply voltage of V_{gb} as gate bias and $V_{sb} = V_{sb0} - V_x$ and $V_{db} = V_{db0} + V_x$, where V_{sb0} and V_{db0} are initial source/bulk and drain/bulk biases, respectively. If these initial values are set to be the same, and V_x is taken as a voltage source, the drain current I_{ds} must be an odd function of V_x

$$I_{ds}(V_x) = -I_{ds}(-V_x) \qquad (3.144)$$

Consequently, all even order derivatives of I_{ds} with respect to V_x must be zero at $V_{ds}=0$, which constitutes an important quality and symmetry requirement for a compact MOSFET model known as the Gummel symmetry test [101]. To preserve the symmetry of the derivatives, I_{ds} must be continuous (known to engineers as "smooth") through $V_{ds}=0$.

The surface-potential-based modeling, which includes both the drift and diffusion currents, in principle preserves the symmetry at $V_{ds}=0$ automatically as shown in Fig. 3.67a for the inversion charge Q_I. However, the fundamental nature of this symmetry may be destroyed by artifacts introduced by the modeling of phenomena observed in advanced MOSFET technologies such as short-channel effects. To avoid the loss of symmetry, model formulation is improved so that the short-channel effects approach zero continuously as V_{ds} approaches zero. Instead of modeling the diminishing magnitude of the modeled effects as V_{ds} reduces, efficient numerical damping is used in HiSIM. Two parameters are introduced: $VZADD0$ and $PZADD0$ [69]. The values of these parameters are fixed, and it is recommended not to change them. A result with the damping at short channel length is shown in Fig. 3.67b for $L_{gate} = 0.13\mu m$ with a $0.13\mu m$ technology. Other modeled phenomena, showing a V_{ds} dependence and included in the surface-potential calculation, cause a similar symmetry problem as the short-channel effects. The same approach is used in these cases to preserve the symmetry of all modeled phenomena.

It has been discussed that the symmetry can be easily preserved by taking the bulk node as a reference [102]. However, the artifacts due to closed-form descriptions of phenomena observed in advanced MOSFETs cause also disturbance of the symmetry. The model parameter BB for describing the high-field mobility relationship with the longitudinal electric field (see Eq. (3.84) in Section 3.4) is often required to be even to preserve the symmetry [103]. However, it is well confirmed that the value of BB is 2 for electrons and 1 for holes. Thus, the symmetry has to be preserved also under the condition of uneven BB, because the overall model accuracy must have higher priority. It has to be realized, that the symmetry can be only disturbed through artifacts introduced during the modeling of additional advanced device phenomena in the surface-potential-based model, and the resultant discontinuities can be eliminated only by numerical damping. The

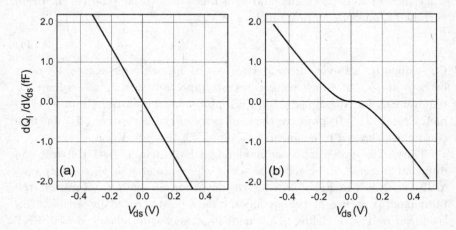

Fig. 3.67   Symmetry test for the inversion-charge derivative of HiSIM at $V_{ds} = 0V$ for (a) $L_{gate} = 10\mu m$ and (b) $L_{gate} = 0.13\mu m$ at $V_{gs} = 3V$.

drain current I_{ds} and its derivatives, all shown in Fig. 3.68, indicate that the damping method used in HiSIM is very effective. It is well known that symmetry needs to be retained in capacitances of the MOSFET, even to the fourth and fifth and higher order derivatives, which is tedious to attain via analytical approximations. Careful treatment of all phenomena occurring in the MOS transistors is therefore required.

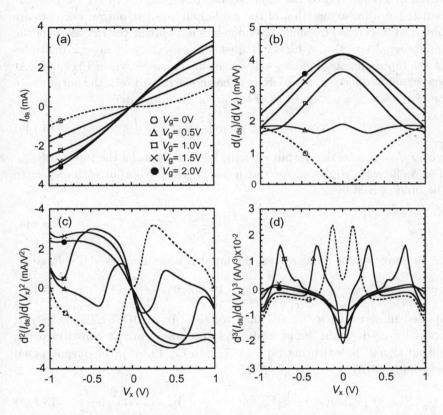

Fig. 3.68 Gummel symmetry test for the drain current I_{ds} of HiSIM up to the 3rd derivative at $V_{\text{ds}} = 0$V for $L_{\text{gate}} = 10\mu$m and $V_{\text{gs}} = 3$V.

3.10   Harmonic Distortions

High performance analog circuit design requires the consideration of all device features explicitly in the compact model for the design. One important feature is the harmonic distortion of MOSFETs [90,105,106] for the output signal in comparison to the input signal. Since the MOSFET output current is not a linear function of the gate-input voltage, output currents are observed to contain frequency components not only at the frequency of the input signal but also at higher frequencies, especially at integer multiples of the input signal frequency called harmonics, as shown in Fig. 3.69. At low signal frequencies, for analysis of harmonic distortions, the output may be expressed as a power series of the input given by [105]

$$S_{\text{out}} = a_1 S_{\text{in}} + a_2 S_{\text{in}}^2 + a_3 S_{\text{in}}^3 + \cdots \tag{3.145}$$

where S_{out} can be the output current, while S_{in} can be the input voltage. The coefficients, a_1, a_2, ... are the derivatives of the output with respect to the input, given by

$$a_{\text{n}} = \frac{1}{n!} \frac{\partial^n S_{\text{out}}}{\partial S_{\text{in}}^n} \tag{3.146}$$

In many analog applications, the input is a small sinusoidal voltage

$$S_{\text{in}} = v_{\text{in}} = V_{\text{p}} \sin(\omega t) \tag{3.147}$$

applied in series with the DC bias voltage. In a MOSFET, the gate is normally used as the input node and the drain current constitutes the output signal. Substituting Eq. (3.147) into Eq. (3.145), the output signal is given by [107]

$$S_{\text{out}} = a_1 V_{\text{p}} \sin(\omega t) + a_2 V_{\text{p}}^2 \sin(\omega t)^2 + a_3 V_{\text{p}}^3 \sin(3\omega t)^3 + \cdots \tag{3.148}$$

$$= a_1 V_{\text{p}} \sin(\omega t) + \frac{a_2 V_{\text{p}}^2}{2} \{1 - \cos(2\omega t)\}$$

$$+ \frac{a_3 V_{\text{p}}^3}{4} \{3 \sin(\omega t) - \sin(3\omega t)\} + \cdots \tag{3.149}$$

$$= \frac{a_2 V_{\text{p}}^2}{2} + \left\{ a_1 V_{\text{p}} + \frac{3 a_3 V_{\text{p}}^3}{4} \right\} \sin(\omega t) - \frac{a_2 V_{\text{p}}^2}{2} \cos(2\omega t)$$

$$- \frac{a_3 V_{\text{p}}^3}{4} \sin(3\omega t) + \cdots \tag{3.150}$$

where the output term at the input frequency, $f = \omega/2\pi$, is known as the fundamental. The terms at multiples of the input frequency, known as the harmonics or harmonic distortions, come exactly from the nonlinear output-input relationship of the MOSFET.

The origin of the harmonics is therefore the nonlinearity of the MOSFET response, and the phenomenon is called as mentioned harmonic distortion. In analog applications, usually only the second and third order harmonic distortions need to be considered, because the input voltage amplitude is made normally quite small. Furthermore, the transistor nonlinearity is mostly also small or can be linearized by negative feedback circuits, and the bandpass filters used in many circuits are narrow or sharply tuned to the fundamental frequency. In analog applications, the harmonic distortions are specified as percentage distortions, thus, $HD1$, $HD2$, and $HD3$ are given by

$$HD1 = a_1 V_{\text{p}} \tag{3.151}$$

$$HD2 = a_2 V_{\text{p}}^2 \tag{3.152}$$

$$HD3 = a_3 V_{\text{p}}^3 \tag{3.153}$$

Measured results are plotted in Fig. 3.70 together with the calculated results by the HiSIM compact model. These measurement data are performed for the DC gate bias voltage V_{gs} as the independent variable, while the frequency f is fixed to 1kHz and the drain voltage V_{ds} is fixed to 0.1V. The theory in Fig. 3.70 was computed using the DC compact model equations and their parameters obtained from fitting to the measured DC characteristics. No additional adjustable parameters were introduced [108]. The excellent agreement of the computed curves to the experiments, to third order, is a most stringent proof of the excellence or accuracy of the DC model. Although the data in Fig. 3.70 covers only an infinitesmal range of V_{ds}, it can be expanded by making the same correlation at other V_{ds} values to cover a wide V_{ds} in order to verify the accuracy of the DC model over a large 2-dimensional domain or map of (V_{gs}, V_{ds}). A specific feature of the harmonic distortion are the zero's at certain DC gate-source voltages, V_{gs}, as shown in Fig. 3.70. These zero's do almost not depend on channel length but depend on the carrier dynamics governed by the electric field, which is discussed in the next paragraphs. The harmonic distortions are in principle derivatives of I_{ds} (see Eq. (3.145)), and thus the calculation results reflect exactly the accuracy of the I-V characteristics. The high reliability of the higher-order derivatives of the drain current is exactly a feature of the surface-potential-based modeling and turns out to be very beneficial for the harmonic-distortion modeling.

Figures 3.71a and b compare the harmonic distortions with derivatives of the mobility extracted from the measured *I-V* characteristics of the device used for the harmonic distortion measurements. The reason for applying small V_{ds} in this analysis is to eliminate the high-field effects, which influence all device features in complicated ways. At low longitudinal channel electric fields, the zero voltages of the harmonic current and the mobility distortions coincide as seen in Fig. 3.71b [108]. Thus the zero's come from the DC gate voltage dependence of the mobility. This is a very powerful method to study the fundamental physics of scattering, in addition to gauge the goodness of the mobility approximation formulas. The derivative method has been used traditionally, for many decades, in the precision measurement of the fundamental properties of materials and space, such as the energy band features of semiconductors, and is also known as the modulation spectroscopy. An example is the modulation reflectance spectroscopy for the determination of the details of the Energy-Wave Vector(k) diagram of silicon and other semiconductors. Most recently, derivative method has been used to detect the geometric quantization of nanometer MOSFET at very low tempertures.

The harmonic distortion analysis is conventionally made at low frequencies. At higher frequencies, the harmonic distortions are influenced by both the external circuits, which can be eliminated, and by the delay of the electron or hole current which carries the signal through the channel. This has been approximated by a lumped charge analysis, given by [108, 109] the charge control, Quasi-Static(QS) equation

$$I_{ds}(t) = I_{ds}(V(t)) + \frac{dQ_D(t)}{dt}; \tag{3.154}$$

which neglects the diffusion and transit time delay through the channel, known as the distributed effects or Non-Quasi-Static(NQS) model, needed for RF modeling of MOSFET for GHz aplications [110]. Fig. 3.72 compares the computed high-frequency harmonic distortions, obtained with the NQS-extended HiSIM model explained in detail in Section 6.2, to the measurements at 1GHz. It can be seen that the zero's or singulurities, observed at low frequency, nearly disappear and that the small V_{gs} distortion amplitudes become much larger than at low frequency. The agreement between the NQS-extended of HiSIM and the measurement is quite good. Discrepancies for the higher order harmonics come mainly from the fact that the data amount of the time-domain computation was in sufficient for numerical stability of the Fourier analysis of the higher order coefficients. The high-frequency properties will be further discussed in Section 6.2.

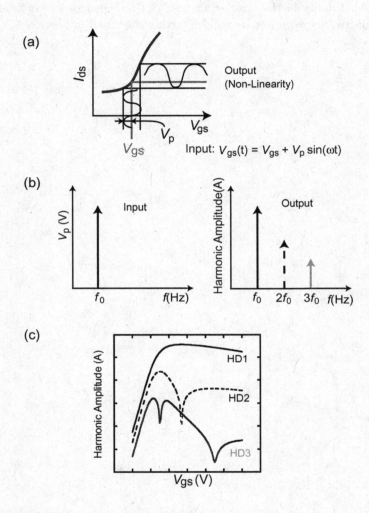

Fig. 3.69 Mechanism of the harmonic distortions. If a small signal voltage of $V_p \sin(\omega t)$ is applied on top of a DC voltage (a), output signals are observed not only at the frequency of $f_0 = \omega/(2\pi)$ but also at $n(n = 2, 3, 4, \cdots)$ times higher frequencies then f_0 (b). This phenomenon is called harmonic distortion and the amplitudes at the harmonic-distortion frequencies can be plotted as a function of the DC voltage, as in (c) for the case of V_{gs}.

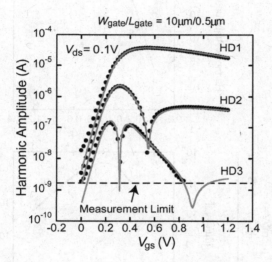

Fig. 3.70 Comparison of calculated harmonic distortions (lines) as a function of V_{gs} with measurements (symbols): $HD1 \propto \frac{dI_{\mathrm{ds}}}{dV_{\mathrm{gs}}}$, $HD2 \propto \frac{d^2 I_{\mathrm{ds}}}{dV_{\mathrm{gs}}^2}$, and $HD2 \propto \frac{d^3 I_{\mathrm{ds}}}{dV_{\mathrm{gs}}^3}$.

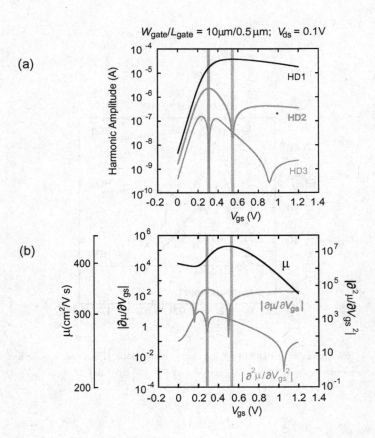

Fig. 3.71   (a) Calculated harmonic distortion with HiSIM, and (b) derivatives of the mobility of the MOSFET used for the calculation of the harmonic distortion.

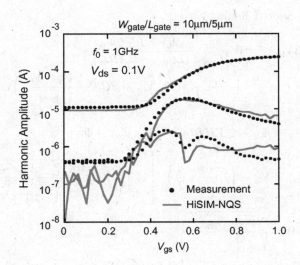

Fig. 3.72 Comparison of calculated harmonic distortions with the Non-Quasi-Static(NQS) extension of HiSIM to measurements at the frequency of 1GHz. The gate length of the measured MOSFET is 5μm and the applied drain-source voltage is $V_{ds}=0.1$V. The cut-off frequency under this condition is about 200MHz.

3.11 Summary of Equations and Model Parameters Appearing in Chapter 3 for Advanced MOSFET Phenomena Modeling

3.11.1 *Section 3.1: Threshold Voltage Shift*

Following threshold voltage shifts are considered in the effective gate voltage V_G' as

$$V_G' = V_{gs} - V_{fb} + \Delta V_{th,SC} - \Delta V_{th,R} + \Delta V_{th,PSC}$$

[$\Delta V_{th,SC}$: Short-Channel Effect]

$$\Delta V_{th,SC} = W_d \frac{\epsilon_{Si}}{C_{ox}} \cdot \frac{2(VBI - 2\Phi_B')}{(L_{gate} - PARL2)^2}$$
$$\left(SC1 + SC2 \cdot V_{ds} + SC3 \cdot \frac{2\Phi_B - V_{bs}}{L_{gate}} \right)$$

$$\Phi_B' = \Phi_B + PTHROU \cdot \left(\Phi_B''(V_{gs}) - \Phi_B \right)$$

$$\Phi_B''(V_{gs}) = V_G' + \left(\frac{const0}{C_{ox}} \frac{\beta}{2} \right)^2 \left[1 - \sqrt{1 - \frac{4\beta(V_G' - V_{bs} - 1}{\beta^2 \left(\frac{const0}{C_{ox}} \right)^2}} \right]$$

$$const0 = \sqrt{\frac{2\epsilon_{Si}qN_{sub}}{\beta}}$$

[$\Delta V_{th,R}$: Reverse-Short-Channel Effect (Pileup)]

$$\Delta V_{th,R} = \frac{Q_b}{C_{ox}} - \frac{Q_b(long)}{C_{ox}}$$
$$Q_b(long) = \sqrt{2\epsilon_{Si}qN_{sub}}\sqrt{2\Phi_B - V_{bs}}$$

$$Q_b = QDEPCC + \frac{QDEPCL}{L_{gate}^{QDEPCS}} + \left(QDEPBC + \frac{QDEPBL}{L_{gate}^{QDEPBS}} \right) \sqrt{2\Phi_B - V_{bs}}$$

[$\Delta V_{th,PSC}$: Reverse-Short-Channel Effect (Pocket Implant)]

$$\Delta V_{th,P} = (V_{th,P} - V_{th0})W_d \frac{\epsilon_{Si}}{C_{ox}} \frac{dE_{y,P}}{dy} - \frac{SCP22}{(SCP21 + V_{ds})^2}$$
$$V_{th,P} = V_{fb} + 2\Phi_B + \frac{Q_{bmod}}{C_{ox}} + \log \left(\frac{N_{subb}}{NSUBC} \right)$$

$$N_{\text{subb}} = 2 \cdot NSUBP - \frac{(NSUBP - NSUBC) \cdot L_{\text{gate}}}{LP} - NSUBC$$

$$V_{\text{th0}} = V_{\text{fb}} + 2\Phi_{\text{BC}} + \frac{\sqrt{2qNSUBC\epsilon_{\text{Si}}(2\Phi_{\text{BC}} - V_{\text{bs}})}}{C_{\text{ox}}}$$

$$2\Phi_{\text{BC}} = \frac{2}{\beta} \ln\left(\frac{NSUBC}{n_{\text{i}}}\right)$$

$$\frac{dE_{y,\text{P}}}{dy} = \frac{2(VBI - 2\Phi_{\text{B}})}{LP^2}\left(SCP1 + SCP2 \cdot V_{\text{ds}} + SCP3 \cdot \frac{2\Phi_{\text{B}} - V_{\text{bs}}}{LP}\right)$$

$$Q_{\text{bmod}} = \sqrt{2q \cdot N_{\text{sub}} \cdot \epsilon_{\text{Si}} \cdot \left(2\Phi_{\text{B}} - V_{\text{bs}} - \frac{BS1}{BS2 - V_{\text{bs}}}\right)}$$

Table 3.1 HiSIM model parameters for threshold voltage shift. ∗ indicates minor parameters.

VBI	built-in potential V_{bi}
$PARL2$	depletion width of channel/contact junction
$SC1$	magnitude of short-channel effect
$SC2$	V_{ds} dependence of short-channel effect
∗$SC3$	V_{bs} dependence of short-channel effect
∗$PTHROU$	correction for subthreshold swing
$NSUBC$	substrate-impurity concentration
$NSUBP$	maximum pocket concentration
LP	pocket penetration length
$SCP1$	magnitude of short-channel effect due to pocket
$SCP2$	V_{ds} dependence of short-channel effect due to pocket
∗$SCP3$	V_{bs} dependence of short-channel effect due to pocket
∗$SCP21$	short-channel-effect modification for small V_{ds}
∗$SCP22$	short-channel-effect modification for small V_{ds}
∗$BS1$	body-coefficient modification due to impurity profile
∗$BS2$	body-coefficient modification due to impurity profile
∗$NPEXT$	maximum concentration of pocket tail
∗$LPEXT$	extension length of pocket tail
$QDEPCC$	depletion charge coefficient 1 for pileup profile
$QDEPCL$	depletion charge coefficient 2 for pileup profile
$QDEPCS$	depletion charge coefficient 3 for pileup profile
$QDEPBC$	Vbs dependence of depletion charge coefficient 1
$QDEPBL$	Vbs dependence of depletion charge coefficient 2
$QDEPBS$	Vbs dependence of depletion charge coefficient 3

$$2\Phi_{\mathrm{B}} = \frac{2}{\beta} \ln \left(\frac{N_{\mathrm{sub}}}{n_{\mathrm{i}}} \right)$$

$$W_{\mathrm{d}} = \sqrt{\frac{2\epsilon_{\mathrm{Si}}(2\Phi_{\mathrm{B}} - V_{\mathrm{bs}})}{q N_{\mathrm{sub}}}}$$

$$N_{\mathrm{sub}} = \frac{NSUBC(L_{\mathrm{gate}} - LP) + NSUBP \cdot LP}{L_{\mathrm{gate}}} + \frac{NPEXT - NSUBC}{\left(\frac{1}{xx} + \frac{1}{\mathbf{LPEXT}} \right) L_{\mathrm{gate}}}$$

$$xx = 0.5 \times L_{\mathrm{gate}} - LP$$

$$C_{\mathrm{ox}} = \frac{\epsilon_{\mathrm{ox}}}{T_{\mathrm{ox}}} = \frac{\kappa}{T_{\mathrm{ox}}}$$

3.11.2   *Section 3.2: Depletion Effect of the Poly-Silicon Gate*

$$\phi_{\mathrm{Spg}} = PGD1 \left(\frac{N_{\mathrm{sub}}}{NSUBC} \right)^{PGD4} \exp \left(\frac{V_{\mathrm{gs}} - PGD2 - PGD3 \cdot V_{\mathrm{ds}}}{V} \right)$$

Table 3.2   HiSIM model parameters. ∗ indicates minor parameters for depletion effect of the poly-silicon gate.

$PGD1$	strength of poly depletion
$PGD2$	threshold voltage of poly depletion
$PGD3$	V_{ds} dependence of poly depletion
∗$PGD4$	L_{gate} dependence of poly depletion

3.11.3   *Section 3.3: Quatum-Mechanical Effects*

$$T_{\mathrm{ox,eff}} = T_{\mathrm{ox}} + \Delta T_{\mathrm{ox}}$$

$$\Delta T_{\mathrm{ox}} = QME1 \left(V_{\mathrm{gs}} - V_{\mathrm{th}}(T_{\mathrm{ox,eff}} = T_{\mathrm{ox}}) - QME2 \right)^2 + QME3$$

Table 3.3   HiSIM model parameters for quatum-mechanical effects.

$QME1$	V_{gs} dependence
$QME2$	V_{gs} dependence
$QME3$	minimum T_{ox} modification

$$V_{\text{th}} = V_{\text{fb}} + 2\Phi_{\text{B}} + \frac{\sqrt{2qN_{\text{sub}}\epsilon_{\text{Si}}(2\Phi_{\text{B}} - V_{\text{bs}})}}{C_{\text{ox}}} \tag{3.155}$$

$$C_{\text{ox}} = \frac{\epsilon_{\text{ox}}}{T_{\text{ox}}} = \frac{\kappa}{T_{\text{ox}}}$$

3.11.4 *Section 3.4: Mobility Model*

[Low-Field Mobility]:

$$\frac{1}{\mu_0} = \frac{1}{\mu_{\text{CB}}} + \frac{1}{\mu_{\text{PH}}} + \frac{1}{\mu_{\text{SR}}}$$

$$\mu_{\text{CB}}(\text{Coulomb}) = MUECB0 + MUECB1 \frac{Q_{\text{i}}}{q \cdot 10^{11}}$$

$$\mu_{\text{PH}}(\text{phonon}) = \frac{MUE_{\text{ph1}}}{(T/TNOM)^{MUETMP} E_{\text{eff}}^{MUEPH0}}$$

$$\mu_{\text{SR}}(\text{surface roughness}) = \frac{MUESR1}{E_{\text{eff}}^{MUE_{\text{sr0}}}}$$

$$E_{\text{eff}} = \frac{1}{\epsilon_{\text{Si}}} \left[\gamma \cdot Q_{\text{b}} + \eta \cdot Q_{\text{i}} \right]$$

$$\gamma = NDEP \left(1 - \frac{NDEPL}{NDEPL + (L_{\text{gate}} \cdot 10^4)^{NDEPLP}} \right)$$

$$\eta = NINV \left(1 - \frac{NINVL}{NINVL + (L_{\text{gate}} \cdot 10^4)^{NINVLP}} \right)$$

$$MUE_{\text{ph1}} = MUEPH1 \cdot \left(1 + \frac{MUEPHL}{(L_{\text{gate}} \cdot 10^4)^{MUEPLP}} \right)$$

$$MUE_{\text{sr0}} = MUESR0 \cdot \left(1 + \frac{MUESRL}{(L_{\text{gate}} \cdot 10^4)^{MUESLP}} \right)$$

[High-Field Mobility]:

$$\mu = \frac{\mu_0}{\left(1 + \left(\frac{\mu_0 E_{\text{y}}}{V_{\text{max}}} \right)^{BB} \right)^{\frac{1}{BB}}}$$

$$V_{\text{max}} = VMAX \cdot \left(1 + \frac{VOVER}{(L_{\text{gate}} \cdot 10^4)^{VOVERP}} \right)$$

Table 3.4   HiSIM model parameters. ∗ indicates minor parameters for mobility model.

T	device temperature in Kelvin
$TNOM$	nominal device temperature in Kelvin, usually room temperature
$MUECB0$	Coulomb scattering
$MUECB1$	Coulomb scattering
$MUEPH0$	phonon scattering
$MUEPH1$	phonon scattering
∗$MUEPHL$	L_{gate} dependence of phonon scattering
∗$MUEPLP$	L_{gate} dependence of phonon scattering
$MUESR0$	surface-roughness scattering
$MUESR1$	surface-roughness scattering
∗$MUESRL$	L_{gate} dependence of surface-roughness scattering
∗$MUESLP$	L_{gate} dependence of surface-roughness scattering
$NDEP$	depletion-charge contribution on effective-electric field
$NDEPL$	L_{gate} dependence of depletion-charge contribution
$NDEPLP$	L_{gate} dependence of depletion-charge contribution
$NINV$	inversion-charge contribution on effective-electric field
BB	high-field-mobility degradation
$VMAX$	maximum saturation velocity
$VOVER$	velocity overshoot effect
$VOVERP$	L_{eff} dependence of velocity overshoot

3.11.5   *Section 3.5: Channel-Length Modulation*

$$\Delta L = \frac{1}{2}\left[-\frac{1}{L_{\text{eff}}}\left(2\frac{I_{\text{dd}}}{\beta Q_{\text{i}}}z + 2\frac{N_{\text{sub}}}{\epsilon_{\text{Si}}}(\phi_{\text{S}}(\Delta L) - \phi_{\text{SL}})z^2 + E_0 z^2 \right) \right.$$

$$+ \left\{ \frac{1}{L_{\text{eff}}^2}\left(2\frac{I_{\text{dd}}}{\beta Q_{\text{i}}}z + 2\frac{N_{\text{sub}}}{\epsilon_{\text{Si}}}(\phi_{\text{S}}(\Delta L) - \phi_{\text{SL}})z^2 + E_0 z^2 \right)^2 \right.$$

$$\left. \left. -4\left(2\frac{N_{\text{sub}}}{\epsilon_{\text{Si}}}(\phi_{\text{S}}(\Delta L) - \phi_{\text{SL}})z^2 + E_0 z^2 \right) \right\}^{\frac{1}{2}} \right]$$

$$z = \frac{\epsilon_{Si}}{CLM2 \cdot Q_{\text{b}} + CLM3 \cdot Q_{\text{i}}}$$

$$\phi_{\text{S}}(\Delta L) = CLM1(\phi_{\text{S0}} + V_{\text{ds}}) + (1 - CLM1)\phi_{\text{SL}}$$

Table 3.5   HiSIM model parameters for channel-length modulation.

$CLM1$	hardness coefficient of channel/contact junction
$CLM2$	coefficient for depletion-charge contribution
$CLM3$	coefficient for inversion-charge contribution

3.11.6 *Section 3.6: Narrow-Channel Effects*

[Threshold Voltage Shift]

$$\Delta V_{\text{th,W}} = \left(\frac{1}{C_{\text{ox}}} - \frac{1}{C_{\text{ox}} + 2C_{\text{ef}}/(L_{\text{eff}}W_{\text{eff}})} \right) qN_{\text{sub}}W_{\text{d}} + \frac{WVTH0}{W_{\text{gate}} \cdot 10^4}$$

$$C_{\text{ef}} = \frac{WFC}{2}L_{\text{eff}}$$

$$W_{\text{d}} = \sqrt{\frac{2\epsilon_{\text{Si}}(2\Phi_{\text{B}} - V_{\text{bs}})}{qN_{\text{sub}}}}$$

$$2\Phi_{\text{B}} = \frac{2}{\beta} \ln\left(\frac{N_{\text{sub}}}{n_{\text{i}}} \right)$$

[Mobility Modification]

$$MUE_{\text{PH1}} = MUE_{\text{ph1}} \left(1 + \frac{MUEPHW}{(W_{\text{gate}} \cdot 10^4)^{MUEPWP}} \right)$$

$$MUE_{\text{SR0}} = MUE_{\text{sr0}} \left(1 + \frac{MUESRW}{(W_{\text{gate}} \cdot 10^4)^{MUESWP}} \right)$$

[Leakage Current due to STI]

$$I_{\text{ds,STI}} = 2\frac{W_{\text{STI}}}{L_{\text{eff}} - \Delta L}\mu\frac{Q_{\text{i,STI}}}{\beta}\left[1 - \exp(-\beta V_{\text{ds}}) \right]$$

$$W_{\text{STI}} = WSTI \left(1 + \frac{WSTIL}{(L_{\text{gate}} \cdot 10^4)^{WSTILP}} \right)$$

$$Q_{\text{i,STI}} = \sqrt{\frac{2\epsilon_{\text{Si}}qNSTI}{\beta}\left[\beta(\phi_{\text{S,STI}} - V_{\text{bs}}) - 1 + \frac{n_{\text{p0}}}{p_{\text{p0}}}\left\{ \exp(\beta\phi_{\text{S,STI}})) \right\} \right]^{\frac{1}{2}}}$$
$$- \sqrt{\beta(\phi_{\text{S,STI}} - V_{\text{bs}}) - 1}$$

$$\phi_{\text{S,STI}} = \phi_{\text{S,STI}}(2) - 0.5 \cdot \left(\phi_{\text{S,STI}}(12) + \sqrt{\phi_{\text{S,STI}}(12)^2 + 4\delta\phi_{\text{S,STI}}(2)} \right)$$

$$\phi_{\text{S,STI}}(12) = \phi_{\text{S,STI}}(2) - \phi_{\text{S,STI}}(1) - \delta$$

$$\phi_{\text{S,STI}}(1) = V'_{\text{gs,STI}} + \frac{\epsilon_{\text{Si}} Q_{\text{STI}}}{C_{\text{ox}}^2} \left[1 - \sqrt{1 + \frac{2C_{\text{ox}}^2}{\epsilon_{\text{Si}} Q_{\text{i,STI}}} \left(V'_{\text{gs,STI}} - V_{\text{bs}} - \frac{1}{\beta} \right)} \right]$$

$$\phi_{\text{S,STI}}(2) = \frac{\ln\left[\left(\frac{C_{\text{ox}}}{const0'}\right)^2 \frac{n_{\text{p0}}}{n_{\text{p0}}} V'^2_{\text{gs,STI}} \right]}{\beta + \frac{2}{V'_{\text{gs,STI}}}}$$

$$\frac{n_{\text{p0}}}{p_{\text{p0}}} = \left(\frac{n_{\text{i}}}{NSTI} \right)^2$$

$$V'_{\text{gs,STI}} = V_{\text{gs}} - V_{\text{fb}} + VTHSTI + \Delta V_{\text{th,SCSTI}}$$

$$\Delta V_{\text{th,SCSTI}} = \frac{\epsilon_{\text{Si}}}{C_{\text{ox}} const0'} W_{\text{d,STI}} \frac{dE_{\text{y}}}{dy}$$

$$\frac{dE_{\text{y}}}{dy} = \frac{2(V_{\text{bi}} - 2\Phi_{\text{B,STI}})}{(L_{\text{gate}} - PARL2)^2}$$
$$\cdot \left(SCSTI1 + SCSTI2 \cdot V_{\text{ds}} + SCSTI3 \cdot \frac{2\Phi_{\text{B,STI}} - V_{\text{bs}}}{L_{\text{gate}}} \right)$$

$$2\Phi_{\text{B,STI}} = \frac{2}{\beta} \ln\left(\frac{N_{\text{STI}}}{n_{\text{i}}} \right)$$

$$const0' = \sqrt{\frac{2\epsilon_{\text{Si}} q N_{\text{STI}}}{\beta}}$$

$$Q_{\text{STI}} = q \cdot N_{\text{STI}} \cdot W_{\text{d,STI}}$$

$$W_{\text{d,STI}} = \sqrt{\frac{2\epsilon_{\text{Si}}(2\Phi_{\text{B,STI}} - V_{\text{bs}})}{qN_{\text{STI}}}}$$

[Small Geometry Effects]

$$L_{\text{eff}} = L_{\text{eff}} + \frac{WL2}{WL^{WL2P}}$$

$$WL = (W_{\text{gate}} \cdot 10^4)(L_{\text{gate}} \cdot 10^4)$$

$$MUE_{\text{PH1}} = MUEPH1 \left(1 + \frac{MUEPHS}{WL^{MUEPSP}}\right)$$

$$V_{\text{max}} = V_{\text{max}} \cdot \left(1 + \frac{VOVERS}{(L_{\text{gate}} \cdot 10^4)^{VOVERSP}}\right)$$

Table 3.6 HiSIM model parameters for narrow-channel effects. ∗ indicates minor parameters.

WFC	threshold voltage change due to capacitance change
∗*WVTH0*	threshold voltage shift
∗*NSUBP0*	modification of pocket concentration
∗*NSUBWP*	modification of pocket concentration
∗*MUEPHW*	W_{gate} dependence of phonon scattering
∗*MUEPWP*	W_{gate} dependence of phonon scattering
∗*MUESRW*	W_{gate} dependence of surface roughness scattering
∗*MUESWP*	W_{gate} dependence of surface roughness scattering
∗*VTHSTI*	threshold voltage shift due to STI
SCSTI1	the same effect as *SC*1 but at STI edge
SCSTI2	the same effect as *SC*2 but at STI edge
SCSTI3	the same effect as *SC*3 but at STI edge
NSTI	substrate-impurity concentration at the STI edge, N_{STI}
WSTI	width of the high-field region at STI edge
WSTIL	channel-length dependence of *WSTI*
WSTILP	channel-length dependence of *WSTI*
WL2	magnitude of small size effect
WL2P	magnitude of small size effect
∗*MUEPHS*	mobility modification
∗*MUEPSP*	mobility modification
∗*VOVERS*	modification of maximum velocity
∗*VOVERSP*	modification of maximum velocity

3.11.7 Section 3.7: Effect of the Source/Drain Diffusion Length for Shallow-Trench Isolation (STI) Technologies

$$N_{\text{subp}} = N_{\text{subp}} \times N_{\text{substi}}$$

$$N_{\text{substi}} = 1 + \frac{1}{1 + NSUBPSTI2} \cdot \left(\frac{NSUBPSTI1}{LOD}\right)^{NSUBPSTI3}$$

$$M_{\text{uephonon}} = M_{\text{uephonon}} \times M_{\text{uesti}}$$

$$M_{\text{uesti}} = 1 + \frac{1}{1 + MUESTI2} \cdot \left(\frac{MUESTI1}{LOD}\right)^{MUESTI3}$$

Table 3.7   HiSIM model parameters for DLSTI technologies.

LOD	length of diffusion between gate and STI
$NSUBPSTI1$	pocket concentration change due to LOD
$NSUBPSTI2$	pocket concentration change due to LOD
$NSUBPSTI3$	pocket concentration change due to LOD
$MUESTI1$	mobility change due to LOD
$MUESTI2$	mobility change due to LOD
$MUESTI3$	mobility change due to LOD

3.11.8   *Section 3.8: Temperature Dependences*

$$\frac{kT}{q} = \beta^{-1}$$

$$E_{\text{g}} = EG0 - BGTMP1 \cdot T - BGTMP2 \cdot T^2$$

$$n_{\text{i}} = n_{\text{i0}} \cdot T^{\frac{3}{2}} \cdot \exp\left(-\frac{E_{\text{g}}}{2q}\beta\right)$$

$$\mu_{\text{PH}}(\text{phonon}) = \frac{MUE_{\text{PH1}}}{(T/TNOM)^{MUETMP} \times E_{\text{eff}}^{MUE_{\text{PH0}}}}$$

$$V_{\text{max}} = \frac{V_{\text{max}}}{1.8 + 0.4(T/TNOM) + 0.1(T/TNOM)^2 - VTMP \cdot (1 - T/TN}$$

Table 3.8   HiSIM model parameters for temperature dependence.

EG0	bandgap
BGTMP1	temperature dependence of bandgap
BGTMP2	temperature dependence of bandgap
MUETMP	temperature dependence of phonon scattering
T	device temperature in Kelvic
TNOM	nominal device temperature in Kelvic, usually room temperature
∗VTMP	temperature dependence of the saturation velocity

Bibliography

[1] Chih-Tang Sah, "Characteristics of the metal-oxide-semiconductor transistors," *IEEE Trans. Electron Devices*, vol. 11, no. 7, pp. 324–345, 1964.

[2] Chih-Tang Sah and Henry C. Pao, "The effect of fixed bulk charge on the characteristics of metal-oxide-semiconductor transistors," *IEEE Trans. Electron Devices*, vol. 13, no. 4, pp. 393–409, 1966.

[3] H. C. Pao and C. T. Sah, "Effects of diffusion current on characteristics of metal-oxide (insulator)-semiconductor transistors," *Solid-State Electron.*, vol. 9, no. 10, pp. 927–937, 1966.

[4] Chih-Tang Sah, "A history of MOS transistor compact modeling," *Proc. Workshop on Compact Modeling*, pp. 347–390, Anaheim, May, 2005.

[5] N. Arora, "MOSFET models for VLSI circuit simulation: theory and practice," World Scientific Publishing Company, 2007. Reprinted from Springer-Verlag, 1993.

[6] M. Tusno, M. Suga, M. Tanaka, K. Shibahara, M. Miura-Mattausch, and M. Hirose, "Physically-based threshold voltage determination for MOSFET's of all gate lengths," *IEEE Trans. Electron Devices*, vol. 46, no. 7, pp. 1429–1434, 1999.

[7] L. A. Akers and J. J. Sanchez, "Threshold voltage models of short, narrow and small geometry MOSFET's," *Solid-State Electron.*, vol. 25, no. 7, pp. 621–641, 1982; L. A. Akers, "The inverse-narrow-width effect," *IEEE Electron Device Lett.*, vol. EDL-7, no. 7, pp. 419–421, 1986.

[8] R. R. Troutman and A. G. Fortino, "Simple model for threshold voltage in short-channel IGFET," *IEEE Trans. Electron Devices*, vol. ED-24, no. 10, pp. 1266–1268, 1977.

[9] B. El-Kareh, W. R. Tonti, and S. L. Titcoms, "A submicron MOSFET parameter extraction technique," *IBM J. Res. Develop.*, vol. 34, no. 2/3, pp. 243–249, 1990.

[10] K. Aoyama, "A method for extracting the threshold voltage of MOSFET's based on current components," *Proc. SISPAD*, pp. 118-121, Erlangen, Sept., 1995.

[11] H. S. Wong, M. H. White, T. J. Krutsick, and R. V. Booth, "Modeling of transconductance degradation and extraction of threshold voltage in thin oxide MOSFET's," *Solid-State Electron.*, vol. 30, no. 9, pp. 953–968, 1987.

[12] M. Miura-Mattausch and H. Jacobs, "Analytical model for circuit simulation with quarter micron metal oxide semiconductor field effect transistors: Subthreshold characteristics," *Jpn. J. Appl. Phys.*, vol. 29, no. 12, pp. L2279–L2282, 1990.

[13] M. Suetake, K. Suematsu, H. Nagakura, M. Miura-Mattausch, H. J. Mattausch, S. Kumashiro, T. Yamaguchi, S. Odanaka, and N. Nakayama, "HiSIM: A drift-diffusion-based advanced MOSFET model for circuit simulation with easy parameter extraction," *Proc. Simulation of Semiconductor Processes and Devices*, pp. 261–264, Seattle, Sept., 2000.

[14] M. Miura-Mattausch, M. Suetake, H. J. Mattausch, S. Kumashiro, N.

Shigyo, S. Odanaka, and N. Nakayama, "Physical modeling of the reverse-short-channel effect for circuit simulation," *IEEE Electron Devices*, vol. 48, no. 10, pp. 2449-2452, 2001.

[15] R. R. Troutman, "VLSI limitation from drain-induced barrier lowering," *IEEE Trans. Electron Devices*, vol. ED-26, no. 4, pp. 461-469, 1979.

[16] L. D. Yau, "A simple theory to predict the threshold voltage of short-channel IGFET's," *Solid-State Electron.*, vol. 17, pp. 1059–1063, 1974.

[17] T. Toyabe and S. Asai, "Analytical models of threshold voltage and break-down voltage of short-channel MOSFET's derived from two dimensional analysis," *IEEE Trans. Electron Devices*, vol. ED-26, no. 4, pp. 453–461, 1979.

[18] K. N. Ratnakumar "Short-channel MOST threshold voltage model," *IEEE J. Solid-State Circuits*, vol. SC-17, no. 5, pp. 937–948, 1982.

[19] K. Nishi, H. Matsuhashi, T. Ochiai, M. Kasai, and T. Nishikawa, "Evidence of channel profile modification due to implantation damage studied by a new method, and its implication to reverse short channel effects of nMOSFETs," in *IEDM Tech. Dig.*, pp. 993–996, Washington DC, Dec., 1995.

[20] C. S. Rafferty, H. -H. Vuong, S. A. Eshraghi, M. D. Giles, M. R. Pinto, and S. J. Hillenius, "Explanation of reverse short channel effect by defect gradient," *IEDM Tech. Dig.*, pp. 311–314, Washington DC, Dec, 1993.

[21] P. M. Rousseau, S. W. Crowder, P. B. Griffin, and J. D. Plummer, "Arsenic deactivation enhanced diffusion and the reverse short-channel effect," *IEEE Electron Device Lett.*, vol. 18, no. 2, pp. 42–44, 1997.

[22] H. Sakamoto, M. Hiroi, M. Hane, and H. Matsumoto, "Simulation of reverse short channel effects wit a consistent point-defect diffusion model," *Proc. SISPAD*, pp. 137–140, Boston, Sept., 1997.

[23] S. Kumashiro, H. Sakamoto, and K. Takeuchi, "Modeling of channel boron distribution in deep sub-0.1μm n-MOSFETs," *IEICE Trans. Electro.*, vol. E82-C, no. 6, pp. 813–820, 1999.

[24] H. J. Mattausch, M. Suetake, D. Kitamaru, M. Miura-Mattausch, S. Kumashiro, N. Shigyo, S. Odanaka, and N. Nakayama, "Simple nondestructive extraction of the vertical channel-impurity profile of small-size metal-oxide-semiconductor field-effect transistors," *Appl. Phys. Lett.*, vol. 80, no. 16, pp. 299–2996, 2002.

[25] T. Hori, "A 0.1-μm CMOS technology with Tilt-Implanted Punch through Stopper (TIPS)," *IEDM Tech. Dig.*, pp. 75–78, San Francisco, Dec., 1994.

[26] K. -I. Goto, M. Kase, Y. Momiyama, H. Kurata, T. Tanaka, M. Deura, Y. Sanbonsugi, and T. Sugii, "A study of ultra shallow junction and tilted channel implantation for high performance 0.1μm pMOSFETs," *IEDM Tech. Dig.*, pp. 631–634, San Francisco, Dec., 1998.

[27] Y. Taur and E. J. Nowak, "CMOS devices below 0.1μm: How high will go?" *IEDM Tech. Dig.*, pp. 215–218, Washington DC, Dec., 1997.

[28] D. Buss, "Device issues in the integration of analog/RF functions in deep submicron digital CMOS," *IEDM Tech. Dig.*, pp. 423–426, Washington D.C., Dec., 1999.

[29] B. Yu, H. Wang, O. Milic, Q. Xiang, W. Wang, J. X. An, and M. -R. Lin, "50nm gate-length CMOS transistor with super-Halo: Design, Process, and Reliability," *IEDM Tech. Dig.*, pp. 653–656, Washington DC, Dec., 1999.

[30] K. Miyashita, H. Yoshimura, M. Takayanagi, N. Fujiwara, and Y. Toyoshima, "Optimized Halo structure for 80nm physical gate CMOS technology with Indium and Antimony highly angled ion implantation," *IEDM Tech. Dig.*, pp. 645–648, Washington DC, Dec., 1999.

[31] D. Kitamaru, H. Ueno, K. Morikawa, M. Tanaka, M. Miura-Mattausch, H. J. Mattausch, S. Kumashiro, T. Yamaguchi, K. Yamashita, and N. Nakayama, "V_{th} Model of Pocket-Implant MOSFETs for Circuit Simulation," *Proc. SISPAD*, Athens, Sept., 2001.

[32] H. Ueno, D. Kitamaru, K. Morikawa, M. Tanaka, M. Miura-Mattausch, H. J. Mattausch, S. Kumashiro, T. Yamaguchi, K. Yamashita, and N. Nakayama, "Impurity-profile-based threshold-voltage model of pocket-implanted MOSFETs for circuit simulation," *IEEE Trans. Electron Devices*, vol. 49, no. 10, pp. 1783–1789, 2002.

[33] M. Miura-Mattausch, U. Feldmann, A. Rahm, M. Bollu, and D. Savignac, "Unified complete MOSFET model for analysis of digital and analog circuits," *IEEE Trans. CAD/ICAS*, vol. 15, no. 1, pp. 1–7, 1996.

[34] A. S. Grove, O. Leistiko, Jr., and C. T. Sah, "Redistribution of acceptor and donor impurities during thermal oxidation of silicon," *J. Appl. Phys.*, vol. 35, no. 9 , pp. 2695–2701, 1964.

[35] S. J. Wang, I. C. Chen, and H. L. Tigelaar, "Effects of poly depletion on the estimate of thin dielectric lifetime," *IEEE Electron Device Lett.*, vol. 12, no. 11 , pp. 617–619, 1991.

[36] S. Matsumoto, K. Hisamitsu, M. Tanaka, H. Ueno, M. Miura-Mattausch, H. J. Mattausch, S. Kumashiro, T. Yamaguchi, S. Odanaka, and N. Nakayama, "Validity of mobility universality for scaled metal-oxide-semiconductor field-effect transistors down to 100nm gate length," *J. Appl. Phys.*, vol. 92, no. 9, pp. 5228–5232, 1997.

[37] Chih-Tang Sah, "Fundamentals of Solid-State Electronics," *World Scientific Publishing Company*, 1991.

[38] C. -Y Lu, J. M. Sung, H. C. Kirsch, S. J. Hillenius, T. E. Smith, and L. Manchanda, "Anomalous $C - V$ characteristics of implanted poly MOS structure in n^+/p^+ dual-gate CMOS technology," *IEEE Electron Device Lett.*, vol. 10, no. 5, pp. 192–194, 1989.

[39] P. Habas and S. Selberherr, "On the effect of non-degenerate doping of polysilicon gate in thin oxide MOS-devices–Analytical modeling," *Solid-State Electron.*, vol. 33, no. 12, pp. 1539–1544, 1990.

[40] S. -W. Lee, C. Liang, C. -S. Pan, W. Lin, and J. B. Mark, "A study on the physical mechanism in the recovery of gate capacitance to C_{ox} in implanted polysilicon MOS structure," *IEEE Electron Device Lett.*, vol. 13, no. 1, pp. 2–4, 1992.

[41] F. Stern and W. E. Howard, "Properties of semiconductor surface-inversion layers in the electric quantum limit," *Phys. Rev.*, vol. 163, no. 3, pp. 816–

835, 19.

[42] F. Stern, "Self-consistent results for n-type Si inversion layers," *Phys. Rev.*, vol. B5, no. 12, pp. 4891–4899, 1972.

[43] T. Ando, A. B. Fowler, and F. Stern, "Properties of two-dimensional systems," *Review of Modern Physics*, vol. 54, no. 2, pp. 437–621, 1982.

[44] A. Hartstein, "Quantum interference in ultrashort channel length silicon metal-oxide-semiconductor field-effect transistors," *Appl. Phys. Lett.*, vol. 59, no. 16, pp. 2028–2030, 1991.

[45] T. Takahashi, "Modeling of current oscillation in SOI-MOSFET and development of its simulation program," Dissertation for Master of Engineering (Japanese), Hiroshima University, February, 2000.

[46] T. Takahashi, M. Miura-Mattausch, and Y. Omura, "Transconductance oscillations in metal-oxide-semiconductor field-effect transistors with thin silicon-on-insulator originated by quantized energy levels," *Appl. Phys. Lett.*, vol. 75, no. 10, pp. 1458–1460, 1999.

[47] H. Ueno, M. Tanaka, K. Morikawa, T. Takahashi, M. Miura-Mattausch, and Y. Omura, "Origin of transconductance oscillations in silicon-on-insulator metal-oxide-semiconductor field-effect transistors with an ultra-thin 6-nm-thick active layer," *J. Appl. Phys.*, vol. 91, no. 8, pp. 5360–5364, 2002.

[48] C. Jumgemann, A. Edmunds, and W. L. Engle, "Simulation of linear and nonlinear electron transport in homogeneous silicon inversion layers," *Solid-State Electron.*, vol. 36, no. 11, pp. 1529–1540, 1993.

[49] M. Rudan, "The $R - \Sigma$ method for nanoscale-device analysis," *Proc. SISPAD*, pp. 13–18, Tokyo, 2005.

[50] D. Querlioz, J. Saint-Martin, V.-N, Do, A. Bournel, and P. Dollfus, "Filly Quantum Self-Consistent Study of Ultimate DG-MOSFETs Including Realistic Scattering Using a Wigner Monte-Carlo Approach," *IEDM Tech. Dig.*, pp. 941–944, San Francisco, Dec., 2006.

[51] F. Stern, "Quantum properties of surface space-charge layers," *CRC Crit Rev. Solid State Sci.*, pp. 499–514, 1974.

[52] Z. Yu, R. W. Dutton, and R. A. Kiehl, "Circuit device modeling at the quantum level," *Proc. IWCE-6*, pp. 222–229, Osaka, Oct., 1998.

[53] R. Rios and N. D. Arora, "Determination of ultra-thin gate oxide thickness for CMOS structure using quantum effects," *IEDM Tech. Dig.*, pp. 613–616, San Francisco, Dec., 1994.

[54] M. G. Ancona and G. J. Iafrate, "Quantum correction to the equation of state of an electron gas in a semiconductor," *Phys. Rev. B*, vol. 39, no. 13, pp. 9536–9540, 1989.

[55] K. Morikawa, H. Ueno, D. Kitamaru, M. Tanaka, T. Okagaki, M. Miura-Mattausch, and H. J. Mattausch, "Quantum effect in sub-0.1μm MOSFET with pocket technologies and its relevance for the on-current condition," *Jpn. J. Appl. Phys.*, vol. 41, no. 4B, pp. 2359–2362, 2002.

[56] S. M. Sze, "Physics of Semiconductor Device (Second Edition)", *New York, John Wiley & Sons, Inc.*, 1981.

[57] S. Selberherr, "Analysis and Simulation of Semiconductor Devices",

Springer-Verlag, Wien New York, 1984.

[58] N. D. Arora and G. Gildenblat, "A semi-empirical model of the MOSFET inversion layer mobility for low-temperature operation," *IEEE Trans. Electron Devices*, vol. ED-34, no. 1, pp. 89–93, 1987.

[59] C. Lombardi, S. Manini, A. Spaporiti, and M. Vanzl, "A physically based mobility model for numerical simulation of nonplanar devices," *IEEE Tran. Computer-Aided Design*, vol. 7, no. 11, pp. 1164–1170, 1988.

[60] H. Shin, A. F. Tasch, Jr., C. M. Maziar, and S. K. Banerjee, ",", *IEEE Trans. Electron Devices*, vol. 36, no. 6, pp. 1117–1124, 1989.

[61] A. G. Savinis and J. T. Clemens, "Characterization of electron mobility in the inverted (100) surface," *IEDM Tech. Dig.*, pp. 18–21, Washington DC, Dec., 1979.

[62] S. Takagi, M. Iwase, and A. Toriumi, "On the universality of inversion-layer mobility in n- and p-channel MOSFETs," *IEDM Tech. Dig.*, pp. 398–401, San Francisco, Dec., 1988.

[63] K. Lee, J.-S. Choi, S.-P. Sim, and C.-K. Kim, "Physical understanding of low-field carrier mobility in silicon MOSFET inversion layer," *IEEE Trans. Electron Devices*, vol. 38, no. 8, pp. 1905–1911, 1991.

[64] C.-L. Huang and G. Sh. Gildenblat, "Measurements and Modeling of the n-channel MOSFET inversion layer mobility and device characteristics in the temperature range 60-300K," *IEEE Tran. Electron Devices*, vol. 37, no. 5, pp. 1289–1300, 1990.

[65] G. Mazzoni, A. L. Lacaita, L. M. Perron, and A. Pirovano, "On surface roughness-limited mobility in highly doped n-MOSFET's," *IEEE Tran. Electron Devices*, vol. 46, no. 7, pp. 1423–1428, 1999.

[66] C. T. Sah, T. H. Ning, and L. L. Tschopp, "The scattering of electrons by surface oxide charges and by lattice vibrations at the silicon-silicon dioxide interface," *Surface Science*, vol. 32, pp. 561–575, 1972.

[67] Y. Matsumoto and Y. Uemura, "Scattering mechanism and low temperature mobility of MOS inversion layers," *Jpn. J. Appl. Phys.*, Suppl. 2, Pt 2, pp. 367–370, 1974.

[68] J. T. Watt and J. D. Plummer, "Universal mobility-field curves for electrons and holes in MOS inversion layers," *Proc. VLSI Tech*, pp. 81–82, Karuizawa, May, 1987.

[69] *HiSIM1.1.0 User's Manual*, Hiroshima University and STARC, 2002.

[70] T. J. Krutsick and M. H. White, "Consideration of doping profiles in MOSFET mobility modeling," *IEEE Tran. Electron Devices*, vol. 35, no. 7, pp. 1153–1155, 1988.

[71] V. Vasileska and D. K. Ferry, "Scaled silicon MOSFET's: Universal mobility behavior," *IEEE Tran. Electron Devices*, vol. 44, no. 4, pp. 577–583, 1997.

[72] D. M. Caughey and R. E. Thomas, "Carrier mobilities in Silicon Empirically Related to Doping and Field," *Proc. Inst. Electr. Eng.*, vol. 55, no. 12, pp. 2192–2193, Dec., 1967.

[73] W. Shockley, "A unipolar field-effect transistor," *Proc. IRE*, vol. 40, no. 11, pp. 1365–1376, 1952.

[74] J. A. Cooper, Jr., and D. F. Nelson, "High-field drift velocity of electrons at the Si-SiO$_2$ interface as determined by a time-of flight technique," *J. Appl. Phys.*, vol. 53, no. 3, pp. 1445–1456, 1983.

[75] M. Tanaka, H. Ueno, O. Matsushima, and M. Miura-Mattausch, "High-electric-field electron transport at silicon/silicon-dioxide interface inversion layer," *Jpn. J. Appl. Phys.*, vol. 53, no. 3, pp. 1445–1456, 1983.

[76] M. Miura-Mattausch, H. Ueno, H. J. Mattausch, K. Morikawa, S. Itoh, A. Kobayashi, and H. Masuda, "100nm-MOSFET model for circuit simulation: Challenges and solutions," *IEICE Trans. Electron.*, vol. E86-C, no. 6, pp. 1009–1021, 2003.

[77] Private communication with C. T. Sah, and will be published at CJS in 2008.

[78] V. G. K. Reddi and C. T. Sah, "Source to drain resistance beyond pinch-off in metal-oxide-semiconductor transistor (MOSFET)," *IEEE Trans. Electron Devices*, vol. 12, no. 3, pp. 139–141, 1965.

[79] D. Frohman-Bentchkowsky and A. S. Grove, "Conductance of MOS transistors in saturation," *IEEE Trans. Electron Devices*, vol. ED-16, no. 1, pp. 108–113, 1969.

[80] Y. A. El-Mansy and A. R. Boothroyd, "A simple two-dimensional model of IGFET operation in the saturation region," *IEEE Trans. Electron Devices*, vol. ED-24, no. 3, pp. 241–253, 1977.

[81] M. E. Benna and M. E. Nokali, "A pseudo-two-dimensional analysis of short-channel MOSFETs," *Solid-State Electron.*, vol. 31, no. 2, pp. 269–274, 1988.

[82] M. Miura-Mattausch, "Analytical MOSFET model for quarter micron technologies," *IEEE Trans. Computer-Aided Design*, vol. 13, no. 5, pp. 610–615, 1994.

[83] *BSIM4.0.0 MOSFET Model, User's Manual*, Department of Electrical Engineering and Computer Science, University of California, Berkeley CA, 2000.

[84] Y. Sambonsugi, T. Maruyama, K. Yano, H. Sakaue, H. Yamamoto, E. Kawamura, S. Ohkubo, Y. Tamura, and T. Sugii, "A Perfect Process Compatible 2.49μm^2 Embedded SRAM Cell Technology for 0.13μm-Generation CMOS Logic LSIs," *VLSI Dig. Tech. Papers*, pp. 62–63, Hawaii, June, 1998.

[85] H. Nicollian and J. R. Brews, "MOS Physics and Technology," *John Wiley & Sons*, 1982.

[86] B. Davari, C. Koburger, T. Furukawa, Y. Tauer, W. Noble, A. Megdanis, J. Warnock, and J. Mauer, "A Variable-Size Shallow Trench Isolation (STI) Technology with Diffused Sidewall Doping for Submicron CMOS," *IEDM Tech. Dig.*, pp. 92–95, San Francisco, Dec., 1988.

[87] C.-R. Ji and C. T. Sah, "Two dimensional numerical analysis of the narrow gate effect in MOSFET," *IEEE Trans. Electron Devices*, vol. 30, no. 6, pp. 635–646, 1983.

[88] C.-R. Ji and C. T. Sah, "Analysis of the narrow gate effect in submicron MOSFETs," *IEEE Trans. Electron Devices*, vol. 30, no. 12, pp. 1672–1182,

1983.

[89] S. Chung and C. T. Sah, "A subthreshold model of the narrow-gate effect in MOSFETs," *IEEE Trans. Electron Devices*, vol. 34, no. 12, pp. 2521–2529, 1987.

[90] Y. P. Tsividis, "Operation and Modeling of the MOS Transistor," *McGraw-Hill*, 1999.

[91] G. Scott, J. Lutze, M. Rubin, F. Nouri, and M. Manley, "NMOS drive current reduction caused by transistor layout and trench isolation induced stress," *IEDM Tech. Dig.*, pp. 827–830, Washington DC, Dec., 1999.

[92] T. Ghani, M. Armstrong, C. Auth, M. Bost, P. Charvat, G. Glass, T. Hoffmann, K. Johnson, C. Kenyon, J. Klaus, B. McIntyre, K. Mistry, A. Murthy, J. Sandford, M. Silberstein, S. Sivakumar, P. Smith, K. Zawadzki, S. Thompson, and M. Bohr, "A 90nm High Volume Manufacturing Logic Technology Featuring Novel 45nm Gate Length Strained Silicon CMOS Transistors," *IEDM Tech. Dig.*, pp. 978–981, Washington DC, Dec., 2003.

[93] M. V. Fischetti and S. E. Laux, "Band Structure, Deformation Potentials, and Carrier Mobility in Strained Si, Ge and SiGe Alloys," *J. Appl. Phys.*, vol. 80, no. 4, pp. 2234–2252, 1996.

[94] N. Shigyo and R. Dang, "Analysis of anomalous subthreshold current in a fully recessed oxide MOSFET using a three-dimensional device simulator," *IEEE Trans. Electron Devices*, vol. ED-32, no. 2, pp. 441–445, 1985.

[95] N. Shigyo and T. Hiraoka, "A review of narrow-channel effects for STI MOSFET's: A difference between surface- and buried-channel cases," *Solid-State Electron.*, vol. 43, no. 11, pp. 2061–2066, 1999.

[96] S. Sivakumar, "Lithography Challenges for 32nm Technologies and Beyond," *IEDM Tech. Dig.*, pp. 985–988, San Francisco, Dec., 2006.

[97] W. Shockley, "Problems related to $p-n$ junctions in silicon," *Solid-State Electronics*, vol. 2, no. 1, pp. 35–67, 1961.

[98] W. Shockley, "Hot electrons in Germanium and Ohm's law," *Bell System Technical Journal*, vol. 30, pp. 990–1034, Oct., 1951.

[99] K. Hisamitsu, H. Ueno, M. Tanaka, D. Kitamaru, M. Miura-Mattausch, H. J. Mattausch, S. Kumashiro, T. Yamaguchi, K. Yamashita, and N. Nakayama, "Temperature-independence-point properties for 0.1μm-scale pocket-implant technologies and the impact on circuit design," *Proc. ASP-DAC*, pp. 179–183, Kita-Kyushu, Jan., 2003.

[100] F. H. Gaensslen and R. C. Jaeger, "Temperature Dependent Threshold Behavior of Depletion Mode MOSFETs," *Solid-State Electron.*, vol. 22, no. 4, pp. 423–430, 1979.

[101] IEEE Recommended Practices #P1485 on: Test Procedures for Microelectronic MOSFET Circuit Simulator Model Validation (http://ray.eeel.nist.gov/modval/database/contents/reports/micromosfet/standard.html).

[102] C. C. Enz, F. Krummenacher, and E. A. Vittoz, "An Analytical MOS Transistor Model Valid in All Regions of Operation and Dedicated to Low-Voltage and Low-Current Applications," *Special Issue of the Analog Integrated Circuits and Signal Processing Journal on Low-Voltage and*

Low-Power Design, vol. 8, pp. 83–114, July, 1995.

[103] K. Joardar, K. K. Gullapalli, C. C. McAndrew, M. E. Burnham, and A. Wild, "An improved MOSFET model for circuit simulation," *IEEE Trans. Electron Devices*, vol. 45, no. 1, pp. 134–148, 1998.

[104] C. C. McAndrew, "Validation of MOSFET Model Source-Drain Symmetry," *IEEE Trans. Electron Devices*, vol. 53, no. 9, pp. 2202–2206, 2006.

[105] E. Fong and R. Zeman, "Analysis of harmonic distortion in single-channel MOS integrated circuits," *IEEE J. Solid-State Circuits*, vol. SC–17, no. 1, pp. 83–86, 1982.

[106] Q. Huang, "A MOSFET-only continuous-time bandpass filter," *IEEE J. Solid-State Circuits*, vol. 32, no. 2, pp. 147–158, 1997.

[107] B. Razavi, "RF Microelectronics," *Prentice Hall PTR*, 1998.

[108] D. Navarro, N. Nakayama, K. Machida, Y. Takeda, H. Ueno, H. J. Mattausch, M. Miura-Mattausch, T. Ohguro, T. Iizuka, M. Taguchi, T. Kage, and S. Miyamoto, "Modeling of carrier transport dynamics at GHz-frequencies for RF circuit simulation," *Proc. SISPAD*, pp. 259-262, Munich, Sept., 2004.

[109] Y. Takeda, D. Navarro, S. Chiba, M. Miura-Mattausch, H. J. Mattausch, T. Ohguro, T. Iizuka, M. Taguchi, S. Kumashiro, and S. Miyamoto, "MOS-FET harmonic distortion analysis up to the non-quasi-static frequency regime," *CICC*, pp. 827–830, San Jose, Sept., 2005.

[110] D. Navarro, Y. Takeda, M. Miura-Mattausch, H. J. Mattausch, T. Ohguro, T. Iizuka, M. Taguchi, S. Kumashiro, and S. Miyamoto, T. Okagaki, M. Tanaka, H. Ueno, and M. Miura-Mattausch, "On the validity of conventional MOSFET nonlinearity characterization at RF switching," *IEEE Microwave and Wireless Components Lett.*, vol. 16, no. 3, pp. 125–127, 2006.

Chapter 4

Capacitances

As indicated in Fig. 4.1, four different capacitances (C_{int}, C_{ov}, C_{Qy}, and C_{f}) can be distinguished. Among them the intrinsic capacitance, C_{int}, is the main concern in the MOSFET capacitance [1–4], but the other three capacitances [1–4] the overlap capacitance C_{ov}, the longitudinal (lateral)-electric-field-induced capacitance C_{Qy}, and the fringing capacitance C_{f}, are also becoming increasingly important with reduced L_{gate} [5,6]. These four capacitances are described in the following four subsections 4.1, 4.2, 4.3 and 4.4. In the HiSIM model, we use the differential capacitance, defined as to the change of charge by the change of voltage applied. The voltage change will be a voltage step from a DC voltage level 1 increased slightly to a DC voltage level 2. The change of the DC charge distribution inside the transistor will be the total positive charge flowing into the respective node. This differential capacitance is also known as the quasi-static capacitance. It differs from the sinusoidal small-signal capacitance that appears in high-frequency circuits which take into account the delay time for the charge change to flow into the node from a change of the voltage. An example for such a delay is the transit delay of the charge flowing from the source to the drain through the MOSFET channel in response to a change of the applied voltage.

4.1 Intrinsic Capacitances

The intrinsic capacitance between two node J and K is defined as the change of the charge flowing out the node J and into of the node K due to the change of the potential applied to the node K as schematically shown in Fig. 4.2 for C_{gs} as an example. The intrinsic capacitance is defined by

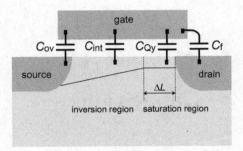

Fig. 4.1   Components of the MOSFET's gate capacitance included in the HiSIM model. In the saturation case the capacitance C_{Q_y} appears and is treated as a part of the gate-drain capacitance.

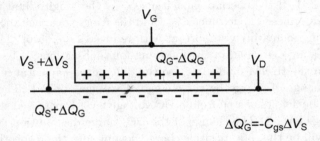

Fig. 4.2   Definition of differential capacitance C_{jk} schematically shown for the C_{gs} case.

$$C_{jk} = \delta \frac{\partial Q_J}{\partial V_K} \qquad (4.1)$$
$$\delta = -1 \text{ for } J \neq K$$
$$\delta = 1 \text{ for } J = K$$

where Q_J is the integrated DC charge distribution on node J at V_{jk} which is the DC potential applied to the node J relative to node K.

For the four-terminal or four-node MOSFET, with nodes S=Source, G=Gain, D=Drain and B=Bulk, there are six pairs of nodes or 12 capacitive elements given by the following four expressions:

$$C_{ss} = C_{sg} + C_{sd} + C_{sb} = C_{gs} + C_{ds} + C_{bs} \qquad (4.2)$$

$$C_{gg} = C_{gs} + C_{gd} + C_{gb} = C_{sg} + C_{dg} + C_{bg} \qquad (4.3)$$

$$C_{dd} = C_{ds} + C_{dg} + C_{db} = C_{sd} + C_{gd} + C_{bd} \qquad (4.4)$$

$$C_{bb} = C_{bs} + C_{bg} + C_{bd} = C_{sb} + C_{gb} + C_{db} \qquad (4.5)$$

These expressions also define the $C_{jk}(j{=}k){=}C_{jj}$ values which are the capacitances measured at the nodes j with all other nodes tied to the referrence. The C_{jj} capacitances are known as the short-circuit input capacitances, i. e., AC-short-circuit all other nodes to the reference node. Equations (4.2) – (4.5) assume $C_{jk}{=}C_{kj}(j{\neq}k)$, that is, the 4-terminal device is completely symmetrical or it is a reciprocal 4-terminal network. This assumption is valid for the MOSFET only when there is no DC current flowing into or out of the 4 terminals, which means at electrical equilibrium [12] with $V_{ds} = 0$V. When $V_{ds} \neq 0$V, the 4-terminal MOSFET is no longer reciprocal and $C_{jk} \neq C_{kj}$. The capacitances are computed as a function of V_{gs} at $V_{ds}{=}1.0$V for an nMOSFET, covering the three current ranges (subthreshold, saturation, linear), and are shown in Figure 4.3. They are also computed as a function of V_{ds} at $V_{gs}{=}1.0$V, covering the linear and saturation current ranges, and are shown in Figure 4.4. The non-reciprocity is a universal property of active networks such as the transistor, and the difference is shown as a function of V_{ds} for $V_{gs}{=}1.0$V in Fig. 4.5 for the common source connection, defined by the following three mutual or transcapacitances

$$C_m = C_{dg} - C_{gd} \qquad (4.6)$$

$$C_{mb} = C_{db} - C_{bd} \qquad (4.7)$$

$$C_{mx} = C_{bg} - C_{gb} \qquad (4.8)$$

They have been called transcapacitances or mutual capacitances, from the short-circuit admittance matrix representation of a two-port network. The curves of Figs. 4.3 – 4.5 can be computed readily using the surface-potential-based analytical solutions given by Eqs. (2.32) – (2.36) in Section 2.4, by taking their partial derivatives with respect to the specific applied voltage at a given node. As mentioned earlier, these voltages have to be taken relative to the source node which is selected as the reference node. In principle, there is no additional model parameter needed to compute these intrinsic capacitances. The area, which has to be considered, is the channel

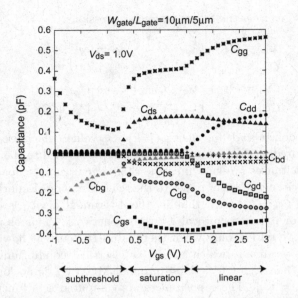

Fig. 4.3 Compact modeling of the 9 independent intrinsic capacitances of the MOSFET with HiSIM as a function of V_{gs}.

width times the channel length, and the channel length is the gate length minus the two overlap lengths between the gate electrode and the diffused drain and souce regions.

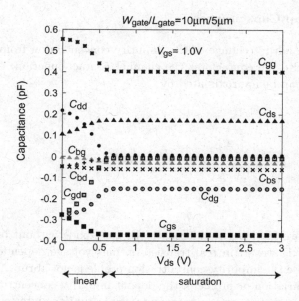

Fig. 4.4 Calculated 9 independent intrinsic capacitances with HiSIM as a function of V_{ds}.

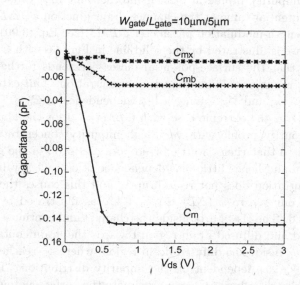

Fig. 4.5 Calculated transcapacitances with HiSIM as a function of V_{ds}. At $V_{ds} = 0$V all capacitances converge to zero.

4.2   Overlap Capacitances

Assuming an adrupt change of the impurity concentration from substrate to source and drain contacts at the channel/contact junction, the overlap capacitance can be approximated by

$$C_{\text{ov}} = -\frac{\epsilon_{\text{ox}}}{T_{\text{ox}}} L_{\text{over}} W_{\text{eff}} \qquad (4.9)$$

L_{over} is the overlap length underneath the gate oxide without bias voltage dependence. However, in real transistors, bias voltage dependence is observed because the impurity-concentration profile is not abrupt. Therefore, the spatial variation or profile of the dopant impurity concentration in the semiconductor surface at the gate-oxide/semiconductor interface must be considered. For the channel-length modulation model (see Section 3.5), the effects from impurity profile are taken into account in the magnitude of the surface-potential value at the channel/drain junction $\phi_{\text{S}}(\Delta L)$ through the channel-length-modulation parameter $CLM1$ (see Eq. (3.90)). A typical general case is illustrated by the solid line in Fig. 4.6 with $CLM1=0.7$ which shows that the surface-potential increase extends further into the drain contact region with high impurity concentration. This extension depends on V_{ds}, V_{gs} and V_{bs}, giving a bias dependence of C_{ov}. The dotted line in Fig. 4.6 is the extreme case with $CLM1=1.0$ for the abrupt channel/drain impurity profile with very high impurity concentration in the drain contact so that there is little or no potential drop in the gate/drain overlap region and hence little bias dependence of C_{ov}. Generally, the impurity concentration does not rise abruptly and this causes the potential to increase from ϕ_{SL} to $\phi_{\S}(\Delta L) \simeq \phi_{\text{S0}} + V_{\text{ds}}$ as illustrated by the solid line in Fig. 4.6. The rise starts inside the channel and continues deep into the implanted and diffused drain region towards the drain contact. Thus, the position y_{n} inside the diffused drain region, where ϕ_{S} reaches its maximum $\phi_{\text{S0}} + V_{\text{ds}}$, is dependent on the impurity distribution. To account for the impurity distribution, we approximate the surface potential distribution (the solid line in Fig. 4.6) in the n$^+$drain by a quadratic function $\phi_{\text{S}}(y) = a(y - y_{\text{n}})^2 + \phi_{\text{S0}} + V_{\text{ds}}$.

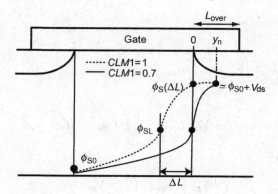

Fig. 4.6 Schematic diagram of the surface-potential distribution along the channel for an abrupt junction ($CLM1=1$) and a non-abrupt junction ($CLIM1=0.7$) between channel and the drain region.

Boundary conditions of the potential distribution are

i) at $y = 0$
$$\phi_S(\Delta L) = a \times y_n^2 + \phi_{S0} + V_{ds}$$
$$E(\Delta L) = -2a \times y_n$$

ii) at $y = y_n$
$$\phi_S(y_n) \simeq \phi_{S0} + V_{ds}$$
$$E(y_n) = 0 \tag{4.10}$$

where a is the steepness of the potential increase along the drain-region surface and $y=0$ is the position of the junction. Once we know the surface potential distribution along the overlap region caused by the impurity gradient, the overlap charge in the drain contact (Q_{gdo}) can be calculated under the assumption that the injected carriers into the drain contact can be ignored as

$$\frac{Q_{gdo}}{W_{eff}C_{ox}} = \int_0^{L_{over}} \{V_{gs''} - (\phi_s(y) - \phi_{s0})\} dy \tag{4.11}$$

resulting in

i) $y_n \leq L_{over}$

$$\frac{Q_{gdo}}{WC_{ox}} = (V''_{gs} - V_{ds})L_{over} - \frac{a}{3}y_n^3$$

ii) $y_n > L_{over}$

$$\frac{Q_{gdo}}{WC_{ox}} = (V''_{gs} - V_{ds})L_{over} - \frac{a}{3}\{(L_{over} - y_n)^3 + y_n^3\} \qquad (4.12)$$

where V''_{gs} is given

$$V''_{gs} = V_{gs} - V'_{fb} \simeq V_{gs} \qquad (4.13)$$

and V'_{fb} is the flat-band voltage in the drain contact, which can approximated to be zero. L_{over} is the overlap length, and y_n is derived as a function of the surface potentials in the form

$$y_n = \left(-\frac{\phi_{S0} + V_{ds} - \phi_S(\Delta L)}{a}\right)^{\frac{1}{2}} \qquad (4.14)$$

The steepness value $a \simeq -1 \times 10^{11}$ can be approximately obtained from a 2D-process simulation result. If the impurity profile is abrupt and the contact is heavily doped, then a is very large ($\sim$ infinity) and y_n reduces to zero. Then, the overlap capacitance is equal to that given in Eqs. (4.9). The drain overlap capacitances has a major influence on the circuit performances as a part of the Miller capacitance.

The gate charge above the overlapped source-side contact (Q_{gso}) can be written as

$$\frac{Q_{gso}}{WC_{ox}} = V''_{gs} \cdot L_{over} \qquad (4.15)$$

Finally, the overlap capacitances at source and drain are determined by

$$C_{gxo} = \frac{dQ_{go}}{dV_x}; \quad Q_{go} = Q_{gs0} + Q_{gdo}. \qquad (4.16)$$

where x stands for d (drain) or s (source).

Fig. 4.7 compares calculated C_{gdo} results of the developed model with those of a 2D-device simulator. It verifies also that the channel-length modulation influences significantly the overlap capacitance C_{ov}. Another feature to be aware of is that C_{ov} is nearly independent of bias conditions for small V_{ds}, where it reaches its maximum value. Fig. 4.8 explains the reason for this fact. Under the non-saturation condition the surface potential ϕ_{SL} is written as

$$\phi_{SL} \simeq \phi_{S0} + V_{ds} \tag{4.17}$$

It must be remembered that under the gradual-channel approximation, the

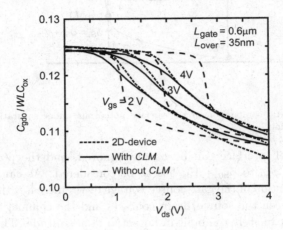

Fig. 4.7 Calculated drain-overlap capacitance C_{gdo} with the derived compact model in comparison to a 2D-device-simulation result.

surface potential ϕ_{SL} gives the value at the end of the region for which this approximation is valid. Thus the position where ϕ_{SL} reaches $\phi_{S0} + V_{ds}$ is approximated to be at the junction of the channel/drain contact, resulting in $y_n = 0$ for the non-saturation condition. On the other hand, under the saturation condition a steep potential increase starts beyond the pinch-off point, and continues further into the drain contact region. This causes an increase of y_n, which means $y_n > 0$, resulting in a reduction of C_{ov}. However, the reduction of C_{ov} is in real technologies not so drastical as could be expected. The reason is attributed to the fact that advanced technologies tend to have a relatively abrupt junction profile, avoiding high contact resistances different from older technologies with the lightly-doped drain (LDD) contact.

An alternative pragmatic overlap-capacitance model is discussed in the following. The most important device features determining circuit performances are those under the normal operation condition, namely beyond the threshold condition. Therefore the potential drop which occurs at the

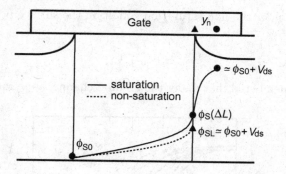

Fig. 4.8 Schematic comparison of the surface potentials under saturation and non-saturation condition.

contact/channel interface can be usually ignored, and the potential distribution from ϕ_{S0} to $\phi_{S0} + V_{ds}$ has to be considered. As can be seen in Fig. 4.9, a reduction of the gate-source voltage of V_{gs} increases the potential difference between the source/drain contacts and the channel from about 60mV to approximately the built-in potential in magnitude. The small potential barrier of 60mV, remaining at the junction for large V_{gs}, protects against free carrier injection from the source contact into the channel. The potential difference $V_{bi} - \phi_{S0}$ between source/drain and channel becomes larger as V_{gs} decreases, resulting in the reduction of the surface potential at the junction. Thus we separate the overlap region into two parts to derive a simple equation, $L_{over} - L'$ and L'. The equation for the overlap capacitance at the source end is then written in the form

$$V''_{gs} \cdot L'_{over} = V''_{gs} \cdot (L_{over} - L') - \frac{(V_{bi} - \phi_{S0})}{2} \cdot L' \qquad (4.18)$$

$$\simeq V''_{gs} \cdot L_{over} - \frac{(V_{bi} - \phi_{S0})}{2} \cdot L' \qquad (4.19)$$

where V_{bi} is fixed to 1.2V so that $(V_{bi} - \phi_{S0})$ can never become negative. Since L' is expected to be V_{gs} dependent, the final model equation is written as

$$V''_{gs} \cdot L'_{over} \simeq V''_{gs} \cdot L_{over} - OVSLP\,(V_{bi} - \phi_{S0})\,(OVMAG + V''_{gs}) \qquad (4.20)$$

with two model parameter $OVSLP$ and $OVMAG$. The model parameter $OVSLP$ describes the V_{gs} dependence of the overlap length, and $OVMAG$ determines the starting value of the V_{gs} dependence. A noticeable effect due to the V_{gs} dependence is in practice only observed for very short-channel devices, when the overlap capacitance dominates over the intrinsic capacitance. Since it is difficult to know the length L_{over} a model parameter named $XLDOV$ is introduced to adjust the overlap capacitance

$$L_{over} = XLDOV \tag{4.21}$$

For the overlap capacitance at the drain end, V_{gs}'' in Eq. (4.20) is replaced

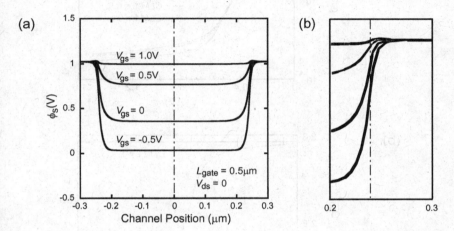

Fig. 4.9 (a) Surface potential magnitude from source to drain as a function of V_{gs} for $V_{ds} = 0$V. (b) An enlargement of (a) at the drain junction. The vertical dotted-dashed line depicts the junction.

with $V_{gs}'' - V_{ds}$.

A big advantage of the developed overlap-capacitance model is that the complicated bias dependence of C_{gdo} can be automatically calculated without fitting parameters as an initial test. The overlap capacitances are included as extensions of the channel-length-modulation model as can be seen in Fig. 4.7 [9]. Therefore, the surface potential at the drain junction $\phi_S(\Delta L)$ influences on the capacitance values at the same time. Fig. 4.10a shows calculated channel conductance g_{ds} for different $CLM1$ values (see Section 3.5). Here, $CLM1=1$ means that the contact profile is abrupt and that no potential increase occurs in the overlap region. In this case, a strong channel-length-modulation effect occurs in the MOSFET saturation, resulting in a large g_{ds} and at the same time in a large overlap capacitance, as demonstrated in Fig. 4.10b.

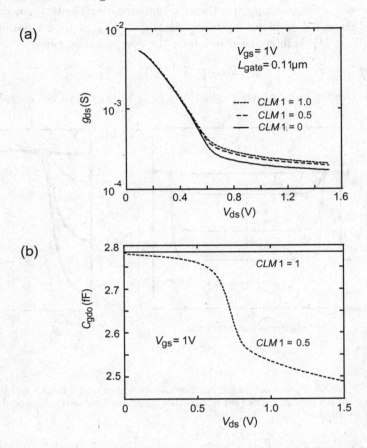

Fig. 4.10   (a) Calculated channel conductance g_{ds}, and (b) gate-drain overlap capacitance C_{gdo} as a function of V_{ds} for different contact profiles, represented by the channel-length-modulation parameter $CLM1$.

The intrinsic part of the Miller capacitance C_{gd} reduces to zero under the saturation condition. However, the overlap capacitances remains finite and give a dominating contribution, as shown in Fig. 4.11, influencing strongly on circuit performances. Importance of accurate modeling of the overlap capacitance is therefore obvious.

HiSIM includes an option for the case that no information about the overlap region is available. The default overlap capacitances (Model Flag: $COOVLP = 0$) are calculated as bias-independent drain and source overlap

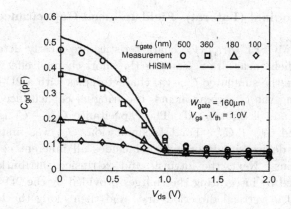

Fig. 4.11   Gate-drain capacitances for different L_{gate} as a function of V_{ds}.

capacitances. In this case user-defined user-defined values can be specified by using the input parameters $CGDO$ and $CGSO$. If these values are not specified, the overlap capacitances are calculated using the equation

$$C_{\text{ov}} = -\frac{\epsilon_{\text{ox}}}{T_{\text{ox}} \cdot L_{\text{over}} \cdot W_{\text{eff}}} \qquad (4.22)$$

The gate-to-bulk overlap capacitance C_{gbo} is calculated only with a user-defined value $CGBO$ using the equation

$$C_{\text{gbo}} = -CGBO \cdot L_{\text{gate}} \qquad (4.23)$$

independent of the model flag $COOVLP$.

4.3   Longitudinal (Lateral) -Field-Induced Capacitance

In small-size MOSFETs, device characteristics are mostly determined by the field gradient near drain, causing the short-channel effects [6]. Figure 4.12 shows the simulated C_{gd}-V_{ds} characteristics with a 2D-device simulator. Among the C_{gd} components, the fringing capacitance C_f is bias-independent, and is subtracted. The capacitance $C_{gd} - C_f$ is compared with simulated $C_{int} + C_{ov}$. For this simulation, C_{gd} was first calculated using the 2D-device simulator, which includes all inherent contributions in 2 dimensions. Next, the intrinsic and extrinsic contributions to C_{gd} were calculated by integrating the charges provided by the 2D-device simulator along the vertical direction first, and then along the longitudinal direction bounded by L_{gate}, separately. The integration includes all the bias-dependent components but excludes the coupling effect between the vertical and the longitudinal charge distribution. This coupling component is coming from the charges induced by the field gradient along the channel, and is called longitudinal-field-induced capacitance C_{qy} in HiSIM. BSIM models this contribution as a so-called inner-fringing capacitance, neglecting the carrier existence in the pinch-off region [10]. However, a detailed investigation with 2D-device simulations verifies the important influence of the velocity saturation on C_{gd} [11], which can be understood as a high-field effect, and is thus based on the same origin as the modeling approach used in HiSIM. In the above described analysis a significant difference between the C_{gd}-C_f and $C_{int}+C_{ov}$ calculations, i.e. a large longitudinal-field-induced capacitance C_{Qy}, was found. The difference as shown in Fig. 4.12 is mainly visible in the saturation region and is explained by the longitudinal electric field (E_y) gradient along the channel.

The space charge Q_y caused by the electric field gradient is calculated according to Eq. (4.24)

$$Q_y = \epsilon_{Si} W_{eff} \int W_d(y) \frac{dE_y(y)}{dy} dy \qquad (4.24)$$

where W_{eff} is the channel width in the high field region, $W_d(y)$ is the depletion width and y is the direction along the channel. The MOSFET condition used in deriving Eq. (4.24) is depicted in Fig. 4.13 with $W_d(y)$ treated as constant in the pinch-off region, the region where the longitudinal field is greater than the field perpendicular to the channel. The longitudinal field increases with V_{ds} as shown in the lower part of Fig. 4.13. The capacitance C_{Qy} due to the charge Q_y is given by $C_{Qy}=dQ_y/dV$, and is

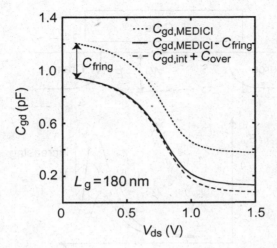

Fig. 4.12 Comparison of the gate-drain capacitance calculated by a 2D-device simulator ($C_{\text{gd,2D}}$) and as calculated by charge integration ($C_{\text{int}}+C_{\text{ov}}$). A difference is visible in the saturation region and is owing to the longitudinal field gradient. The fringing capacitance C_{f} is constant and bias-independent, and thus subtracted from $C_{\text{gd,2D}}$.

added to the conventional intrinsic and extrinsic components as shown in Fig. 4.1. The quantitative magnitude of C_{Qy} for $L_{\text{gate}} = 0.18\mu\text{m}$, calculated with a 2D-device simulator, is demonstrated in Fig. 4.14. As expected, this C_{gd} component becomes significant under the saturation condition. More importantly in short-channel MOSFETs, E_{y} becomes larger and thus C_{Qy} influences the C_{gd} capacitance stronger.

Since HiSIM is surface-potential based, implementation of the developed C_{gd} model is straightforward. HiSIM knows all the potential values necessary to describe the induced capacitance, especially near the drain junction, where the maximum longitudinal field occurs. The surface-potential distribution as described by HiSIM is schematically shown in Fig. 4.15. By applying the Gauss law in the region where the longitudinal field steeply increases [80], Q_{y} is expressed as

$$Q_{\text{y}} = \epsilon_{\text{Si}} W_{\text{eff}} W_{\text{d}}(E_{\text{max}} - E_{\text{c}}) \qquad (4.25)$$

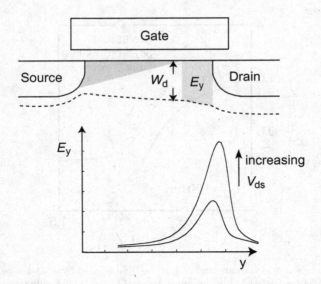

Fig. 4.13 A MOSFET in saturation showing the effect of the longitudinal field. The Gauss law is applied in the shaded region near the drain contact. The lower graph shows the calculated longitudinal field with a 2D-device simulator. This field increases with drain-source bias, penetrating the overlap region at higher bias voltages.

with

$$E_{\max} = \frac{\phi_{s0} + V_{ds} - \phi_{S}(\Delta L)}{X_{Qy}} \tag{4.26}$$

$$W_{d} = \sqrt{\frac{2\epsilon_{Si}}{qN_{sub}}(\phi_{S}(\Delta L) - V_{bs})} \tag{4.27}$$

Here, $\phi_{s0} + V_{ds}$ is the surface potential in the drain region, $\phi_s(\Delta L)$ is the surface potential at the channel/drain junction [9], and E_c is the electric field at the end of the gradual-channel approximation. W_d is approximated as constant near the drain region. A parameter X_{Qy}, independent of L_{gate}, is introduced indicating the position of the maximum longitudinal field relative to the channel/drain junction. The induced capacitance C_{Qy} is then calculated as

$$C_{Qy} = \epsilon_{Si} W_{eff} W_d \left(\frac{\dfrac{d\phi_{s0}}{dV_{ds}} + 1 - \dfrac{d\{\phi_S(\Delta L)\}}{dV_{ds}}}{X_{Qy}} - \frac{dE_c}{dV_{ds}} \right) \tag{4.28}$$

where the second term of the right-hand-side equation is expected to be

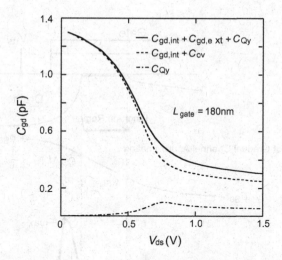

Fig. 4.14 $C_{\text{gd,int}}+C_{\text{gd,ext}}$ and C_{Qy} contributions to the gate-drain capacitance as simulated in a 2D-device simulator. C_{Qy} becomes a significant contribution to C_{gd} in the saturation region.

small and is neglected under the gradual-channel approximation. In the saturation condition, C_{Q_y} together with the overlap capacitance dominates the gate-drain capacitance C_{gd}. This effect is more visibly observed as the gate-length reduces. Therefore, in the C_{gd} modeling, C_{Q_y} is added to the conventional components as depicted in Fig. 4.14.

4.4 Fringing Capacitance

To derive a simple equation for calculating the fringing capacitance we consider two dielectric plates, the edge of the gate-poly electrode and the surface of the contact which are facing each other with an angle α as schematically shown in Fig. 4.16. Here the lines of the electric force between the two electrodes are assumed to be parallel. For further simplification the angle α is fixed to $\pi/2$. The resulting fringing capacitance is bias independent and is written as [5]

$$C_{\text{f}} = \frac{\epsilon_{\text{ox}}}{\pi/2} W_{\text{gate}} \ln\left(1 + \frac{TPOLY}{T_{\text{ox,eff}}}\right) \tag{4.29}$$

where $TPOLY$ is the gate-poly thickness.

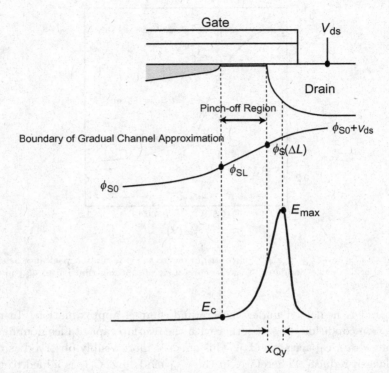

Fig. 4.15 Schematic representation of the surface potential distribution of a MOSFET in saturation as modeled by HiSIM. Also depicted are the field points significant in the C_{gd} modeling.

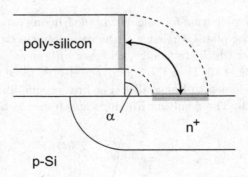

Fig. 4.16 Schematic for the modeling approach of the fringing capacitance.

4.5 Summary of Equations and Model Parameters Appearing in Chapter 4 for Capacitances

[Overlap Capacitances]
Bias Dependent-1

i) $y_n \leq L_{\text{over}}$

$$\frac{Q_{\text{gxo}}}{WC_{\text{ox}}} = V'_{\text{gs}} \cdot L_{\text{over}} - \frac{a}{3} y_n^3$$

ii) $y_n > L_{\text{over}}$

$$\frac{Q_{\text{gxo}}}{WC_{\text{ox}}} = V'_{\text{gs}} \cdot L_{\text{over}} - \frac{a}{3} \{ (L_{\text{over}} - y_n)^3 + y_n^3 \}$$

$$y_n = \left(-\frac{\phi_{\text{S0}} + V_{\text{ds}} - \phi_{\text{S}}(\Delta L)}{a(\simeq -1 \times 10^{11})} \right)^{\frac{1}{2}} \qquad (V_{\text{ds}}=0 \text{ for x=S})$$

Bais Dependent-2

$$\frac{Q_{\text{gxo}}}{WC_{\text{ox}}} = \frac{XLDOV - (OVMAG + V'_{\text{gs}})OVSLP(V_{\text{bi}} - \phi_{\text{S0}})}{V'_{\text{gs}}}$$

Bias independent

$$\frac{Q_{\text{gxo}}}{WC_{ox}} = V'_{\text{gs}} \cdot XLDOV$$

i) x=S

$$V'_{\text{gs}} = V_{\text{gs}}$$

ii) x=D

$$V'_{\text{gs}} = V_{\text{gs}} - V_{\text{ds}}$$

$$C_{\text{gxo}} = \frac{dQ_{\text{go}}}{dV_{\text{x}}}; \quad Q_{\text{go}} = Q_{\text{gso}} + Q_{\text{gdo}}.$$

[Longitudinal-Field-Induced Capacitance]

$$C_{\mathrm{Qy}} = \epsilon_{\mathrm{Si}} W_{\mathrm{eff}} W_{\mathrm{d}} \left(\frac{\dfrac{d\phi_{\mathrm{S0}}}{dV_{\mathrm{ds}}} + 1 - \dfrac{\{d\phi_{\mathrm{S}}(\Delta L)\}}{dV_{\mathrm{ds}}}}{XQY} \right)$$

$$W_{\mathrm{d}} = \sqrt{\frac{2\epsilon_{\mathrm{Si}}}{qN_{\mathrm{sub}}}(\phi_{\mathrm{S}}(\Delta L) - V_{\mathrm{bs}})}$$

$$\phi_{\mathrm{S}}(\Delta L) = CLM1(\phi_{\mathrm{S0}} + V_{\mathrm{ds}}) + (1 - CLM1)\phi_{\mathrm{SL}}$$

[Fringing Capacitance]

$$C_{\mathrm{f}} = \frac{\epsilon_{\mathrm{ox}}}{\pi/2} W_{\mathrm{gate}} \ln \left(1 + \frac{TPOLY}{T_{\mathrm{ox,eff}}} \right)$$

$$T_{\mathrm{ox,eff}} = T_{\mathrm{ox}} + \Delta Tox$$

Table 4.1   HiSIM model parameters for capacitances.

XQY	distance from drain junction to maximum electric field point
XLDOV	overlap length L_{over}
OVSLP	model parameter for overlap capacitance
OVMAG	model parameter for overlap capacitance
CGSO	gate-to-source overlap capacitance
CGDO	gate-to-drain overlap capacitance
CGBO	gate-to-bulk overlap capacitance
TPOLY	height of the gate poly-Si

Bibliography

[1]   Y. P. Tsividis, "Operation and Modeling of the MOS Transistor (Second Edition)," *McGraw-Hill*, 1999.

[2]   B. J. Sheu and P.-K. Ko, "Measurement and modeling of short-channel MOS transistor gate capacitances," *IEEE J. Solid-State Circuits*, vol. SC-22, no. 3, pp. 464–472, 1987.

[3]   H.-J. Park, P.-K. Ko, and C. Hu, "A charge sheet capacitance model of short channel MOSFET's for SPICE," *IEEE Trans. Computer-Aided Design*, vol. 10, no. 3, pp. 376–389, 1991.

[4] W. Budde and W. H. Lamfried, "A charge-sheet capacitance model based on drain current modeling," *IEEE Trans. Electron Devices*, vol. 37, no. 7, pp. 1678–1687, 1990.

[5] R. Shrivastava and K. Fitzpatrik, "A simple model for the overlap capacitance of a VLSI MOS device," *Proc. IEEE*, vol. ED-29, no. 12, pp. 1870–1875, 1982.

[6] D. Navarro, H. Kawano, K. Hisamitsu, T. Yamaoka, M. Tanaka, H. Ueno, M. Miura-Mattausch, H. J. Mattausch, S. Kumashiro, T. Yamaguchi, K. Yamashita, and N. Nakayama, "Circuit-simulation model of C_{gd} changes in small-size MOSFETs due to high channel-field gradients," *IEICE Trans. Electron.,*, vol. E-86-C, no. 3, pp. 474–480, 2003.

[7] D. E. Ward and R. W. Dutton, "A charge-oriented model for MOS transistor capacitances," *IEEE J. Solid-State Circuits*, vol. SC-13, no. 5, pp. 703–708, 1978.

[8] C. Turchetti, G. Masetti and Y. Tsividis, "On the small-signal behavior of the MOS transistor in quasistatic operation," *Solid-State Electron.*, vol. 26, no. 10, pp. 941–949, 1983.

[9] D. Navarro, T. Mizoguchi, M. Suetake, K. Hisamitsu, H. Ueno, M. Miura-Mattausch, H. J. Mattausch, S. Kumashiro, T. Yamaguchi, K. Yamashita, and N. Nakayama, "A compact model of the pinch-off region of 100nm MOSFETs based on the surface potential," *IEICE Trans. Electron.*, vol. E88-C, no. 5, pp. 1079-1086, 2005.

[10] *BSIM4.0.0 MOSFET Model, User's Manual*, Department of Electrical Engineering and Computer Science, University of California, Berkeley CA, 2000.

[11] H. Iwai, M. R. Pintp, C. S. Rafferty, J. E. Oristian, and R. W. Dutton, "Analysis of velocity saturation and other effects on short-channel MOS transistor capacitances," *IEEE Trans. Computer-Aided Design*, vol. CAD-6, no. 2, pp. 173–184, 1987.

[12] J. Meyer, "MOS models and circuit simulation," *RCA Review*, vol.32, pp. 42-63, 1971.

Chapter 5

Leakage Currents and Junction Diode

The DC current in MOSFET is observed not only to flow into the drain electrode but into all electrodes or terminals in the entire applied voltage or bias voltage range. The origins of the currents are different under different bias voltage configurations. The differences arise from the increasing electric fields at different locations inside the transistor as transistor size scales down. The MOSFET (nMOST to be used as the example in this chapter) consists of two or three p/n junctions: the mandatory n^+Source/p-Base and n^+Drain/p-Base junctions, and the Low/High/Low p-Base/p^+Well/p-Substrate junction or the optional p-Base/n^+Well/p-Substrate electrical isolation junction. Here + and - indicate the very high and moderately low dopant impurity concentration in each of the regions or layers. In addition, the Gate-conductor/Oxide-insulator/p-Base is a M/O/S junction. DC currents pass through all of these p/n, $p^-/p^+/p$, $p^-/n^+/p^-$ and M/O/S junctions. Some of the currents are undesirable because they are not switching or amplifying the signals, while they are dissipating power and generating heat. We shall focus on the undesirable currents, as leakage currents and stray capacitances.

5.1 Leakage Currents

5.1.1 *Substrate Current*

The substrate current, I_{sub}, in a MOS transistor at hight voltage is generated via interband impact generation of electron-hole pairs by the energetic or hot electrons and holes [1], known historically and habitually by the jargon "impact ionization" although there is no ionization at all [2]. The energetic or hot electrons are first created by acceleration of the thermal electrons (or holes) injected into the high-electric-field depleted space-charge region (or depletion region) of the reversed-biased n^+ Drain/p-Base junction, labeled by a length from $y=0$ to $y = \Delta L'$ as shown in Fig. 5.1. The lighter shaded region is the high-field space-charge region on the p-Basewell side and the darker shaded region is the high-field space-charge region on the n^+ Drain side of the p-Basewell/n^+ Drain junction. This p/n junction leakage current can be expressed by

$$I_{\mathrm{sub}} = \alpha I_{\mathrm{ds}} \Delta L' \tag{5.1}$$

This formula is applicable to all regions of high electric field, such as the $\Delta L'$ in the channel-length-modulation modeling in Section 3.5, with $\Delta L'$ replaced by an effective impact-generation length. The coefficient α in Eq. (5.1) is the interband e-h pair generation coefficicent (also known as the impact ionization coefficient) which was derived by Shockley [2] using his lucky electron model for hot electrons and phonon emission [1, 2]. His formula gives [1–4]

$$\alpha = A \exp\left(-\frac{B}{E_{\mathrm{y}}}\right). \tag{5.2}$$

where A is a power law of the electric field E_{y} and B is an electron-phonon collision parameter. In HiSIM, A and B are taken as model parameters. Because $E_{\mathrm{y}} = E_{\mathrm{y}}(y)$, Eq. (5.1) needs to be integrated over the high-field region

$$I_{\mathrm{sub}} = \int_0^{\Delta L'} I_{\mathrm{ds}} A \exp\left(-\frac{B}{E_{\mathrm{y}}(y)}\right) dy. \tag{5.3}$$

So, in HiSIM, we approximate

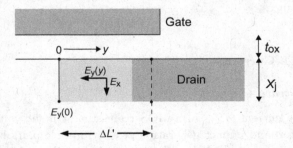

Fig. 5.1   Schematics of the high field region at the n^+ Drain/p-Base junction of the drain end of the MOS transistor.

$$E_y(y) = E_0 + \frac{(E_{\max} - E_0)}{\Delta L'} y \tag{5.4}$$

and we get the Arora solution [5] assuming the maximum field occurs at $y = \Delta L'$ and the field at the pinch-off point is much smaller than $E_{\max}$

$$I_{\text{sub}} = \frac{A}{B} \big(\phi(\Delta L') - \phi(0)\big) I_{\text{ds}} \exp\left(-\frac{\lambda B}{\phi(\Delta L') - \phi(0)}\right) \tag{5.5}$$

where

$$\lambda^2 = \frac{\epsilon_{\text{Si}} X_j T_{\text{ox}}}{\epsilon_{\text{ox}}} \tag{5.6}$$

and X_j is the n$^+$Drain/p-Base junction depth. Note that Eq. (5.5) is just a function of the total surface potential drop through the p/n junction with two model parameters A and B, thus, it can be simplified by using the HiSIM model parameters defined by $SUB1 = A/B$ and $SUB2 = \lambda B$, then Eq. (5.5) becomes

$$I_{\text{sub}} = SUB1 \big(\phi(\Delta L') - \phi(0)\big) I_{\text{ds}} \exp\left(-\frac{SUB2}{\phi(\Delta L') - \phi(0)}\right) \tag{5.7}$$

The surface potentials $\phi(0)$ and $\phi(\Delta L')$ are further modeled as

$$\phi(0) = \phi_{\text{SL}} \tag{5.8}$$
$$\phi(\Delta L') = \phi_{\text{S0}} + V_{\text{ds}} \tag{5.9}$$

The simple description of Eqs. (5.7) – (5.9) is insufficient to reproduce substrate-current measurements in MOSTs of different channel lengths. The reason is the 2-dimensional variation of the electric field at the n$^+$Drain/p-Basewell junction region and the n$^+$Drain contact [6]. One additional parameter is introduced in HiSIM to account for the 2D electric field variation [7] by replacing the constant $SUB1$ by adding a channel length dependence as follows [8]

$$SUB1 = SUB1 + \frac{SUB4}{L_{\text{eff}}} \tag{5.10}$$

The Fowler-Nordheim tunneling mechanism is included in order to reproduce a wide range of measurements, as shown in Fig. 5.2.

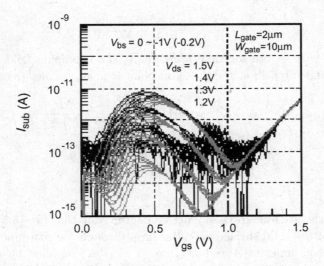

Fig. 5.2   Measured substrate currents and their reproduction with HiSIM. $V_{\mathrm{bs}} = 0\mathrm{V}$ to $-1.0\mathrm{V}$ $(-0.2\mathrm{V})$ and $V_{\mathrm{ds}} = 1.5\mathrm{V}$ to $1.2\mathrm{V}$ $(-0.1\mathrm{V})$ with $L_{\mathrm{gate}} = 2\mu\mathrm{m}$ and $W_{\mathrm{gate}} = 10\mu\mathrm{m}$.

5.1.2   *Gate Current*

All possible gate leakage currents of the MOS transistor due to tunneling are schematically shown in Fig. 5.3.

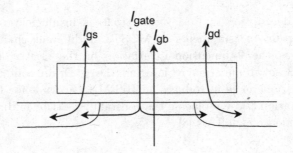

Fig. 5.3   Possible gate leakage currents of the MOS transistor.

(i) Between Gate and Channel, I_{gate}

As for the gate current (I_{gate}) to the MOS transistor channel the direct-tunneling mechanism has to be considered [9]. Since measured I_{gate} shows nearly linear L_{gate} dependence, the tunneling is concluded to occur at all positions of the whole channel. Consequently, the basic description implemented in HiSIM is [10]

$$I_{\text{gate}} = q \cdot GLEAK1 \frac{E^2}{E_{\text{gp}}^{\frac{1}{2}}} \exp\left(-\frac{E_{\text{gp}}^{\frac{3}{2}} GLEAK2}{E}\right) \sqrt{\frac{Q_{\text{i}}}{const0}} W_{\text{eff}} L_{\text{eff}} \quad (5.11)$$

where E is defined as

$$E = \frac{V_{\text{G}} - GLEAK3 \times \phi_{\text{S}}(\Delta L)}{T_{\text{ox}}} \quad (5.12)$$

$$V_{\text{G}} = V_{\text{gs}} - V_{\text{fb}} + GLEAK4 \times (\Delta V_{\text{th}})(L_{\text{gate}} \cdot 10^4) \quad (5.13)$$

Altogether 4 model parameters, namely $GLEAK1, 2, 3, 4$ are introduced.

(ii) Between Gate and Bulk, I_{gb}

The DC leakage current from tunneling between gate and bulk, I_{gb}, is important under the accumulation condition and is modeled as

$$I_{\text{gb}} = GLKB1 \cdot E_{\text{gb}}^2 \cdot \exp\left(-\frac{GLKB2}{E_{\text{gb}}}\right) W_{\text{eff}} L_{\text{eff}} \quad (5.14)$$

$$E_{\text{gb}} = \frac{-V_{\text{gs}} - V_{\text{fbc}}}{T_{\text{ox}}} \quad (5.15)$$

where $GLKB1$ and $GLKB2$ are the model parameters.

(iii) Between Gate and Source/Drain, $I_{\text{gs}}/I_{\text{gd}}$

The tunneling current between the gate and the source/drain overlap region is model as

$$I_{\text{gs}} = sign \cdot GLKSD1 \cdot E_{\text{gs}}^2$$
$$\exp\left(T_{\text{ox,eff}}(-GLKDS2 \cdot V_{\text{gs}} + GLKSD3)\right) W_{\text{eff}} \quad (5.16)$$

$$E_{\text{gs}} = \frac{V_{\text{gs}}}{T_{\text{ox}}} \quad (5.17)$$

$$I_{gd} = sign \cdot GLKSD1 \cdot E_{gd}^2$$
$$\exp\left(T_{ox,eff}(GLKSD2 \cdot (-V_{gs} + V_{ds}) + GLKSD3)\right) W_{eff} \qquad (5.18)$$

$$E_{gd} = \frac{V_{gs} - V_{ds}}{T_{ox}} \qquad (5.19)$$

$$sign = -1 \text{ for } E \leq 0$$
$$sign = 1 \text{ for } E \geq 0$$

where $GLKSD1, 2, 3$ are the used model parameters.

Schematical graphs of the different gate currents as a function of V_{gs} are shown in Fig. 5.4. It is noted here that the signs of I_{gs} and I_{gd} are opposite for the left and right parts of the respective current curves.

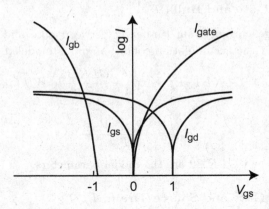

Fig. 5.4   Schematic plots of the various gate currents as a function of V_{gs}.

(iv) Partitioning of the gate-channel current I_{gate} between source/drain

As the oxide thickness decreases in advanced MOSFETs, the gate/basewell-channel component I_{gate} increasingly dominates [11]. Accurate division of I_{gate} between the source and drain is required [12, 13]. We adopted the division scheme that was used for dividing or partitioning the stored charge for capacitance computations. Thus,

$$Q_{\text{gate,D}} = W_{\text{eff}} \int_0^{L_{\text{eff}}} \frac{y}{L_{\text{eff}}} j_{\text{gate}}(y) dy \qquad (5.20)$$

$$Q_{\text{gate,S}} = Q_{\text{gate}} - Q_{\text{gate,D}}. \qquad (5.21)$$

The current density, j_{gate}, given in Eq. (5.20), contains an exponential term with a variable in the denominator, it does not have a closed form solution. After some simplifications, the approximate analytical solution can be expressed in terms of error functions with different arguments. Figure 5.5 shows the computed results. However, since the contribution of I_{gate} is not so drastically affected by the division, as reported in [14], a 50:50 partitioning is used as the default in HiSIM. More accurate division can be implemented as a user option.

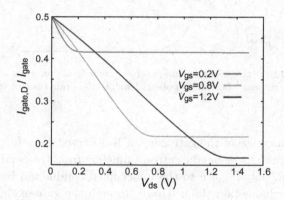

Fig. 5.5 Partitioning of the gate-channel leakage current into drain and source components.

5.1.3 *GIDL (Gate-Induced Drain Leakage) Current*

As the gate voltage V_{gs} is lowered, the n$^+$Drain silicon surface under the Gate/Drain overlap interface turns from electron accumulation to electron depletion. At the same time, the potential difference between the p-substrate or the p-Basewell-body and the n$^+$Drain V_{db} increases, which induces a high longitudinal (lateral) electric field at the Gate/n$^+$Drain interface. As a consequence, the equi-potential contour at the n$^+$Drain/p-

Basewell junction extends into the n^+ Drain contact region, forming a very narrow potential well, as shown by the shaded region in Fig. 5.6. The electric field in this region is very high, causing substantial impact generation of electron-hole pairs [15]. It is measured as an enhancement of the drain current for negative V_{gs} in the basewell-channel surface accumulation range of the nMOSFET. It is known as the Gate-Induced-Drain-Leakage (GIDL) current. Its strong dependence on V_{ds} is illustrated by the experimental data shown in Fig. 5.7.

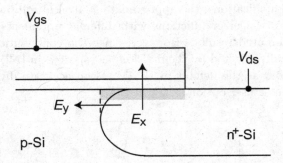

Fig. 5.6 Schematical graph of the equi-potential contour at the drain contact for low gate voltage V_{gs}, showing the narrow potential well in the drain contact underneath the gate.

A further increase of the gate current is observed from further reduced V_{gs}. This is due to intra-band electron tunneling from the conduction band states of the n^+ poly-Si-gate to the unoccupied conduction band states of the induced p-channel on the n^+ Drain through the gate-oxide trapezoidal potential barrier. This is known as intra-band tunneling [1] (so-called "direct" tunneling).

As the V_{gs} decreases further to larger negative values, the trapezoidal barrier abruptly changes to the trianglar barrier (known as Fowler-Nordheim tunneling) at a given negative V_{gs}, which abruptly changes the slope of tunneling rate and current. In scaled down MOS transistors, the main tunneling is the intraband electron tunneling from n^+ Gate to the overlapped and inverted n^+ Drain of a narrow induced potential well of length ΔY

$$I_{\mathrm{GIDL}} = \alpha I_{ds} \Delta Y \qquad (5.22)$$

The oxide electric field in this region is determined by the potential differ-

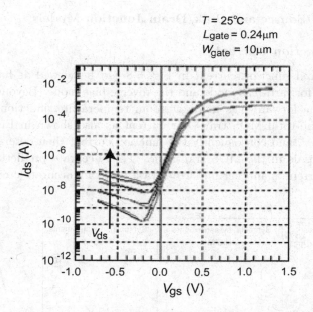

Fig. 5.7   Measured drain current data and their reproduction with HiSIM, showing the strong V_{ds} dependence of the GIDL current contribution.

ence $V_{ds} - V_{gs}$. The final equation is

$$I_{GIDL} = q \cdot GIDL1 \cdot \frac{E^2}{E_g^{\frac{1}{2}}} \cdot \exp\left(-GIDL2 \cdot \frac{E_g^{\frac{3}{2}}}{E}\right) \cdot W_{eff} \qquad (5.23)$$

where

$$E = \frac{GIDL3 \cdot (V_{ds} + GIDL4) - V_G'}{T_{ox,eff}} \qquad (5.24)$$

and

$$V_G' = V_{gs} - V_{fb} + \Delta V_{th} \qquad (5.25)$$
$$\Delta V_{th} = \Delta V_{th,SC} + \Delta V_{th,R} + \Delta V_{th,P} + \Delta V_{th,W} - \phi_{Spg} \qquad (5.26)$$

All the threshold voltage component terms in Eq. (5.26) were described in Chapter 3. The model parameters $GIDL3$ and $GIDL4$ are optional, if they are needed to fit the data. The default values are $GIDL3=1.0$ and $GIDL4=0$.

5.2　Bulk/Source and Bulk/Drain Junction Models

5.2.1　*Junction Current*

The measured junction current in Fig. 5.8 can be viewed as having two modes, the forward-bias mode and the reverse-bias mode. Beyond $V_{junc} \simeq |0.5|$V in the forward-bias mode at room temperature, junction currents dominate the total MOS transistor current as also shown in Fig. 5.9 for $V_{bd} < -0.5$V, and consequently the junction current then determines the node potentials in the MOS transistor. Thus accurate prediction of the junction current is another important task to attain accurate circuit simulation.

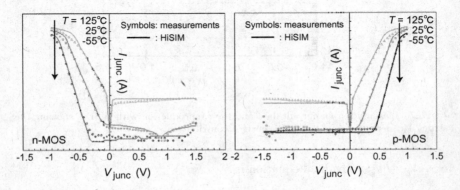

Fig. 5.8　Typical measured junction currents (symbols) of n-type and p-type MOSFETs as a function of applied bias V_{junc}. Three temperatures are shown. Simulation results with HiSIM are shown as lines.

The basic equation for the junction current I_{junc} has been derived by Shockley [1,2,5] as a function of the applied voltage V_{junc}

$$I_{junc} = I_{d0}\left[\exp(\beta V_{junc}) - 1\right] \tag{5.27}$$

$$I_{d0} = Area \cdot q \cdot n_i^2 \left(\frac{D_p}{L_p N_D} + \frac{D_n}{L_n N_A}\right) \tag{5.28}$$

$$\beta = \frac{q}{kT} \tag{5.29}$$

where the p/n junction area is *Area*, the intrinsic carrier concentration, n_i, is given by Eq. (2.27) and k is the Boltzmann constant. $\beta^{-1} = kT/q$ is known as the 'thermal voltage' and is approximately 26mV at room

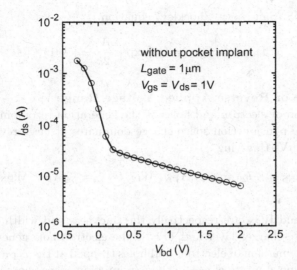

Fig. 5.9 A 2D-device simulation result of the junction current as a function of the junction voltage V_{bd} ($=V_{ds} - V_{bs}$), where the bulk voltage V_{bs} is varied from -1V to 1.3V with V_{gs} and V_{ds} fixed to 1V.

temperature. D_p & D_n, L_p & L_n, and N_D & N_A are intrinsic carrier concentration, diffusion constant of hole and electron, diffusion length of hole and electron, impurity concentration of donor and acceptor, respectively. Shockley's p/n diode equation (5.27) shows the exponential current rise with voltage in the forward-bias mode, $V_{JUNC} = V_{PN} > 0$ and the diminished and voltage-independent saturation current in the reverse-bias mode, $V_{JUNC} = V_{PN} < 0$. However, measured $I - V$ characteristics deviate from the Shockley ideal diode of Eqs. (5.27)- (5.29), due to three major factors in real p/n junctions that are not taken into account by Shockley in (5.27). They are explained briefly as follows.

1) Small Applied Voltage Range $-400\text{mV} < V_{PN} \lesssim +400\text{mV}$

Recombination of electrons and holes at recombination-generation (r-g) centers (such as gold and other metals from contamination and lattice vacancies and vacancy-impurity-and-oxygen pairs from high-temperature diffusion cooling rate stress during fabrication) in the p/n junction space-charge regions dominates in silicon p/n junctions in the room temperature ranges at low V_{PN} ($\approx -400\text{mV}$ to $+400\text{mV}$). The diode current is given by [1] the SNS (Sah-Noyce-Shockley) equation,

$$I_{\text{JUNC}} = I_{\text{SNS}} = I_{20}\{\exp\left(\frac{qV_{\text{PN}}}{2kT}\right) - 1\} \tag{5.30}$$

2) Negative or Reverse Applied Voltage Range $V_{\text{PN}} < -400\text{mV}$

Generation of electron and holes at the generation-recombination (g-r) centers in the p/n junction space charge dominates at large reverse biases $V_{\text{PN}} < -400\text{mV}$ [1], giving

$$I_{\text{JUNC}} = I_{\text{SNS}} = -I_{20} = -qn_iX_{\text{PN}} \cdot W_{\text{PN}}/\tau_g \;\;= -qX_{\text{PN}} \cdot W_{\text{PN}}e_{\text{PN}}N_{\text{TT}} \tag{5.31}$$

where X_{PN} and W_{PN} are respectively the thickness and width of the p/n junction space-charger layer and $1/\tau_g$ is the electron-hole generation rate from thermal emission of electrons and holes trapped at the N_{TT} g-r centers per unit volume at a rate of e_{PN} in the p/n junction space-charge layer with energy level at silicon midgap. At higher reverse biases, interband impact generation of electron-hole pairs in the p/n junction space charge region will also take place, giving a electron-hole pair multiplication factor

$$\alpha_{\text{SNS}} = \alpha_2 = A_2 \exp\left(-\frac{B_2}{E_{\text{PN}}}\right) \tag{5.32}$$

similar to Eq. (5.2) for the Shockley diode current.

3) High-Forward Bias - High Injection Level $V_{\text{PN}} > +700\text{mV}$

Recombination of electrons and holes at high-carrier injection occurs for large Vjunc, which has been ignored in the derivation of these equations, and also occurs due to leakage currents caused by the generation-recombination process of carriers for small Vjunc. Therefore several improvements have to be undertaken to reproduce real device measurements [16]. Three different regions are normally distinguished in the modeling and are treated separately according to their origins. These three regions which are denoted (a), (b) and (c) in the schematic diagram of Fig. 5.10, correspond to the forward-bias current saturation, forward-bias linear region, and the backward-bias region, respectively.

In MOS transistors, the source and drain p/n junction currents can be divided into two geometric pathways, the bottom and the perimeter of the drain and source diffused regions, shown in Fig. 5.11. The model equations for the forward-biased current densities, describing the area and periphery

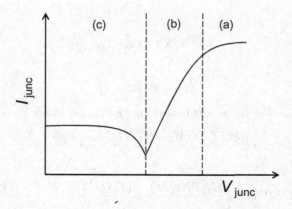

Fig. 5.10 The two I_{junc} currents (between bulk and drain I_{bd} and between bulk an source I_{bs}) are modeled separately in the three different operating regions (a), (b) and (c).

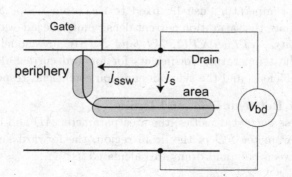

Fig. 5.11 Periphery and bottom area components of the junction current flow between bulk and drain.

components of the source/drain region, are given by

$$j_{\text{s}} = JS0 \exp \left\{ \frac{E_{\text{g}}(TNOM) \cdot \beta(TNOM) - E_{\text{g}}\beta + Xtil}{NJ} \right\} \quad (5.33)$$

$$j_{\text{ssw}} = JS0SW \exp \left\{ \frac{E_{\text{g}}(TNOM) \cdot \beta(TNOM) - E_{\text{g}}\beta + Xtil}{NJSW} \right\} \quad (5.34)$$

where

$$Xtil = XTI \cdot \log(T_{\text{tnom}})$$

$$T_{\text{tnom}} = \frac{T}{TNOM} \tag{5.35}$$

The corresponding backward-biased current densities are given by

$$j_{s2} = JS0 \exp\left\{ \frac{E_{\text{g}}(TNOM) \cdot \beta(TNOM) - E_{\text{g}}\beta + Xti2l}{NJ} \right\} \tag{5.36}$$

$$j_{\text{ssw2}} = JS0SW \exp\left\{ \frac{E_{\text{g}}(TNOM) \cdot \beta(TNOM) - E_{\text{g}}\beta + Xti2l}{NJSW} \right\} \tag{5.37}$$

where

$$Xti2l = XTI2 \cdot \log(T_{\text{tnom}})$$

In Eqs. (5.33) – (5.35), T is the temperature investigated and $TNOM$ is the nominal temperature usually fixed to the room temperature. JSO and $JS0SW$ are the saturation current density and the sidewall saturation current density. XTI & $XTI2$, NJ, and $NJSW$ are model parameters describing the temperature coefficients for forward current densities, the emission coefficient, and the sidewall emission coefficient, respectively.

(i) Current Between Bulk and Drain

With these current densities, the area parameter AD and the side-wall periphery parameter PD of the drain region, the forward and backward currents between bulk and drain are calculated as

$$I_{\text{sbd}} = AD \cdot j_{\text{s}} + PD \cdot j_{\text{ssw}} \tag{5.38}$$

$$I_{\text{sbd2}} = AD \cdot j_{s2} + PD \cdot j_{\text{ssw2}} \tag{5.39}$$

The resulting bulk-drain junction-current equations in the 3 operating regions (a), (b) and (c) are derived as follows.

a) $V_{\text{bd}} \geq T_1$

$$I_{\text{bd}} = I_{\text{sbd}} \left\{ \exp\left(\frac{T_1}{N_{\text{vtm}}} \right) - 1 \right\} + \frac{I_{\text{sbd}}}{N_{\text{vtm}}} \left\{ \exp\left(\frac{T_1}{N_{\text{vtm}}} \right) - 1 \right\} (V_{\text{bd}} - T_1) \tag{5.40}$$

where

$$T_1 = N_{\text{vtm}} \cdot \log\left\{ \frac{VDIFFJ}{I_{\text{sbd}}} \cdot (T_{\text{tnom}})^2 + 1 \right\}$$

$$N_{\text{vtm}} = \frac{NJ}{\beta}$$

b) $T_1 \geq V_{\text{bd}} \geq 0$

$$I_{\text{bd}} = I_{\text{sbd}} \left\{ \exp\left(\frac{V_{\text{bd}}}{N_{\text{vtm}}} \right) - 1 \right\}$$

$$+ I_{\text{sbd2}} \cdot CISB \cdot \left\{ \exp\left(-\frac{V_{\text{bd}} \cdot CVB}{N_{\text{vtm}}} \right) - 1 \right\} Expt$$

$$+ CISBK \cdot \left\{ \exp\left(-\frac{V_{\text{bd}} \cdot CVBK}{N_{\text{vtm}}} \right) - 1 \right\} \tag{5.41}$$

where

$$Expt = \exp\left\{ (T_{\text{tnom}} - 1)CTEMP \right\}$$

c) $V_{\text{bd}} \leq 0$

$$I_{\text{bd}} = I_{\text{sbd2}} \cdot CISB \left\{ \exp\left(-\frac{V_{\text{bd}} \cdot CVB}{N_{\text{vtm}}} \right) - 1 \right\} Expt$$

$$+ CISBK \cdot \left\{ \exp\left(-\frac{V_{\text{bd}} \cdot CVBK}{N_{\text{vtm}}} \right) - 1 \right\} \tag{5.42}$$

Finally, the sidewall current is added to the calculated bottom-area current

$$I_{\text{bd}} = I_{\text{bd}} + DIVX \cdot I_{\text{sbd2}} \cdot V_{\text{bd}} \tag{5.43}$$

$CISB, CVB, CTEMP, CISBK, CVBK$, and $DIVX$ are reverse saturation current, bias dependence coefficient of $CISB$, temperature coefficient of reverse currents, saturation current of reverse bias at low temperature, bias dependence coefficient of $CISB$, and reverse current coefficient, respectively. $VDIFFJ$ is the junction diode threshold voltage between bulk substrate and source/drain.

(ii) Current Between Bulk and Source

The area parameter AS and the side-wall perimeter PS of the source region are used to calculate the forward and backward currents between bulk and source. They are similar to those for the current between bulk and drain given in (i) with proper change of symbols, namely D to S and 1 to 2.

$$I_{\text{sbs}} = AS \cdot j_{\text{s}} + PS \cdot j_{\text{ssw}} \tag{5.44}$$

$$I_{\text{sbs2}} = AS \cdot j_{\text{s2}} + PS \cdot j_{\text{ssw2}} \tag{5.45}$$

The resulting bulk-source junction current equation in the three operating ranges (a), (b) and (c) are given as follows.

a) $V_{\text{bs}} \geq T_2$

$$I_{\text{bs}} = I_{\text{sbs}} \left\{ \exp\left(\frac{T_2}{N_{\text{vtm}}}\right) - 1 \right\} + \frac{I_{\text{sbs}}}{N_{\text{vtm}}} \left\{ \exp\left(\frac{T_2}{N_{\text{vtm}}}\right) - 1 \right\} (V_{\text{bs}} - T_2) \tag{5.46}$$

where

$$T_2 = N_{\text{vtm}} \cdot \log \left\{ \frac{VDIFFJ}{I_{\text{sbs}}} \cdot \{T_{\text{tnom}}\}^2 + 1 \right\}$$

b) $T_2 \geq V_{\text{bs}} \geq 0$

$$\begin{aligned} I_{\text{bs}} = & I_{\text{sbs}} \cdot \left\{ \exp\left(\frac{V_{\text{bs}}}{N_{\text{vtm}}}\right) - 1 \right\} \\ & + I_{\text{sbs2}} \cdot CISB \cdot \left\{ \exp\left(-\frac{V_{\text{bs}} \cdot CVB}{N_{\text{vtm}}}\right) - 1 \right\} Expt \\ & + CISBK \cdot \left\{ \exp\left(-\frac{V_{\text{bs}} \cdot CVBK}{N_{\text{vtm}}}\right) - 1 \right\} \end{aligned} \tag{5.47}$$

c) $V_{\text{bs}} \leq 0$

$$\begin{aligned} I_{\text{bs}} = & I_{\text{sbs2}} \cdot CISB \left\{ \exp\left(-\frac{V_{\text{bs}} \cdot CVB}{N_{\text{vtm}}}\right) - 1 \right\} Expt \\ & + CISBK \cdot \left\{ \exp\left(-\frac{V_{\text{bs}} \cdot CVBK}{N_{\text{vtm}}}\right) - 1 \right\} \end{aligned} \tag{5.48}$$

Finally the total bulk-source junction current is given by

$$I_{\text{bs}} = I_{\text{bs}} + DIVX \cdot I_{\text{sbs2}} \cdot V_{\text{bs}} \tag{5.49}$$

5.2.2 Junction Capacitance

The p/n junction is treated with the simplified charge-control formulas for reverse bias condition. The capacitance per unit junction area is given then as below by [5]

$$C_{\text{junc}} = \frac{dQ_{\text{dep}}}{dV_{\text{junc}}} \tag{5.50}$$

$$Q_{\text{dep}} = qN_A W_p = qN_D W_n \tag{5.51}$$

where W_p and W_n are the depletion thicknesses ("width") describing the thicknesses from the p/n metallurgical boundary to the edge of depletion layer on the p-Si and n-Si sides, respectively. From the above equations, the junction capacitance per unit junction area for an abrupt transition of the impurity concentration from p-Si, N_A, to n-Si, N_D, in a step is given by

$$C_{\text{junc}} = \sqrt{\frac{q\epsilon_{\text{Si}}}{2(V_{\text{bi}} - V_{\text{junc}})} \left(\frac{N_A N_D}{N_A + N_D}\right)} \tag{5.52}$$

which is an approximation to the real diffused, hence non-abrupt, p/n junction. By further assuming that the impurity concentration on one side of the junction is much higher than on the other side, to be denoted by N_{sub}, Eq. (5.52) can be simplified to

$$C_{\text{junc}} = \sqrt{\frac{q\epsilon_{\text{Si}} N_{\text{sub}}}{2(V_{\text{bi}} - V_{\text{junc}})}} \tag{5.53}$$

which can be written in the form

$$C_{\text{junc}} = \frac{C_{\text{junc},0}}{\sqrt{1 - \frac{V_{\text{junc}}}{V_{\text{bi}}}}} \tag{5.54}$$

where $C_{\text{junc},0}$ is a bias independent constant capacitance

$$C_{\text{junc},0} = \sqrt{\frac{q\epsilon_{\text{Si}} N_{\text{sub}}}{2V_{\text{bi}}}} \tag{5.55}$$

For applications to real devices, the exponent $1/2$ in the voltage dependence of Eq. (5.54) is generalized to become a fitting parameter MJ, giving

$$C_{\text{junc}} = \frac{C_{\text{junc},0}}{\left(1 - \frac{V_{\text{junc}}}{V_{\text{bi}}}\right)^{MJ}} \tag{5.56}$$

We follow the previous junction current modeling to account for the capacitance from the perimeter and the bottom areas. This is further improved by dividing the perimeter contributions into two parts, the side-wall at the gate edge and bottom area intercepted by the isolation edge [15, 16] as indicated in Fig. 5.12. Then the total junction capacitance is given by

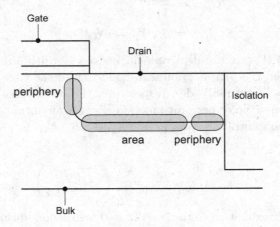

Fig. 5.12   Schematic explanation of the two peripheral components and the bottom-area component in the junction capacitance between bulk and drain.

$$
\begin{aligned}
C_{\text{junc}} =& Area \cdot CJUN \left(1 - \frac{V_{\text{junc}}}{V_{\text{bi}}}\right)^{MJ} \\
&+ W_{\text{eff}} \cdot CJSWG \left(1 - \frac{V_{\text{junc}}}{V_{\text{bi}}}\right)^{MJSWG} \\
&+ (PX - W_{\text{eff}}) \cdot CJSWI \left(1 - \frac{V_{\text{junc}}}{V_{\text{bi}}}\right)^{MJSWI}
\end{aligned}
\tag{5.57}
$$

where PX is either PS or PD. The built-in potential V_{bi} is here also different for these three regions. In principle the model parameters considered in a simplified model like that of Eq. (5.56) are $C_{\text{junc},0}$ and MJ. However, for applications, the junction models include many more fitting parameters due to the unknown physics which occurs at the diode junction [5, 7]. For example, the built-in potential is not accurately known due to the uncertain impurity diffusion or implantation gradient and MJ can vary from 1 in the Shockley ideal p/n-junction diode to 2 or more in the Sah-Noyce-Shokley p/n-junction diode [1] from electron-hole generation at the impurity and defect traps or generation-centers in the depletion-layer of the p/n junction. Nevertheless, the minimum number of parameters introduced in the compact model is found sufficient to account for all these dependences on process conditions during the fabrication of the p/n junctions and the MOSFET.

5.3 Summary of Equations and Model Parameters Appeared in Chapter 5 for Leakage Currents and Junction Diode

5.3.1 *Section 5.1: Leakage Currents*

[Substrate Current]

$$I_{\text{sub}} = SUB1\big(\phi(\Delta L') - \phi(0)\big) I_{\text{ds}} \exp\left(-\frac{SUB2}{\phi(\Delta L') - \phi(0)}\right)$$

$$SUB1 = SUB1 + \frac{SUB4}{L_{\text{eff}}}$$

[Gate Current]

$$I_{\text{gate}} = q \cdot GLEAK1 \frac{E^2}{E_{\text{gp}}^{\frac{1}{2}}} \exp\left(-\frac{E_{\text{gp}}^{\frac{3}{2}} GLEAK2}{E}\right) \sqrt{\frac{Q_i}{const0}} W_{\text{eff}} L_{\text{eff}}$$

$$E = \frac{V_{\text{G}} - GLEAK3 \times \phi_{\text{S}}(\Delta L)}{T_{\text{ox}}}$$
$$V_{\text{G}} = V_{\text{gs}} - V_{\text{fb}} + GLEAK4 \times (\Delta V_{\text{th}})(L_{\text{gate}} \cdot 10^4)$$

$$I_{\text{gb}} = GLKB1 \cdot E_{\text{gb}}^2 \cdot \exp\left(-\frac{GLKB2}{E_{\text{gb}}}\right) W_{\text{eff}} L_{\text{eff}}$$
$$I_{\text{gs}} = sign \cdot GLKSD1 \cdot E_{\text{gs}}^2$$
$$\qquad \exp\left(T_{\text{ox,eff}}(-GLKDS2 \cdot V_{\text{gs}} + GLKSD3)\right) W_{\text{eff}}$$
$$I_{\text{gd}} = sign \cdot GLKSD1 \cdot E_{\text{gd}}^2$$
$$\qquad \exp\left(T_{\text{ox,eff}}(GLKSD2 \cdot (-V_{\text{gs}} + V_{\text{ds}}) + GLKSD3)\right) W_{\text{eff}}$$

[GIDL Current]

$$I_{\text{GIDL}} = q \cdot GIDL1 \cdot \frac{E^2}{E_{\text{g}}^{\frac{1}{2}}} \cdot \exp\left(-GIDL2 \cdot \frac{E_{\text{g}}^{\frac{3}{2}}}{E}\right) \cdot W_{\text{eff}}$$

$$E = \frac{GIDL3 \cdot (V_{\text{ds}} + GIDL4) - V_{\text{G}}'}{T_{\text{ox,eff}}}$$
$$V_{\text{G}}' = V_{\text{gs}} - V_{\text{fb}} + \Delta V_{\text{th}}$$
$$\Delta V_{\text{th}} = \Delta V_{\text{th,SC}} + \Delta V_{\text{th,R}} + \Delta V_{\text{th,P}} + \Delta V_{\text{th,W}} - \phi_{\text{Spg}}$$

Table 5.1   HiSIM model parameters for leakage currents.

$SUB1$	magnitude of substrate current
$SUB2$	coefficient of electric field
$SUB3$	modification of electric field
$SUB4$	L_{gate} dependence of substrate current
$GLEAK1$	magnitude of gate to channel current
$GLEAK2$	coefficient of electric field for gate to channel current
$GLEAK3$	modification of electric field for gate to channel current
$GLEAK4$	modification of electric field for gate to channel current
$GLKB1$	magnitude of gate to bulk current
$GLKB2$	coefficient of electric field for gate to bulk current
$GLKSD1$	magnitude of gate to source/drain current
$GLKSD2$	coefficient of electric field for gate to source/drain current
$GLKSD3$	modification of electric field for gate to source/drain current
$GIDL1$	magnitude of GIDL current
$GIDL2$	coefficient of electric field for GIDL current
$GIDL3$	modification of electric field for GIDL current

5.3.2   Section 5.2: Junction Diode

Table 5.2   HiSIM model parameters for junction diode

$JS0$	saturation current density
$JS0SW$	sidewall saturation current density
NJ	emission coefficient
$NJSW$	sidewall emission coefficient
XTI	temperature coefficient for forward-current densities
$XTI2$	temperature coefficient for reverse-current densities
$CISB$	reverse biased saturation current
CVB	bias dependence coefficient of **CISB**
$CTEMP$	temperature coefficient of reverse currents
$CISBK$	reverse biased saturation current (at low temperature)
$CVBK$	bias dependence coefficient of **CISB** (at low temperature)
$DIVX$	reverse current coefficient
$VDIFFJ$	junction diode threshold voltage between source/drain and substrate
MJ	bottom junction capacitance grading coefficient
$CJUN$	bottom area junction capacitance per unit area at zero bias
$CJSWG$	side-wall peripheral junction capacitance per unit length
$CJSWI$	peripheral junction capacitance at isolation per unit length

Bibliography

[1] Chih-Tang Sah, "Fundamentals of solid-state electronics," *World Scientific Publishing Co*, Singapore, 1991.

[2] W. Shockley, "Problems related to p-n junctions in silicon," *Solid-State Electron.*, vol. 2, no. 1, pp. 35–67, 1961.

[3] K. K. Thornber, "Applications of scaling to problem in high-field electronic transport," *J. Appl. Phys.*,vol. 52, no. 1, pp. 279–290, 1981.

[4] S. Selberherr, "Analysis and Simulation of Semiconductor Devices," *Springer Verlag*, 1984.

[5] N. Arora, "MOSFET models for VLSI circuit simulation: theory and practice," *Springer Verlag*, 1993. Reprinted World Scientific Publishing Co. Singapore. 2007.

[6] C. Jungemann, S. Yamaguchi, and H. Goto, "On the accuracy and efficiency of substrate current calculation for sub-μm n-MOSFET's," *IEEE Electron Device Lett.*,vol. 17, no. 1, pp. 464–466, 1996.

[7] *HiSIM1.0.0, User's Manual*, Hiroshima University and STARC, 2002.

[8] R. Inagaki, N. Sadachika, D. Navarro, Q. Ngo, C. Y. Yang, M. Miura-Mattausch, and Y. Inoue, "A Substrate–Current Model for Advanced MOSFET Technologies Implemented into HiSIM2," *Proc. Int. 4th Workshop on Compact Modeling*, Yokohama, pp. 89-92, Jan. 2007.

[9] E. O. Kane, "Zener Tunneling in Semiconductors," *J. Phys. Chem. Solids*, vol. 12, no. 1, pp. 181–188, 1959.

[10] Q. Ngo, D. Navarro, T. Mizoguchi, S. Hosokawa, H. Ueno, M. Miura-Mattausch, and C. Y. Yang, "Gate Current Partitioning in MOSFET Models for Circuit Simulation," *Proc. Modeling and Simulation of Microsystems*, vol. 1.2, pp. 322–325, 2003.

[11] C.-H. Choi, K.-H. Oh, J.-S. Goo, Z. Yu, and R. W. Dutton, "Direct tunneling current model for circuit simulation," *IEDM Tech. Dig.*, pp. 735–738, 1999.

[12] W.-K. Shih, R. Rios, P. Packan, K. Mistry, T. Abbott, "A general partition scheme for gate leakage current suitable for MOSFET compact models," in *IEDM Tech. Dig.*, pp. 293–296, 2001.

[13] Y. Yang, "Gate Current Partitioning in MOSFET Models for Circuit Simulation," *Proc. Modeling and Simulation of Microsystems*, vol. 1.2, pp. 322-325, 2003.

[14] R. Inagaki, K. Konno, N. Sadachika, D. Navarro, K. Machida, Q. Ngo, C. Y. Yang, T. Ezaki, H. J. Mattausch, M. Miura-Mattausch, and Y. Inoue, "A Gate–Current Model for Advanced MOSFET Technologies Implemented into HiSIM2," *Proc. Int. 3rd Workshop on Compact Modeling*, Yokohama, pp. 43-46, Jan. 2006.

[15] T. Y. Chen, J. Chen, P. K. Ko, and C. Hu, "The impact of gate-induced leakage current on MOSFET scaling," *IEDM Tech. Dig.*, pp. 718–721, 1987.

[16]   *BSIM4.0.0 MOSFET Model, User's Manual,* Department of Electrical Engineering and Computer Science, University of California, Berkeley CA, 2000.

Chapter 6

Modeling of Phenomena Important for RF Applications

6.1 Noise Models

The microscopic features of current include the influence of random fluctuation of carriers in the channel as shown in Fig. 6.1. The magnitude of the fluctuation Δi is considered to characterize the noise as

$$S_I = \Delta i^2 \qquad (6.1)$$

The noise spectrum density S_I is determined to characterize the different

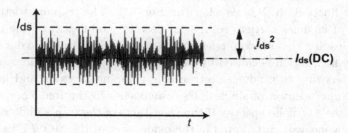

Fig. 6.1 Schematic illustration of the current fluctuations in the drain-source current I_{ds}.

noise features. Advanced MOSFETs are suffering from two dominating noise contributions: the trapping noise (frequently referred as the flicker noise) and the thermal noise [1]. The former noise shows frequency dependence of $1/f$, called therefore the $1/f$ noise, and the latter noise is independent of the frequency. In addition, the frequency dependent induced gate noise and the cross-correlation noise are observed under high-frequency operation of the MOSFET [2] as shown in Fig. 6.2.

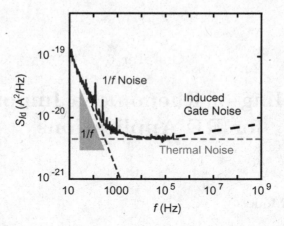

Fig. 6.2   Different noise contributions present in the MOSFET, namely $1/f$ noise, thermal noise and induced gate noise as well as the frequency range of their occurrence.

6.1.1 $1/f$ Noise Model

The origin of the conventional $1/f$ noise in MOSFETs has been understood theoretically as a fluctuation in the number of carriers due to trapping/detrapping processes at the gate oxide interface [3,4], as well as by the mobility fluctuation [5,6] as shown in Fig. 6.3. The trapping/detrapping process of channel carriers results in discrete modulation of the channel current leading to a random telegraph signal (RTS) in the time domain [7]. Mobility fluctuation is caused by Coulomb-scattering fluctuations of channel carriers due to random electron-hole trapping at oxide and interface traps. Superposition of single RTS components in the frequency domain leads to the $1/f$ noise spectra, if time constants of the trap and detrap sites are homogeneously distributed in the oxide layer of the MOSFETs.

Figure 6.4 compares measured $1/f$ noise intensity S_{I_d} with the simulated number of carriers hitting the oxide interface for various V_{gs} biases [4]. The simulation applies the Monte Carlo method, which considers the dynamics of each carrier individually. Increase of V_{gs}-V_{th} from 0.1V to 0.6V causes an increase of S_{I_d} as well as an increase of the number of carriers hitting the interface by nearly two orders of magnitude. This coincidence in the behavior of the two quantities in Fig. 6.4 concludes that the trap/detrap process at the gate-oxide interface is the main cause of the $1/f$ noise under the normal circuit operation conditions. Since the two types of fluctuations, namely carrier-density and carrier-mobility fluctuations, have the same origin, they

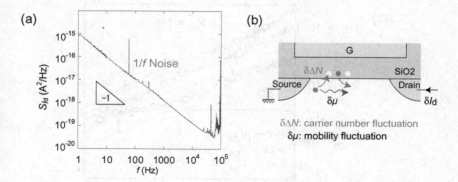

Fig. 6.3 (a) $1/f$ noise measurement and (b) microscopic trap-detrap model for $1/f$ noise explanation.

are correlated with each other [7].

The dotted line in Fig. 6.4a is a Lorentzian noise spectrum which deviates from the $1/f$ characteristics (see the dashed line) written as a function of frequency f

$$S_{I_d} = \frac{A\tau}{1 + (2\pi f\tau)^2} \qquad (6.2)$$

where A is the magnitude of the Lorentzian nois determined by the trap density, and τ is the trapping time constant of the carriers determined by the trap position, z, in the depth direction in the oxide layer [3]

$$\tau = \tau_0 \exp(\eta z) \qquad (6.3)$$

Here τ_0 is the time constant for trapping/detrapping processes at the interface, η is the tunneling attenuation coefficient through the oxide layer. The intensity and the frequency value, at which S_{I_d} starts to reduce, are determined by τ. The $1/f$ noise is a superposition of the Lorentzian noise, coming from different trap sites as schematically shown in Fig. 6.5, and requires that trap energy levels and positions in the oxide are distributed [8–12]

$$S_{I_d} = \int_0^{T_{\text{ox}}} \frac{A\tau}{1 + (2\pi f\tau)^2} dz \qquad (6.4)$$

As the device size decreases, measured low frequency noise departs strongly from the $1/f$ dependence [13] as shown in Fig. 6.6 [14]. Measurement setups for the noise are similar to those commonly used [15] and are illustrated in Fig. 6.7. Figures 6.8a and 6.8b show measured noise spectra under the

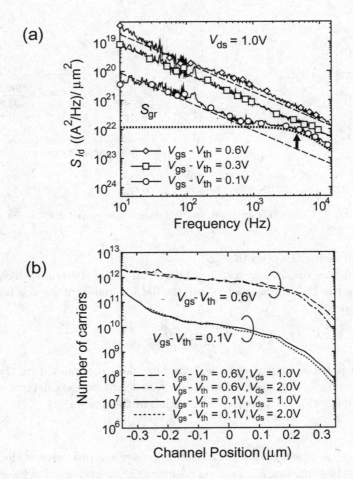

Fig. 6.4   Comparison of (a) the current-noise density S_{I_d} and (b) the number of carriers hitting the gate-oxide interface for different gate-source voltages V_{gs}.

linear and saturation operating conditions of the MOSFET, respectively, for a relatively long gate length (L_{gate}) of $1.0\mu m$ [16]. Measurements with exchanged source (forward) and drain (backward) contacts are also shown in the Figs. 6.8(a) and (b). The carrier density distribution along the channel is sketched in the insets. Under the linear operating condition of the MOSFT the difference in the noise spectra between the forward and backward measurement is hardly visible. On the contrary, a difference between the forward and backward noise spectra becomes obvious under the sat-

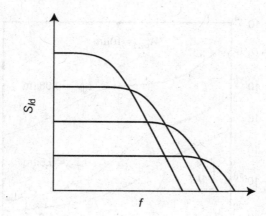

Fig. 6.5 Schematic of the superposition of Lorentzians with different energy and position, resulting in a $1/f$ dependence of the current-noise density S_{I_d} of the MOSFET.

uration condition. With reduced L_{gate} the difference observed under the saturation condition becomes even more obvious. However, no difference in the corresponding measured drain currents (I_{ds}) is observed due to the exchange of source and drain. Also, device degradations could not be confirmed during the measurements. This concludes that the measured noise characteristics of Figs. 6.8a and 6.8b are due to the position dependent trap density and trap energy along the channel direction, which are expected to influence only on the $1/f$ noise but not on the drain current.

Figure 6.9a shows measured noise spectra of 30 different chips located on a wafer with $L_{gate} = 0.46\mu$m [16]. The average of all these noise spectra gives the typical $1/f$ noise characteristics as shown by the thick line. Figure 6.9b shows the histogram of the measured deviation of the noise from its average value at $f = 100$Hz. It is nearly a normal distribution given by the solid curve in the figure. This further proves that the inhomogeneity of the trap density and energy, is randomly distributed across a wafer. For a circuit simulation model it is therefore appropriate to reduce the description to the averaged $1/f$ noise characteristics and to specify two noise boundaries corresponding to the measured worst and best cases. Figure 6.10 summarizes the measured gate voltage dependences (symbols) of the $1/f$ noise S_{I_d} for $L_{gate} = 1.0\mu$m, 0.46μm and 0.12μm at $f = 100$Hz for different drain source voltages V_{ds}. All measured points are average values over 30 chips on a wafer. The logarithmic relationship of S_{I_d} for small V_{gs} saturates at large V_{gs}. By reducing L_{gate}, the V_{gs} dependence is

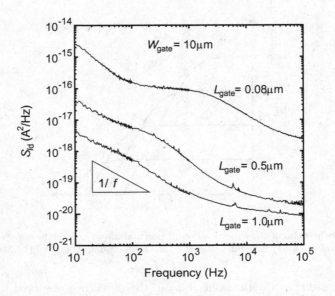

Fig. 6.6   Drain current noise of an n-MOSFET with different gate length (0.08, 0.5 and 1.0μm) under linear operating condition.

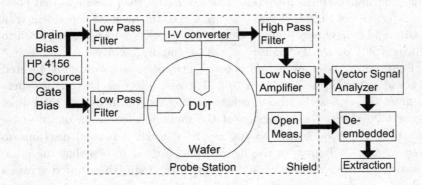

Fig. 6.7 $1/f$ noise measurement system.

enhanced. These features of the $1/f$ noise are also observed in the $I_{ds} - V_{gs}$ characteristics.

The current noise density, under the approximation that most of carriers

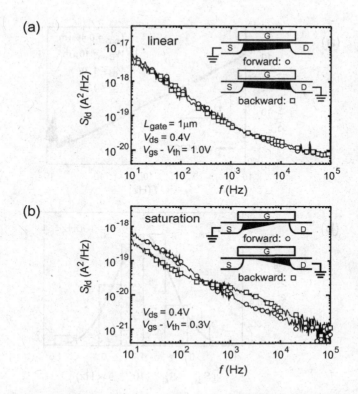

Fig. 6.8 Comparison of drain current noise spectrum density between forward measurement and backward measurement under (a) linear and (b) saturation condition for $L_{\text{gate}} = 1.0\mu m$. The insets show schematics of the inversion charge distribution in the forward and backward measurement.

have energies around the Fermi level E_{f} [6], can be written as

$$S_{I_{\text{d}}} = \frac{I_{\text{ds}}^2 \cdot NFTRP}{\beta f (L_{\text{eff}} - \Delta L)^2 W_{\text{eff}}} \int \left[\frac{1}{(N(x) + N^*)} + NFALP \cdot \mu \right]^2 dy \quad (6.5)$$

$$NFTRP = \frac{N_{\text{t}}(E_{\text{f}})}{\eta} \quad (6.6)$$

where η is the tunneling attenuation coefficient into the oxide, and N_{t} is the trap density. From Eq. (6.5) the compact model equation for the $1/f$ noise is obtained by integrating the carrier density under the approximation of a linearly graded carrier distribution along the channel. The final equation

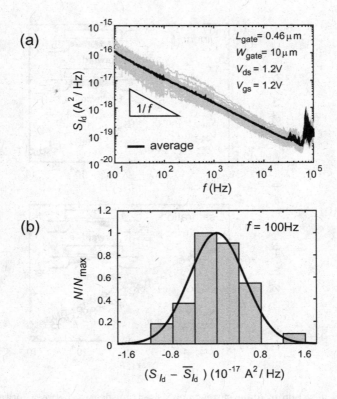

Fig. 6.9   (a) Measured drain current noise spectra of 30 devices with the same size under the same bias conditions on a wafer. The fat curve represents an averaged noise spectrum. (b) Histogram of measured drain current noise spectra at 100Hz. The fat curve shows the normal distribution function. $\overline{S_{I_d}}$ is about $1.0 \times 10^{-17} \text{A}^2/\text{Hz}$.

becomes [16]

$$S_{I_{ds}} = \frac{I_{ds}^2 \cdot NFTRP}{\beta f(L_{eff} - \Delta L)W_g} \left[\frac{1}{(N_0 + N^*)(N_L + N^*)} \right.$$
$$\left. + \frac{2 \cdot NFALP \cdot \mu}{N_L - N_0} \ln\left(\frac{N_L + N^*}{N_0 + N^*}\right) + (NFALP \cdot \mu)^2 \right] \qquad (6.7)$$

Here the trap density, $NFTRP$, and the contribution of the mobility fluctuation due to the traps, $NFALP$, are the compact model parameters. N_0 and N_L are the carrier concentration at source and at drain, respectively, which are calculated by the surface potential values ϕ_{S0} and ϕ_{SL}. N^* is written as

$$N^* = \frac{C_{\text{ox}} + C_{\text{dep}} + CIT}{q\beta} .$$

(6.8)

where q is magnitude of the electron charge, and C_{ox}, C_{dep}, CIT are the oxide capacitance, the depletion capacitance, and the capacitance induced by the trap carriers, respectively. To achieve better fitting to measurements of advanced technologies, the field increase along the channel E_y is also considered. Though the mobility μ is approximated to be constant along the channel to obtain a closed form description, reality is that it varies drastically along the channel as the channel length becomes shorter. This effect is included by considering E_y. The final equation is thus written as

$$S_{I_{\text{ds}}} = \frac{I_{\text{ds}}^2 NFTRP}{\beta f(L_{\text{eff}} - \Delta L)W_{\text{eff}}} \left[\frac{1}{(N_0 + N^*)(N_L + N^*)} \right.$$
$$\left. + \frac{2\mu E_y NFALP}{N_L - N_0} \ln \left(\frac{N_L + N^*}{N_0 + N^*} \right) + (\mu E_y NFALP)^2 \right]$$

(6.9)

We have to note that $NFTRP$ and $NFALP$ are nearly technology independent, if the technology is mature, and that CIT can normally be set to zero.

Calculationed results using the preceding model are shown as continuous lines in Fig. 6.10, where a single model parameter set is used for all three channel lengths. It can be further concluded, that the $1/f$ noise characteristics are mostly determined by the I-V characteristics, because the noise is directly related to the carrier density as can be seen from Eq. (6.9). The measured and calculated $1/f$ noise characteristics are shown as a function of V_{ds} in Fig. 6.11 for $L_{\text{gate}}=0.46\mu$m. From the comparison with I_{ds}^2, it can be seen that the saturation behavior of $S_{I_{\text{d}}}$ is caused exactly by the saturation behavior of I_{ds} [17].

6.1.2 *Thermal Noise Model*

The Nyquist theorem describes the thermal noise of a resistance R at temperature T as [18]

$$S_{\text{v}} = 4kTR$$

(6.10)

where S_v is the voltage spectral dentensity. Van der Ziel extended the theorem to the current spectral intensity of semiconductors [2]

$$S_{I_d} = \frac{4kT}{R} = 4kTg_{ds} \tag{6.11}$$

where k is the Boltzmann constant, g_{ds} is the channel conductance, and S_{I_d} is the current spectral dentensity. Van der Ziel has shown that the carrier-number fluctuation caused by carrier scattering with the lattice is the origin of the thermal noise in semiconductors as schematically shown in Fig. 6.12. The fluctuation in the number of carriers by scattering in a small segment is depicted in schematic form. The thermal noise in the MOSFET can be obtained by integrating the transconductance along the channel direction y, namely over all the small segments along the channel [19]

$$S_{I_d} = \frac{4kT}{L_{\text{eff}}^2} \int g_{ds}(y)dy \tag{6.12}$$

In equation (6.12), $g_{ds}(y)$ is the position-dependent channel conductance.

In HiSIM the integration is performed with respect to the surface potential ϕ_s instead of the channel position as [20]

$$S_{I_d} = \frac{4kT}{L_{\text{eff}}^2 I_{ds}} \int g_{ds}^2(\phi_s)d\phi_s \tag{6.13}$$

$$g_{ds}(\phi_s) = \frac{W_{\text{eff}}}{L_{\text{eff}}\beta} \frac{d\{\mu(\phi_s)f(\phi_s)\}}{d\phi_s} \tag{6.14}$$

where I_{ds} is the drain current and β is the thermal voltage. Here $f(\phi_s)$ is equal to I_{dd} given in Eq. (2.59) and is a characteristic function of HiSIM which is the carrier concentration multiplied by the electric field with the dimension of $CVcm^2$ [21]. To simplify the integration, μ is approximated by a linearly decreasing from source to drain. The validity of the linearly varying approximation of the mobility μ is confirmed by a 2D-device simulation as shown in Fig. 6.13.

The final equations for S_{I_d}, obtained after the integration of Eq. (6.13), becomes a function of the self-consistent surface potential values [22]

$$S_{I_d} = 4kT\frac{W_{\text{eff}}C_{\text{ox}} V_g V_t \mu}{(L_{\text{eff}} - \Delta L)} \frac{1}{15(1+\eta)\mu_{\text{av}}^2} [(1 + 3\eta + 6\eta^2)\mu_d^2$$
$$+ (3 + 4\eta + 3\eta^2)\mu_d\mu_s + (6 + 3\eta + \eta^2)\mu_s] \tag{6.15}$$

where μ_s, μ_d and μ_{av} are mobilities at source end, drain end, and the averaged mobility value, respectively. μ_s and μ_d are calculated by Eq. (3.84) with the electric field E_y, and μ_{av} is simply calculated under the linear approximation as

$$\mu_{av} = \frac{\mu_s + \mu_d}{2} \tag{6.16}$$

The quantity ΔL is the channel length reduction under the saturation condition determined by Eq. (3.101) in Section 3.5. η is a function of the surface potentials and is described by the following equations:

$$\eta = 1 - \frac{(\phi_{SL} - \phi_{S0}) + \chi(\phi_{SL} - \phi_{S0})}{V_g V_t} \tag{6.17}$$

where

$$\chi = 2\frac{const0}{C_{ox}} \left[\left(\frac{2}{3}\frac{1}{\beta}\frac{\{\beta(\phi_{SL} - V_{bs}) - 1\}^{\frac{3}{2}} - \{\beta(\phi_{S0} - V_{bs}) - 1\}^{\frac{3}{2}}}{\phi_{SL} - \phi_{S0}} \right) \right.$$
$$\left. -\sqrt{\beta(\phi_{S0} - V_{bs}) - 1} \right] \tag{6.18}$$

$$const0 = \sqrt{\frac{2\epsilon_{Si} q N_{sub}}{\beta}} \tag{6.19}$$

$V_g V_t$ in Eqs. (6.15) and (6.17) gives the carrier density divided by the oxide capacitance, which is equivalent to the $V_{gs} - V_{th}$ term of a threshold (V_{th}) based model.

A simple closed-form equation can be derived by assuming that the gradient of the mobility is negligible in comparison with the gradients of other physical quantities. The above Eq. (6.15) can then be reduced to simpler equations under the condition of $\mu_d = \mu_s$ as

$$S_{I_d} = 4kT\frac{W_{eff} C_{ox} V_g V_t \, \mu}{(L_{eff} - \Delta L)}; \quad \text{for } \eta = 1 \tag{6.20}$$

$$S_{I_d} = 4kT\frac{W_{eff} C_{ox} V_g V_t \, \mu}{(L_{eff} - \Delta L)}\frac{2}{3}; \quad \text{for } \eta = 0 \tag{6.21}$$

where $\eta=1$ and $\eta=0$ are values appropriate under the linear and saturation conditions for long channel transistors. An important fact derived from the above equations is that no additional model parameters are required for the compact thermal noise model.

Fig. 6.14 and Fig. 6.15 show calculated S_{I_d} characteristics with the derived thermal-noise model as a function of V_{gs} and V_{ds}, respectively, in

The Physics and Modeling of MOSFETs

comparison with measurements. For the calculation only HiSIM model parameters extracted from measured I-V characteristics are used, and no special fitting was made for the noise simulation.

In contrast to the $1/f$ noise, the thermal noise is hard to measure directly due to the required high frequency and its relatively small magnitude. Thus the usually applied AC measurement system requires a considerable experimental effort [23]. To characterize the features of the thermal noise, the noise coefficient γ is normally analyzed

$$S_{I_d} = \frac{4kT}{L_{\text{eff}}^2 I_{ds}} \int g_{ds}(y) dy = 4kT g_{ds0} \gamma \qquad (6.22)$$

$$\gamma = \frac{1}{L_{\text{eff}}^2 I_{ds} g_{ds0}} \int g_{ds}(\phi)^2 d\phi \qquad (6.23)$$

where g_{ds0} is the channel conductance at $V_{ds}=0$. As can be expected from Eqs. (6.20) and (6.21), the noise coefficient γ decreases from 1 to 2/3 as a function of V_{ds} for long-channel transistors, ignoring the mobility variation along the channel. Because of measurement difficulties, the thermal noise characteristics was very controversal especially for short-channel transistors. Figure 6.16 shows the measured thermal noise coefficients γ, normalized to the value of the channel conductance at $V_{ds}=0$, by several authors. Most of these authors have reported a drastic increase at the saturation condition [24–26]. The reasons put forward for explaining this increase were also very contradicting. Our experimental and theoretical evidences suggest that the thermal drain-current noise is neither due to hot electrons nor velocity saturation as argued by other authors, but is caused by the potential gradient in the channel and is therefore determined only by channel position dependent device quantities as can be seen in Eq. (6.15)- (6.19) [20].

Simulation results of γ are shown in Fig. 6.17 as a function of V_{ds}. For long-channel transistors γ reduces from 1 to 2/3 under the saturation condition as expected. The γ increase for short L_{gate} is much less drastic than reported previously. γ values in the linear region i. e. for small V_{ds} show a similar qualitative behavior for long as well as for short L_{gate}. A different L_{gate}-dependent behavior, with a strong increase of γ at short L_{gate}, starts when V_{ds} enters saturation, and is explained by the steepness of the surface-potential gradient, which becomes steeper with decreasing L_{gate}. For very short L_{gate} even the γ minimum at the end of the linear condition in Fig. 6.17 becomes larger than 2/3. However, it has to be noted again that the deviation from long-channel features are not as drastic as predicted previously. The increase of the noise coefficient under the saturation condition with reducing the gate length is as mentioned above attributed to the potential increase along the channel. This potential increase along the channel causes enhanced carrier scattering resulting in a reduction of the mobility along the channel, which is the main origin of the γ increase.

To illustrate the accuracy of the thermal-noise model, calculated results for two different CMOS technologies fabricated by two different manufacturers are compared in Fig. 6.17. The technology results of the first technology are shown in Fig. 6.17 by dotted lines at $V_{gs}=0.8V$ and the other technology results are shown by solid lines at $V_{gs}=1.0V$. The two different technologies show quite similar themal-noise characteristics and agree well with the prediction of our thermal-noise model. A particularly important observation for reduced L_{gate} is the increased gradient of γ with respect to V_{ds} in the saturation region. Fig. 6.18 shows the gradient of the $\gamma - V_{ds}$ characteristic as a function of the gradient of the $\Delta V_{th,SCE} - V_{ds}$ characteristic, where $\Delta V_{th,SCE} = V_{th}(\text{long } L_{gate}) - V_{th}(\text{short } L_{gate})$ represents the pure short-channel effect (SCE) without the reverse-short-channel effect (RSCE). Corresponding symbols in Fig. 6.17 denote the same L_{gate} and the same technology. The γ characteristics of the two different technologies fall on the same theoretical curve in Fig. 6.18, which apparently constitutes a fixed, technology-independent relationship. The technology-independent relationship between the thermal drain-noise coefficient γ and the measured threshold-voltage shift ΔV_{th} relative to a long-channel MOSFET is important, because it is exploitable for predicting γ without direct measurements.

6.1.3 *Induced Gate and Cross-Correlation Noise Model*

At high frequencies the MOSFET can be considered as an RC distributed network with a capacitive coupling between the resistive channel and the gate as schematically shown in Fig. 6.19 [2]. The distributed effect is treated as a transmission line for compact modeling, where the carrier distribution along the channel is included by solving the continuity equation together with the current-density equation [27]

$$j_n = -q\mu_n n \frac{\partial \phi}{\partial y} + qD_n \nabla n \tag{6.24}$$

$$\frac{\partial n}{\partial t} = \frac{1}{q} \nabla j_n \tag{6.25}$$

The final solution can be written in terms of Bessel functions, which is then further simplified by assuming moderately high frequencies to derive a closed form equation. The main factor determining the features of the induced gate noise S_{I_g} written as [28]

$$S_{I_g} = \frac{16}{135} kT \frac{\omega^2 C^2}{g_{ds0}} \frac{4\xi_0 + 20\xi_0^{\frac{3}{2}}\xi_L^{\frac{1}{2}} + 42\xi_0\xi_L + 20\xi_0^{\frac{1}{2}}\xi_L^{\frac{3}{2}} + 4\xi_L}{(\xi_0^{\frac{1}{2}} + \xi_L^{\frac{1}{2}})} \cdot (2\pi f)^2$$

(6.26)

where

$$\xi(y) = V_g V_t^2 - 2 \cdot Q_i(y) y$$ (6.27)

is again the carrier distribution $Q_i(y)$ along the channel. The variable y denotes the channel direction from source (0) to drain (L), and the definition of $V_g V_t$ is the same as in the thermal noise modeling, namely has the same meaning as $V_{gs} - V_{th}$ in a V_{th} based compact model. For the short-channel MOSFETs the equation is extended to include the gradient of the potential distribution in similar way as in the thermal noise case [28]. No additional model parameters are required for the induced gate-noise calculation.

A large difference between the induced-gate-noise mechnism and the thermal-noise mechanism is that the former is induced by the capacitive coupling of the carrier density fluctuation in the channel, whereas the latter is introduced by carrier scattering. Due to the gate insulator the same amount of carriers with opposite sign are collected at the gate side of the MOS capacitor. Therefore, the number of carrier fluctuations in the channel caused by carrier scattering induces the number of carrier fluctuations in the gate. Thus the carrier fluctuations in the channel, which are the origin of the thermal noise, are also the origin of the induced gate noise. It has to be noticed that the capacitive coupling becomes important only under high frequency operations. Simulation results with the induced gate-noise model of HiSIM are shown in Fig. 6.20.

The noise cross-correlation between the thermal noise and the induced gate noise is also expected under high frequencuy operation. The cross-coupling noise $S_{I_d I_g}$ is modeled in the same way as S_{I_g} with g_{ds}. Closed form equations for both S_{I_g} and $S_{I_d I_g}$ are derived by integrating the channel conductance along the channel [28]. This is possible because complete surface-potential-based models such as HiSIM provide the variation of all physical quantities along the channel. These variations along the channel determine exactly the features of MOSFETs under the high frequency operation. Though the final equations are rather complicated, a clear advantage of the used approach is that no extra model parameters are required. However, the equations can be reduced to a simpler form in the case of long-channel transistors under special conditions as can be seen in the van der Ziel description for the induced gate noise [2]

$$S_{I_g} = 4kT g_g \beta \tag{6.28}$$

$$g_g = \frac{\omega C_{gs}^2}{5 g_{ds0}} \tag{6.29}$$

$$\omega = 2\pi f \tag{6.30}$$

which is valid only for the saturation condition.

It is not easy to extend the presented modeling approaches to the nm scale MOSFET case. The reason is that the basic physical quantities influencing the carrier dynamics become complicated due to 2D effects as well as the necessity to include quantum and ballistic carrier transport for reduced L_{gate}. However, the fact that no additional model parameters are required for the noise modeling with HiSIM even in the high-frequency regime, and that only the potential distribution along the channel has to be considered, proves that advanced MOSFET features are still mostly governed by electrostatic featrues and that a quasi-2D modeling approach is still valid. The importance of this conclusion is that the MOSFET development can be undertaken in such a way that the key property of the MOSFET operation is kept, namely by keeping the property of main MOSFET control through the gate voltage.

As a summary of the noise modeling, it is worthwhile to emphasize that model parameters are required only for the $1/f$ noise to describe the trap density and the mobility fluctuation due to random trapping of charges. These parameters are nearly universal, if the technology is mature [16]. Thus most of the microscopic phenomena can be predicted by commonly measured I-V characteristics. This fact concludes that the majority of the carrier dynamics, even in microscopic aspects such as the noise, is still governed by electrostatic effects, which are determined by the Poisson equation [29].

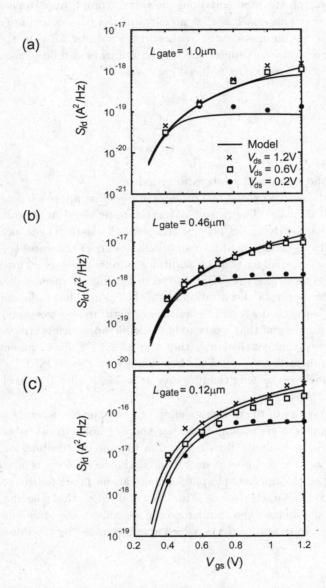

Fig. 6.10 Comparison of the bias dependence of the measured and simulated drain current noise by our model for L_{gate} = (a) 1.0μm, (b) 0.46μm and (c) 0.12μm at a frequency of 100Hz. Model parameter values are the same for all L_{gate} values.

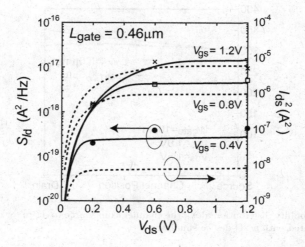

Fig. 6.11 Comparison of the V_{ds} dependence of the measured and simulated drain current noise (solid curves). Additionally, the square of the corresponding drain currents (dotted curves) is also plotted.

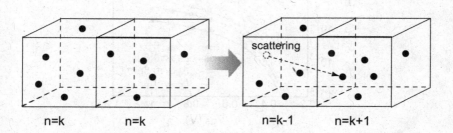

Fig. 6.12 Schematics explaining the origin of the thermal noise, namely carrier-number n fluctuation by scattering.

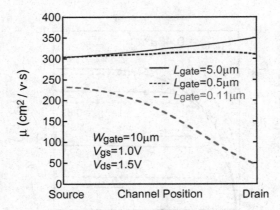

Fig. 6.13   Mobility distribution along the channel from source to drain for different gate length simulated with a 2D-device simulator.

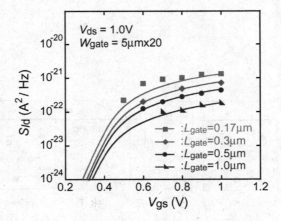

Fig. 6.14   Calculated thermal noise current with the developed compact model (solid lines) in comparison to measurements (symbols), as a function of V_{gs}. Measurements have been performed at five frequencies between 2 and 6 GHz.

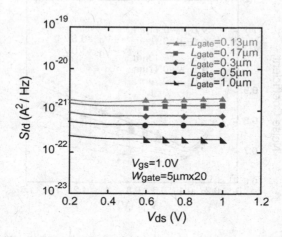

Fig. 6.15 The same figure as Fig. 6.14 but as a function of V_{ds}.

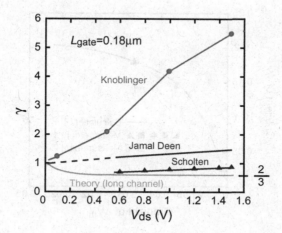

Fig. 6.16 Comparison of different measurements for the thermal noise coefficient γ by Knoblinger et al. [24], Jamal Deen et al. [25], and Scholten et al. [26].

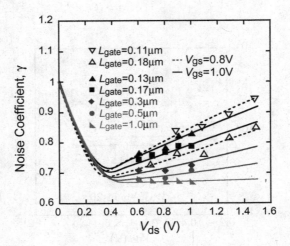

Fig. 6.17 Comparison of calculated γ (lines) with measurements (symbols) for two pocket-implant technologies from different manuafacturers. Solid symbols and solid lines refer to technology 1 ($L_{gate,min}$=130mn) while open symbols and dashed lines refer to technology 2 ($L_{gate,min}$=110mn). Different V_{gs} values were chosen for the comparison in addition, so that V_{ds} values for entering saturation are also different.

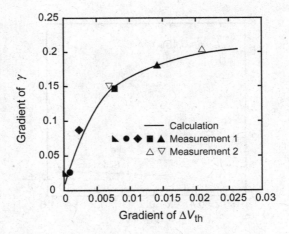

Fig. 6.18 Calculated (lines) and measured (symbols) gradient of the $\gamma - V_{ds}$ characteristic as a function of the gradient of the threshold-voltage shift relative to a long-channel transistor $\Delta V_{th,SCE}$ [$= V_{th}$(short L_{gate}) $- V_{th}$(long L_{gate})]. Lines and symbols refer to those given for the measurements in Fig. 6.17.

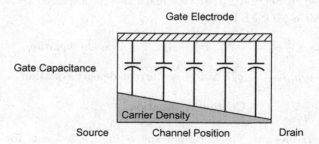

Fig. 6.19 Schematic for explaining the induced gate noise and the corresponding compact model concept.

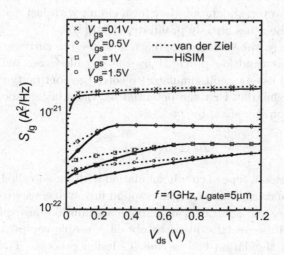

Fig. 6.20 Calculated induced gate noise as a function of V_{ds} for $L_{gate} = 5\mu m$. For comparison results with the van der Ziel equation are depicted together.

6.2   Non-Quasi-Static (NQS) Model

As explained in Section 2.1, device features are described by three basic equations for n-MOSFETs [30]:

- $\nabla^2 \phi = -\dfrac{q}{\epsilon_{Si}} \left(N_D - N_A + p - n \right) :$ Poisson equation

- $j_n = -q\mu_n n \dfrac{\partial \phi}{\partial y} + q D_n \nabla n :$ Current-density equation

- $\dfrac{\partial n}{\partial t} = \dfrac{1}{q} \nabla j_n :$ Continuity equation

Here all variables are four-terminal space-time dependent, for example, the mobile electron and hole concentrations are $n = n(x, y, z, t)$ and $p = p(x, y, z, t)$; the immobile donor and acceptor impurity concentrations are $N_D = N_D(x, y, z)$ and $N_A = N_A(x, y, z)$; the electron current density is the vector $j_n = j_n(x, y, z, t)$, and the electron potential is $\phi = \phi(x, y, z, t)$. To determine the circuit performance during an IC-design process using MOSFETs, ideally transient simulations are performed by solving the above three equations numerically for each type of MOSFET, which is called mixed-mode device simulation (see for example MEDICI [31] or ATLAS [32]). However, simulation time for a circuit with just a small number of transistors becomes already prohibitively large.

Consequently, the common approach is to use the currents and charges at a given bias condtion provided by the so-called compact model like HiSIM, and to use a circuit simulator to solve the continuity equation under the approximation that the potential responds instantaneously to the applied bias voltage, given by

$$I(t) = I(V(y)) + \frac{dQ}{dt} \tag{6.31}$$

In this conventional type of circuit simulation with the so-called quasi-static (QS) approximation, where carriers respond instantaneously to the voltage changes, further and additional simplifying assumptions as explained in the previous chapters are introduced to obtain a usable compact description for the circuit simulation task of the IC-design process. The inaccuracy of the QS approximation is, however, becoming increasingly obvious for applications at very high operating frequencies such as those in the RF-circuits. An improved compact-model approach, which goes beyond the QS approximation by considering the carrier response delays is known as the non-quasi-static (NQS) approach for compact modeling [33] which includes only drift and diffusion delays but not yet the delays due to generation-recombination, trapping or tunneling delays [30].

Validity of model-simulated device and circuit performances during high-frequency operation is often investigated in a frequency-domain analysis. Therefore, compact modeling based on two different approaches, in domains, namely the time-domain and the frequency-domain, will also be discussed separately in this section. Furthermore, a simple method for transforming the time-domain model into a frequency-domain model is presented.

6.2.1 *Time-Domain Analysis*

The carrier behavior during the transit along the MOSFET channel is described by the continuity equation

$$\frac{\partial n}{\partial t} = \frac{1}{q}\nabla j_n \tag{6.32}$$

Under the assumption that the potential responds instantaneously to each voltage change, the equation can be written in the form [34]

$$I_x(t) = I_{x0}(t) + \frac{dQ_{x0}(t)}{dt} \tag{6.33}$$

$$Q_{x0}(t) = Q_x(V(t)) \quad x = S, D, B, \text{ or } G \tag{6.34}$$

and is called Quasi-Static (QS) current equation as explained before. In Eq. (6.34) x=S, D, B and G stand for the source, drain, bulk and gate of the MOSFET, respectively. In particular the amount of charges $Q_x(V(t))$ required for a given voltage $V(t)$ is assumed available instantaneously. Figure 6.21 shows a comparison of the drain current under fast switching calculated by the QS approximation and by a 2D-device simulator. Substantial differences between the two results can be seen in Fig. 6.21, with the correct 2D-device simulation results giving a much smoother switching waveform than that from the QS approximation. The same is true for the two-transistor CMOS inverter circuit in Fig. 6.22. In these two figures of switching voltage applied to the gate/source terminals, $V_{gs}(t)$, Fig. 6.21 for a single linearly rising ramp and Fig. 6.22 for a periodic saw-tooth, we see that the neglected carrier transit delays in the QS approximation give an unphysical output current overshoot and current step at the input discontinuities or abrupt changes of the slope dV_{gs}/dt. This is not unexpected since numerical 2D-device simulators solve the three basic equations simultane-

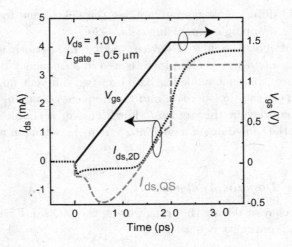

Fig. 6.21   MOSFET drain current calculated under the QS approximation $I_{ds,QS}$ in comparison to 2D-device simulation $I_{ds,2D}$.

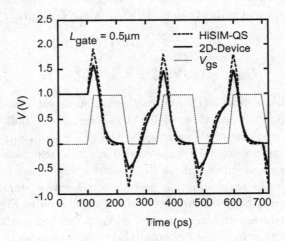

Fig. 6.22   Inverter output voltage under very fast gate switching. Again compact-model results with the QS approximation and 2D-device simulator results are compared.

ously, and thus 2D-device simulators give of course exact Non-Quasi-Static (NQS) solutions. Figure 6.23 shows the "snap shots" at four times for the simulated carrier distributions along the channel at different time steps (t=5ps: V_{gs}=0.375V; t=10ps: V_{gs}=0.75V; t=15ps: V_{gs}=1.125V; t=20ps:

V_{gs}=1.5V) during the switching-on transient of an n-MOSFET with a relatively long channel of 0.5μm length and for a fast input gate-voltage rise time of 20ps. In comparison, the Quasi-Static (QS) results (dotted curves) show the overshoots from neglecting the carrier transit delays. The NQS results verify that carriers in the channel indeed take time to build up and do not follow the fast gate-voltage switching as erroneously assumed by the QS approximation.

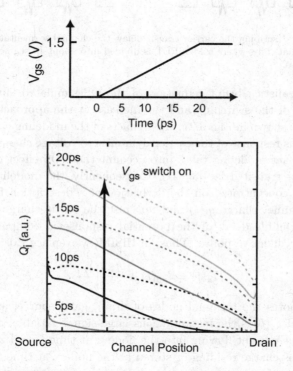

Fig. 6.23 Carrier distribution along the channel for different gate voltages in transient 2D-device simulation (continuous lines) and static 2D-device simulation at the same gate voltages (dotted lines).

The NQS carrier distribution in the channel at time step t_i has been modeled previously either by a polynomial function [35] or a Fourier series [36] of the channel position. Phenomenologically, such previous approaches can be understood as modeling the distributed effect along the channel and are often requiring the devision of the channel into many small segments, as shown in Fig. 6.23 [37]. These so-called channel-segmentation

approaches require huge simulation time and represents a rather artificial
device description.

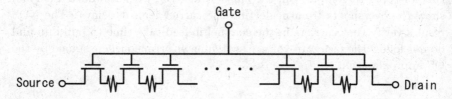

Fig. 6.24 For describing the carrier transit delay, the channel segmentation method is
depicted schematically, where a MOSFET is divided into about 10 segments connected
with resistances.

The unrealistic abrupt reduction of the QS simulated current at the
initial stage of the switching and the increase at the approach to the DC
steady-state, shown in Fig. 6.21, are artifacts of the modeling assumption of
instantaneous response of the carrier dynamics to voltage changes [38]. Our
HiSIM compact modeling takes into account the NQS effect on the basis
of the carrier transit delay [39–41]. Consequently, the modeling approach
in HiSIM is concentrated on the derivation of a description for the total
dynamic channel charge $q_I(t)$ and not just the static charge $Q_{I0}(t)$ that
assumes negligible delay of the potential response in comparison to the
delay of the carrier response. Thus, in HiSIM, we replace Eqs. (6.33) by

$$I_x(t) = I_{x0}(t) + \frac{dz_x(t)_{x0}}{dt} \qquad (6.35)$$

where x denotes the terminal nodes of drain, gate, source, and bulk (D,
G, S, B) and $z_x(t)$ is the instantaneous large-signal variable, $z = i$ for the
instantaneous current flowing into the node x at time t, and $z = q$ for the
instantaneous charge residing or stored at the node x at time t. As can be
seen in Eq. (6.35), the terminal currents consist of the transport (drift and
diffusion) current and the charging (displacement) current. The channel
charge q_I is sum of the source charge q_S and the drain charge q_I. The
transport current is a function of the instantaneous terminal voltages and
is approximated by the steady-state solution. The charging currents are
the time derivatives of the charges q_x including the carrier transit delay.

The Poisson equation relates local potential to the local carrier concentration. However, a compact model normally can consider only the
potential at the source end and the drain end for describing the terminal
currents. These two potentials at the source end and the drain end reach

their expected values at given bias condition nearly instantaneously as can be seen in Fig. 6.25. On the other hand, if we look into the channel-charge distribution, a clear carrier deficit is observed (see Fig. 6.23). Thus in our NQS model, this carrier deficit is considered and the physical reason for the charging delay, namely the carrier transit delay, is explicitly introduced into the modeling.

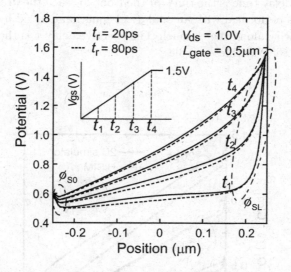

Fig. 6.25 Potential distribution along the MOSFET channel for 20ps (solid lines) and 80ps (dashed lines) gate-voltage switching. Potentials, in particular at source (ϕ_{SO} depicted by a circle) and drain end (ϕ_{SL} depicted by another circle), nearly instantaneously react to the gate voltage changes.

Without solving numerically, we approximate the carrier distribution along the channel from source to drain by a linear variation with distance as illustrated in Fig. 6.26. The validity of the linear approximation can also be seen in the 2D-device simulation results. The modeling of the NQS effect [39] is made by introducing the carrier transit delay, which consists of two possible delay mechanisms, represented by the delay times τ_{sup}, and τ_{tran}. τ_{sup} represents the delay from the carrier supply mechanism which typically occurs at the source side. τ_{tran} represents (i) the finite time it takes for injected carriers (electrons) from the source (n^+ Source) to drain by field-induced drifting (in strong inversion) and/or diffusion (subthreshold) through the length of the channel and (ii) the finite time to form the surface inversion.

The speed with which the carrier-concentration or carrier-density front of the injected carriers moves from the source to the drain through the channel is determined by the saturation velocity in short channels at high V_{ds} or high longitudinal (lateral) electric field, $E_y > 8 \times 10^3 \text{V/cm}$ with a saturation velocity of about $1 \times 10^7 \text{cm/s}$ for electrons in Si and $E_y > 2 \times 10^4 \text{V/cm}$ with a saturation velocity of about $8 \times 10^6 \text{cm/s}$ for holes in Si [30]. The delay time is the time for the front to reach the drain side. The calculation is performed with an analytical equation derived by integrating over the velocity along the channel. The effective delay τ is the mean of all delays using the Matthiessen rule

$$\frac{1}{\tau} = \frac{1}{\tau_{\text{sup}}} + \frac{1}{\tau_{\text{tran}}} \tag{6.36}$$

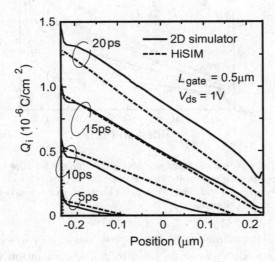

Fig. 6.26 Comparison of the inversion charge Q_i along the MOSFET channel in 2D-device simulation (solid lines) and in the linear approximation of HiSIM-NQS (dashed lines).

The carrier density in the channel q_i is then written as

$$q_i(y, t_j) = q_i(y, t_{j-1}) + \frac{t_j - t_{j-1}}{\tau} [Q_i(y, t_j) - q_i(y, t_{j-1})] \tag{6.37}$$

where $y = 0$ denotes the source end and $y = L$ the drain end. Q_i is the carrier density under the QS approximation and j is the index for numbering the time steps. In Eq. (6.37) the basic modeling idea is that only a part of the carriers required for the static distribution at a given applied voltage can be supplied within the time interval of Δt. The magnitude of this portion is determined by the carrier delay τ, as schematically shown in Fig. 6.27.

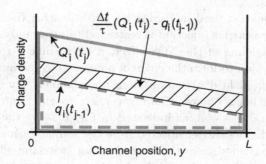

Fig. 6.27   Schematic illustration of the HiSIM model for Non-Quasi-Static (NQS) charge supply to the MOSFET channel. Only a portion of the Quasi-Static charge is supplied in each time step and the magnitude of this portion is determined by the carrier-delay time τ.

Assuming that the carrier density decreases linearly from $y = 0$ to $y = L$, integration of the carriers in the channel becomes [40, 41]

$$q_{\mathrm{I}}(t_{\mathrm{j}}) = \frac{1}{2}W_{\mathrm{eff}}[q_{\mathrm{i}}(0, t_{\mathrm{j}}) + q_{\mathrm{i}}(L, t_{\mathrm{j}})]L = W_{\mathrm{eff}} \int^{L_{\mathrm{eff0}}} q_{\mathrm{i}}(y, t_{\mathrm{j}}) \qquad (6.38)$$

where W_{eff} is the effective channel width. Equation (6.38) is used to rewrite Eq. (6.37) with the integrated charges q_{I} as

$$q_{\mathrm{I}}(t_{\mathrm{j}}) = q_{\mathrm{I}}(t_{\mathrm{j}-1}) + \frac{t_{\mathrm{j}} - t_{\mathrm{j}-1}}{\tau}[Q_{\mathrm{I}}(t_{\mathrm{j}}) - q_{\mathrm{I}}(y, t_{\mathrm{j}-1})] \qquad (6.39)$$

For the case that the carrier front has not reached the drain yet, such as after the time of 5ps in Fig. 6.26, the front position y_{f} of the carriers in the channel can be written as

$$y_{\mathrm{f}}(t_{\mathrm{j}}) = y_{\mathrm{f}}(t_{\mathrm{j}-1}) + \frac{t_{\mathrm{j}} - t_{\mathrm{j}-1}}{\tau_{\mathrm{tran}}}L \qquad (6.40)$$

so that the carrier density in the channel becomes

$$q_{\mathrm{I}}(t_{\mathrm{j}}) = \frac{1}{2}W_{\mathrm{eff}} \cdot q_{\mathrm{i}}(0, t_{\mathrm{j}}) \cdot y_{\mathrm{f}}(t_{\mathrm{j}}) \qquad (6.41)$$

With one more approximation of the carrier-front delay τ_{tran} shorter than the time step Δt, we are then able to write Eq. (6.41) in the same form as Eq. (6.39). Although this further approximation can be expected to cause an inaccuracy, the main reason for modeling of the NQS effect, which is the elimination of the artifacts caused by negelecting the carrier transit delay, is still addressed.

The final equation used in HiSIM for implementing the NQS effect is Eq. (6.39), where carrier injection is restricted to occur only at the source end. In this modeling of the NQS effect, if we assume a relatively large V_{ds}, carrier injection to form the channel remains to come mainly from the source due to the field distribution along the channel. For small V_{ds} the field gradient is not so large, allowing carriers to enter into the channel also from the drain as well, as indicated in Fig. 6.28 by the enhanced Q_{i} at the drain end. The carrier injected from the source during switching-on or the carrier extracted from the source during switching-off is called the source charge Q_{S}. The carrier density injected or extracted at the drain is called the drain charge Q_{D}. The sum of these two charges is equal to the total inversion charge Q_{I}. For calculating terminal currents on the source and drain contacts, charge partitioning (or division) between Q_{S} and Q_{D} must be made. The partition position is normally approximately in the channel middle. For the DC condition the partitioning definition is given by Eqs. (2.36) and (5.20). To reproduce the carrier deficit, Q_{S}:Q_{D}=100:0 is taken as an option for charge partitioning in the BSIM compact model [43], however, this cannot give consistent switching solutions. Our observation is that the ratio of 50:50 for the partitioning gives rather better result after inclusion of the NQS effect [42]. The conventional dynamic partitioning scheme used in HiSIM-QS has been applied also for the present HiSIM-NQS version as the default option. For simplicity the two different transit delays (τ_{sup} and τ_{tran}) are hereafter not distinguished and are treated by a combined τ parameter.

In our modeling approach, the carrier delay τ determines the magnitude of the NQS effect. Correlation with the time interval in the circuit simulation Δt is through a direct factor as can be seen from Eq. (6.39). Up to weak inversion, carriers diffuse through the channel, the transit delay is assumed to be independent of the applied electric field, and written as

$$\tau_{\text{diff}} = DLY1 \qquad (6.42)$$

where $DLY1$ is a modeling parameter normally fixed to a default value 10^{-10}s. In strong inversion, conduction in the surface inversion channel is

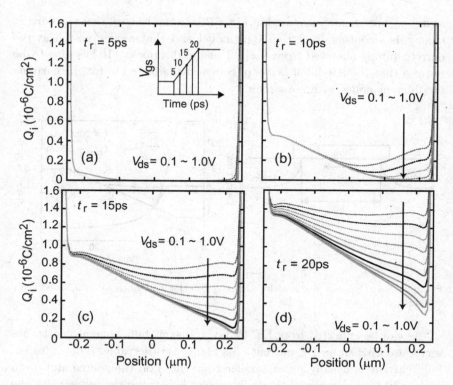

Fig. 6.28 Transient channel charge densities for 20ps gate-voltage switch-on at drain voltages from 0.1V to 1V after (a) 5ps, (b) 10ps, (c) 15ps and (d) 20ps elapsed time.

dominated by drift or electric-field-driven carriers. The transit delay due to such conductive carriers is modeled as

$$\tau_{\text{cond}} = DLY2 \cdot \frac{Q_{\text{I}}}{I_{\text{ds}}} \tag{6.43}$$

where $DLY2$ is a constant coefficient used as a second modeling parameter. The two delay mechanisms of diffusion and drift are then combined using the Matthiessen rule to determine the carrier delay as

$$\frac{1}{\tau} = \frac{1}{\tau_{\text{diff}}} + \frac{1}{\tau_{\text{cond}}} \tag{6.44}$$

The carrier-delay times calculated for the MOSFET switch-on operation with a gate-voltage rise time t_r of 20ps are illustrated in Fig. 6.29. Simulated transient currents during the switch-on and switch-off transients for two switching speeds are compared in Fig. 6.30. It can be seen that the current change is smoother from the NQS model and that artifacts such as the current jumps observed from the QS model disappear. It has also to be noticed that the Kirchhoff law is of course still adhered to for the currents on different nodes within a circuit.

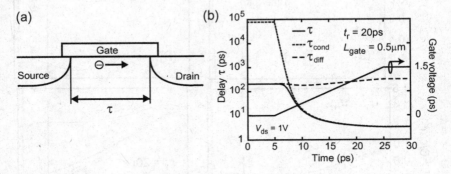

Fig. 6.29   (a) Dynamic carrier-delay times τ, (b) τ_{diff} and τ_{cond} in a 20ps MOSFET switch-on operation

As can be expected from Eq. (6.35), transient bulk current can be observed even when no DC current ($I_{B0} \simeq 0$A) is observable. This transient bulk current is usually much smaller than those on the source and drain nodes. The suppression of the bulk current is because the influence of the carrier dynamics at the surface of the substrate on the bulk can be assumed to be nearly negligible. Applying the same approach as before for the formation of bulk carriers, leads to an approximation of the bulk carrier delay as an RC delay in the form

$$\tau_{\text{B}} = DLY3 \cdot C_{\text{ox}} \tag{6.45}$$

where $DLY3$ is a constant coefficient and C_{ox} is the oxide capacitance.

A very large advantage of the modeling approach presented in this section is that the simulation time of the NQS model is comparable to that of the QS model and sometimes even smaller. The reason is that changes of device operating conditions such as fast changing currents are suppressed and smoothed out due to the carrier-delay mechanisms, which results in an easier convergence for the Newton iterations in circuit simulators. However, since the NQS effect is mainly not so drastic as previously expected, a simulation time increase is observed for most cases, but this simulation time increase is as small as only 3% on the average. Another advantage

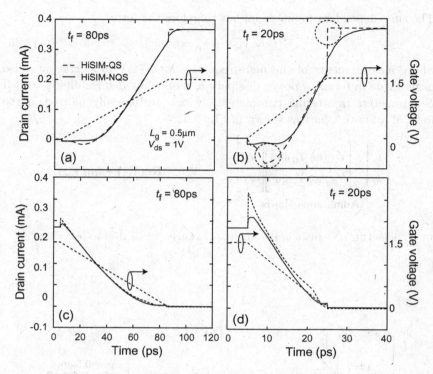

Fig. 6.30 Transient drain currents for 80ps (a) rising and (c) falling gate-voltage input as well as for a 20ps (b) rising and (d) falling gate-voltage input.

of the HiSIM-NQS model is the relatively low user effort for performing an NQS simulation, just by selecting the respective model flag provided in HiSIM, without having to select the specific transistors which are vulnerable to NQS effects, and by using the default parameters which are already about the right values for most practical cases.

6.2.2 *Frequency-Domain Analysis*

The most often used analysis for characterizing the high-frequency response of MOSFETs devices is with the admittance matrix, of which the elements are called Y-parameters [33] as shown in Fig. 6.31. For this type of high-frequency investigation, a small signal with the desired angular frequency ω $(=2\pi f)$ and an amplitude V_p is superimposed on a DC bias voltage V_{DC}.

The time-dependent input signal has therefore the form

$$V(t) = V_{DC} + V_p \exp(i\omega t) \qquad (6.46)$$

Many measurements of the high-frequency MOSFET properties are also often made in terms of power dissipation, namely by determining so-called S-parameters (scattering parameters), which are ususally plotted in the form of a Smith Chart as shown in Fig. 6.32.

$$\begin{bmatrix} i_g \\ i_d \end{bmatrix} = \begin{bmatrix} Y_{gg}(\omega) & Y_{gd}(\omega) \\ Y_{dg}(\omega) & Y_{dd}(\omega) \end{bmatrix} \begin{bmatrix} V_g \\ V_d \end{bmatrix} \qquad Y_{\alpha\beta} = Re(Y_{\alpha\beta}) + Im(Y_{\alpha\beta})$$

Admittance Matrix

Fig. 6.31   Definition of the admittance matrix $[Y]$ for the two-port case.

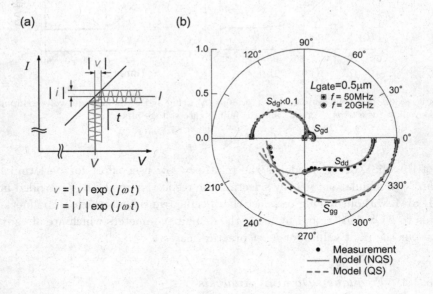

Fig. 6.32   (a) Principle for measuring admittance-matrix elements, and (b) an example of a Smith Chart.

The S-parameters are transformed into the Y-parameters by using the formulas listed in (6.47) for the two-port system.

$$S_{gg} = \frac{(1 - Y_{gg})(1 + Y_{dd}) + Y_{gd}Y_{dg}}{(1 + Y_{gg})(1 + Y_{dd}) - Y_{gd}Y_{dg}}$$

$$S_{gd} = \frac{-2Y_{gd}}{(1 + Y_{gg})(1 + Y_{dd}) - Y_{gd}Y_{dg}}$$

$$S_{dg} = \frac{-2Y_{dg}}{(1 + Y_{gg})(1 + Y_{dd}) - Y_{gd}Y_{dg}}$$

$$S_{dd} = \frac{(1 + Y_{gg})(1 - Y_{dd}) + Y_{gd}Y_{dg}}{(1 + Y_{gg})(1 + Y_{dd}) - Y_{gd}Y_{dg}} \tag{6.47}$$

Each Y-parameter consists of a real part and an imaginary part. A simplified description for the parameters has been derived as [44]

$$Y_{gg}(\omega) = \omega^2 R_g C_{gg}^2 + j\omega C_{gg} \tag{6.48}$$

$$Y_{gd}(\omega) = -\omega^2 R_g C_{gg} C_{gd} - j\omega C_{gd} \tag{6.49}$$

$$Y_{dg}(\omega) = g_m - \omega^2 R_g C_{gg}(C_m + C_{gd}) - j\omega(C_m + C_{gd} + g_m R_g C_{gg}) \tag{6.50}$$

$$Y_{dd}(\omega) = g_{ds} + \omega^2 R_g C_{gg}(C_{bd} + C_{gd}) + j\omega(C_{bd} + C_{gd} - g_{ds} R_g C_{gg}) \tag{6.51}$$

where C_m is the transcapacitance

$$C_m = C_{dg} - C_{gd} \tag{6.52}$$

Conventionally an equivalent circuit with a substrate network is used to reproduce the Y-parameter measurements as shown in Fig. 6.33 [33,44,45]. This approach is commonly used for most applications, however, a shortcoming of such an equivalent-circuit method is that a further parameter extraction is required for the elements included in the equivalent circuit in addition to the normal parameter extraction for the $I - V$ characteristics of the MOSFET. Such a high-frequency parameter extraction is usually quite difficult. Here two approaches are discussed, which aim at improving the conventional frequency-domain analysis:
   i) Bessel-Function Description
   ii) Transformation of Time-Domain NQS model into Frequency-Domain

i) Bessel-Function Description
   The substrate network in the equivalent circuit of Fig. 6.33 describes the carrier delay in a phenomenological way, which theoretically is exactly governed by the continuity equation. To derive a closed form equation based on the origin of the high frequency response, the continuity equation

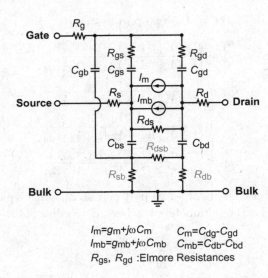

$I_m = g_m + j\omega C_m$ $C_m = C_{dg} - C_{gd}$
$I_{mb} = g_{mb} + j\omega C_{mb}$ $C_{mb} = C_{db} - C_{bd}$
R_{gs}, R_{gd} :Elmore Resistances

Fig. 6.33   Equivalent circuit with a substrate network to reproduce the Y-parameters of MOSFETs

is thus analytically solved simultaneously with the current density equation. Following this method, the final description reduces to an expression of Bessel functions [46, 47]. The surface-potential-based modeling approach includes both the drift and the diffusion contributions thus allowing the Y-parameter calculation at any bias condition for any signal frequency [48,49]. Comparisons of calculated results based on this Bessel-function description and measurements are shown in Fig. 6.34. For the calculation in Fig. 6.34 only the gate resistance is fitted to measurements. Since Y_{gd} and Y_{dd} include effects of the drain junction, additional parameter fitting is needed such as fitting the junction current. Therefore, only the two Y-parameters Y_{gg} and Y_{dg} are plotted in this comparison.

Usually external device features such as the overlap capacitances and resistances influence strongly the Y-parameter characteristics under high frequency operation. To get a better fit to measurements all these resistance and capacitance values have to be also considered, which is not done extensively for the simulation results shown in Fig. 6.34. Simulation results under the QS approximation are also plotted. For these QS results only the linear-order term of the series of Bessel functions is considered. The differences between NQS and QS results are not as drastic as predicted by others [44]. To observe the NQS effect more clearly the gate resistance contribution is omitted from the simulation and the result is shown in Fig. 6.35. The NQS effect can be see to become obvious at frequencies beyond 1/3 of the cut-off frequency.

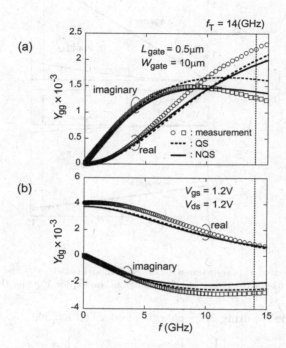

Fig. 6.34 Measured (open symbols) and calculated Y-parameters with the NQS model (solid lines) and the QS model (dashed lines), (a) for Y_{gg} and (b) for Y_{dg}. Model parameters are extracted by normal $I-V$ measurements and no further fitting parameters are introduced. As external contribution only the gate resistance is included.

The Bessel function description is implemented in HiSIM, and is proved to provide good reproducibility of measurements without laborious parameter extraction for the elements included in the equivalent circuit shown in Fig. 6.33. A problem of the Bessel-function approach is that a substantial enhancement of the simulation time cannot be avoided at the circuit level.

ii) Transformation of Time-Domain model into Frequency-Domain

The non-quasi-static (NQS) charge formulation in the time domain can also be transformed into the frequency domain by using the Fourier Trans-

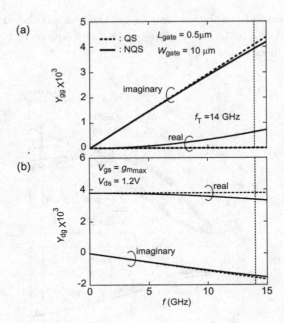

Fig. 6.35   Calculated Y-parameters with the NQS model (solid lines) and the QS model (dashed lines) without the gate-resistance contribution, (a) for Y_{gg} and (b) for Y_{dg}.

formation. The resulting NQS charge $\hat{q}$ is derived as [50]

$$\hat{q}_a(\omega) = \left\{ \frac{1}{1 + (\tau\omega)^2} - i\frac{\tau\omega}{1 + (\tau\omega)^2} \right\} \hat{Q}_a(\omega) \qquad (6.53)$$

where $\hat{Q}_a(w)$ is the QS charge in the frequency domain and τ is the carrier transit delay. The NQS capacitances are consequently calculated as

$$C_{ab} \equiv \frac{\partial q_a}{\partial V_b} = -\frac{2(\tau\omega)^2}{(1 + (\tau\omega)^2)^2} \frac{1}{\tau} \frac{\partial \tau}{\partial V_b} \cdot Q_{a,QS} + \frac{1}{1 + (\tau\omega)^2} \cdot C_{ab,QS}$$

$$-i\left[\frac{\tau\omega(1 - (\tau\omega)^2)}{(1 + (\tau\omega)^2)^2} \frac{1}{\tau} \frac{\partial \tau}{\partial V_b} \cdot Q_{a,QS} + \frac{\tau\omega}{1 + (\tau\omega)^2} \cdot C_{ab,QS} \right] \qquad (6.54)$$

where $C_{ab,QS}$ is the quasi-static capacitance. a and b denote the MOSFET terminals; gate g, drain d, source s and bulk b.

For Y-parameter calculation, conventional circuit simulators implement the quasi-static description as given for example in [33]. In order to obtain the Y-parameters in NQS description, the capacitances in the conventional equations are replaced by the NQS capacitances derived in Eq. (6.54).

In general, Y_{ab} is then obtained as

$$Y_{ab} = \frac{i\omega}{1 + (\tau\omega)^2} \left[C_{ab,QS} + (\tau\omega)^2 A_{ab}(\omega) - i \left[\tau\omega B_{ab}(\omega) + \tau\omega C_{ab,QS} \right] \right]$$

(6.55)

where A_{ab} and B_{ab} are defined in Eqs. (6.56) and (6.57), respectively.

$$A_{ab} = \frac{1}{\tau} \frac{-2}{1 + (\tau\omega)^2} \frac{\partial \tau}{\partial V_b} \cdot Q_{a,QS}$$

(6.56)

$$B_{ab} = \frac{1}{\tau} \frac{1 - (\tau\omega)^2}{1 + (\tau\omega)^2} \frac{\partial \tau}{\partial V_b} \cdot Q_{a,QS}$$

(6.57)

Y_{gg}, for example, is explicitly written as

$$Y_{gg} = \frac{i\omega}{1 + (\tau\omega)^2} \left[C_{gg,QS} + (\tau\omega)^2 A_{gg}(\omega) - i \left[\tau\omega B_{gg}(\omega) + \tau\omega C_{gg,QS} \right] \right]$$

(6.58)

Due to the described approach, the non-quasi static AC-analysis model implemented in HiSIM requires no additional model parameters.

A unified NQS model for both transient analysis and AC analysis is desired by most users to extend the simulation capability of the compact MOSFET model. This unified NQS model would enable high-frequency verification of circuits analyzed for both the time-domain and the frequency-domain characteristics without additional efforts, immediately after the conventional parameter extraction from the $I - V$ characteristics. For this purpose the Fourier transformation of the time-domain NQS model is the most straightforward approach. The result for the accuracy of our unified approach is shown in Fig. 6.36 in comparison with the numerical Fourier transformation of the time-domain result. Good agreement up to frequencies of 2 times the cut-off frequency proves the validity of the just-described unified method for real applications [50]. This can be improved by reducing approximation levels introduced for the calculation. Simulated frequency dependencies of the Y-parameters are compared with measurements in Fig. 6.37 for an example where the cut-off frequency is about 0.5GHz.

In the forgoing HiSIM options have been introduced to select either the described QS or NQS model, furthermore, to select either the time-domain or frequency-domain analysis. It should be emphasized again that the model-parameter extraction with only the measured I-V characteristics is sufficient even for the NQS simulation. Consequently, HiSIM's NQS model allows to describe all MOSFET related RF effects without any addi-

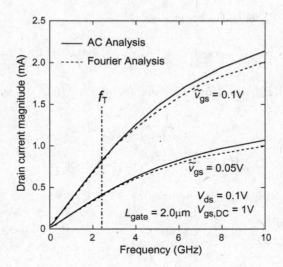

Fig. 6.36   Comparison of the drain current magnitude calculated with the frequency-domain model and the time-domain model. Good agreement until clearly beyond the cut-off frequency f_T is obtained.

tional parameters. Previously, the description of these RF effects was only possible by extracting the elements of a substrate-resistance network from additional Y-parameter measurements.

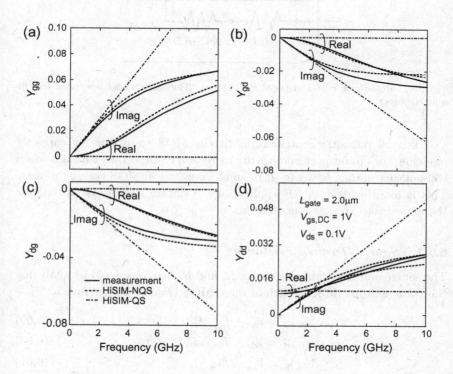

Fig. 6.37 Comparison of the measured Y-parameters with the Y-parameters simulated with the developed NQS model and the conventional QS model. The cut-off frequency of the studied MOSFET device is about 0.5GHz.

6.3   External MOS Transistor Resistances

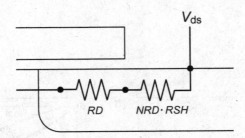

Fig. 6.38   Resistance model of the drain contact. The same model is applied for the source contact.

External parasitic contributions to the MOS transistor for correctly modeling the circuit performance are increasingly important with advanced technologies scaling down to ultra small geometrical MOS transistor sizes. This is more evident in RF applications. Contact resistances are among the important external contributions.

6.3.1   *Source/Drain Resistances*

The source and the drain resistances R_s and R_d are considered by analyzing the voltage drops through each of these MOS transistor terminals as:

$$V_{gs,eff} = V_{gs} - I_{ds} \cdot R_s \tag{6.59}$$

$$V_{ds,eff} = V_{ds} - I_{ds} \cdot (R_s + R_d) \tag{6.60}$$

$$V_{bs,eff} = V_{bs} - I_{ds} \cdot R_s \tag{6.61}$$

where the resistances R_s and R_d consist of two main contributions as shown in Fig. 6.38 for the drain resistance R_d.

$$R_s = RS + NRS \cdot RSH \tag{6.62}$$

$$R_d = RD + NRD \cdot RSH. \tag{6.63}$$

Here the first terms of the right hand sides of Eqs. (6.62) and (6.63) come from the resistances in the low impurity concentration region also called extension region of the contacts. The second terms $NRS \cdot RSH$ and $NRD \cdot RSH$ are the resistances in the conventional diffused regions, which are layout dependent. The parameters NRS and NRD are the number of fingers used, and RSH is the sheet-resistance of the diffusion region. The voltage drops of Eqs. (6.59) − (6.61) are calculated iteratively for given external terminal voltages (V_{gs}, V_{ds}, and V_{bs}) to keep consistency among all device performances. However, R_s and R_d can be also treated as extrinsic resistances, which are not part of the MOS transistor model, and can be included in an equivalent circuit. In HiSIM these parasitic source and drain resistances, R_s and R_d, can be included by two optional approaches. The first approach is to include them as external resistances of HiSIM, so that the circuit simulator has to find the iterative solution for the potential drops across the source and drain resistances. This first approach is commonly used by all MOS transistor models for circuit simulation. The second approach is to include the resistances as internal resistances of HiSIM, so that HiSIM has to solve for the resistance contributions iteratively. The flag $CORSRD$ is provided for the selection of one of the possible approaches. $CORSRD = 0,1,-1$ means "no", "internal", "external" source/drain resistances.

The first approach with external source and drain resistances leads to shorter simulation times for circuits with small to medium transistor numbers, while the second approach with internal source/drain resistances leads to shorter simulation times for circuits with very large transistor numbers. The transistor number, for which both approaches result in approximately equal simulation times (the switching point for the choice between the two approaches) is normally between 10000 and 50000 transistors.

6.3.2 *Gate Resistance*

The gate resistance becomes large as the gate width becomes large in comparison to the length, which is often the case for RF circuits. The equation for the gate-resistance calculation is [43]

$$R_g = \frac{RSHG \cdot \left(XGW + \frac{W_{eff}}{3 \cdot NGCON}\right)}{NGCON \cdot (L_{drawn} - XGL) \cdot NF} \qquad (6.64)$$

where $RSHG$ is the gate sheet resistance, and other parameters are parameters dependent on the layout. The flag **CORG** is provided in HiSIM for the inclusion of the gate resistance. $CORG = 0$ and 1 means "no" and "external" gate resistance, respectively.

6.4 Summary of Equations and Model Parameters Appeared in Chapter 6 for Modeling of Phenomena Important for RF Applications

6.4.1 *Section 6.1: Noise Models*

[$1/f$ Noise Model]

$$S_{I_d} = \frac{I_{ds}^2 NFTRP}{\beta f (L_{eff} - \Delta L) W_{eff}} \left[\frac{1}{(N_0 + N^*)(N_L + N^*)} \right. $$
$$\left. + \frac{2vNFALP}{N_L - N_0} \ln \left(\frac{N_L + N^*}{N_0 + N^*} \right) + (vNFALP)^2 \right]$$

[Thermal Noise Model]

$$S_{I_d} = 4kT \frac{W_{eff} C_{ox} V_g V_t \mu}{(L_{eff} - \Delta L)} \frac{1}{15(1+\eta)\mu_{av}^2} \left[(1 + 3\eta + 6\eta^2)\mu_d^2 \right.$$
$$\left. + (3 + 4\eta + 3\eta^2)\mu_d\mu_s + (6 + 3\eta + \eta^2)\mu_s \right]$$

$$\eta = 1 - \frac{(\phi_{SL} - \phi_{S0}) + \chi(\phi_{SL} - \phi_{S0})}{V_g V_t}$$

$$\chi = 2 \frac{const0}{C_{ox}} \left[\left(\frac{2}{3} \frac{1}{\beta} \frac{\{\beta(\phi_{SL} - V_{bs}) - 1\}^{\frac{3}{2}} - \{\beta(\phi_{S0} - V_{bs}) - 1\}^{\frac{3}{2}}}{\phi_{SL} - \phi_{S0}} \right) \right.$$
$$\left. - \sqrt{\beta(\phi_{S0} - V_{bs}) - 1} \right]$$

$$const0 = \sqrt{\frac{2\epsilon_{Si} q N_{sub}}{\beta}}$$

[Induced Gate Noise]

$$S_{I_g} = \frac{16}{135} kT \frac{\omega^2 C^2}{g_{ds0}} \frac{4\xi_0 + 20\xi_0^{\frac{3}{2}}\xi_L^{\frac{1}{2}} + 42\xi_0\xi_L + 20\xi_0^{\frac{1}{2}}\xi_L^{\frac{3}{2}} + 4\xi_L}{(\xi_0^{\frac{1}{2}} + \xi_L^{\frac{1}{2}})} \cdot (2\pi f)^2$$

$$\xi(y) = V_g V_t^2 - 2 \cdot Q_i(y) y$$

Table 6.1   HiSIM model parameters for noise models.   ∗ indicates minor parameters.

$NFTRP$	ratio of trap density to attenuation coefficient
$NFALP$	contribution of the mobility fluctuation
$*CIT$	capacitance caused by the interface trapped carriers

6.4.2   *Section 6.2: Non-Quasi-Static (NQS) Model*

[Time-Domain Analysis]

$$\frac{1}{\tau} = \frac{1}{\tau_{\text{diff}}} + \frac{1}{\tau_{\text{cond}}}$$

$$\tau_{\text{diff}} = DLY1$$

$$\tau_{\text{cond}} = DLY2 \cdot \frac{Q_{\text{i}}}{I_{\text{ds}}}$$

$$\tau_{\text{B}} = DLY3 \cdot C_{\text{ox}}$$

Table 6.2   HiSIM model parameters for NQS model.

$DLY1$	coefficient for delay due to diffusion of carriers
$DLY2$	coefficient for delay due to conduction of carriers
$DLY3$	coefficient for RC delay of bulk carriers

6.4.3   *Section 6.3: External MOS Transistor Resistances*

[Source/Drain Resistances]

$$V_{\text{gs,eff}} = V_{\text{gs}} - I_{\text{ds}} \cdot R_{\text{s}}$$
$$V_{\text{ds,eff}} = V_{\text{ds}} - I_{\text{ds}} \cdot (R_{\text{s}} + R_{\text{d}})$$
$$V_{\text{bs,eff}} = V_{\text{bs}} - I_{\text{ds}} \cdot R_{\text{s}}$$

$$R_{\text{s}} = RS + NRS \cdot RSH$$
$$R_{\text{d}} = RD + NRD \cdot RSH$$

[Gate Resistance]

$$R_{\text{g}} = \frac{RSHG \cdot \left(XGW + \frac{W_{\text{eff}}}{3 \cdot NGCON}\right)}{NGCON \cdot (L_{\text{drawn}} - XGL) \cdot NF}$$

Table 6.3   HiSIM model parameters for resistances.

RS	source-contact resistance of LDD region
RD	drain-contact resistance of LDD region
RSH	source/drain sheet resistance of diffusion region
NRS	number of source squares
NRD	number of drain squares
RSHG	gate sheet resistance
NGCON	number of gate
NF	number of fingers
XGW	distance from the gate contact to the channel edge
XGL	offset of the gate length

Bibliography

[1] B. Razavi, *IEEE J. Solid-State Circuit*, vol. 34, pp. 268, 1996.

[2] A. van der Ziel, "Noise in Solid Sate Devices and Circuit," New York, John Wiley & Sons, Inc., 1986.

[3] S. Christensson, I. Lundstrom, and C. Svensson, "Low frequency noise in MOS transistors — I: Theory," *Solid-State Electron.*, vol. 11, no. 9, pp. 797–812, 1968.

[4] H. Ueno, T. Kitamura, S. Matsumoto, T. Okagaki, M. Miura-Mattausch, H. Abe, and T. Hamasaki, "Evidence for an additional noise source modifying conventional $1/f$ frequency dependence in sub-μm metal-oxide-semiconductor field-effect transistors," *Appl. Phys. Lett.*, vol. 78, no. 3, pp. 380–382, 2001.

[5] F. N. Hooge, "$1/f$ noise sources," *IEEE Trans. Electron Devices*, vol. 41, no. 11, pp. 1926–1935, 1994.

[6] K. K. Hung, P. K. Ko, C. Hu, and Y. C. Cheng, "A physics-based MOSFET noise model for circuit simulators," *IEEE Trans. Electron Devices*, vol. 37, no. 5, pp. 1323–1333, 1990.

[7] L. D. Yau and C. T. Sah, "Theory and experiments of low-frequency generation-recombination noise in MOS transistors," *IEEE Trans. Electron Devices*, vol. 16, no. 2, pp. 170–177, 1969.

[8] A. Longoni, E. Gatti, and R. Sacco, "Trapping noise in semiconductor-device: A method for determining the noise spectrum as a function of the trap position," *J. Appl. Phys.*, vol. 78, no. 10, pp. 6283–6297, 1995.

[9] R. Brederlow, W. Weber, D. Schmitt-Landsiedel, and R. Thewes, "Fluctuations of the low frequency noise of MOS transistors and their modeling in analog and RF-circuits," *IEDM Tech. Dig.*, pp. 159–162, 1999.

[10] M. Schulz, "Coulomb energy of traps in semiconductor space-charge regions," *J. Appl. Phys.*, vol. 74, no. 4, pp. 2649–2657, 1993.

[11] H. H. Mueller, D. Woerle, and M. Schulz, "Evaluation of the Coulomb energy for single-electron interface trapping in sub-μm metal-oxide-semiconductor field-effect transistors," *J. Appl. Phys.*, vol. 75, no. 6, pp. 2970–2979, 1994.

[12] H. H. Mueller and M. Schulz, "Random telegraph signal: An atomic probe of the local current in field-effect transistors," *J. Appl. Phys.*, vol. 83, no. 3, pp. 1734–1741, 1998.

[13] M.-H. Tsai and T.-P. Ma, "The impact of device scaling on the current fluctuations in MOSFETs," *IEEE Trans. Electron Devices*, vol. 41, no. 11, pp. 2061–2068, 1994.

[14] J. S. Kolhatkar, L. K. J. Vandamme, C. Salm, H. Wallinga, "Separation of random telegraph signals from $1/f$ noise in MOSFETs under constant and switched bias condition," *Proc. of ESSDERC*, pp. 549–552, 2003.

[15] N. Mutoh and N. Teranishi, "New empirical relation for MOSFET $1/f$ noise unified over linear and saturation regions," *Solid-State Electron.*, vol. 31, no. 12, pp. 1675–1680, 1988.

[16] S. Matsumoto, H. Ueno, S. Hosokawa, T. Kitamura, M. Miura-Mattausch, H. J. Mattausch, T. Ohguro, S. Kumashiro, T. Yamaguchi, K. Yamashita, and N. Nakayama, "$1/f$ noise characteristics in 100nm-MOSFETs and its modeling for circuit simulation," *IEICE Trans. Electron.*, vol. E88-C, no. 2, pp. 247–254, 2005.

[17] H. Ueno, S. Matsumoto, S. Hosokawa, M. Miura-Mattausch, H. J. Mattausch, T. Ohguro, S. Kumashiro, T. Yamaguchi, K. Yamashita, and N. Nakayama, "Modeling $1/f$ noise with HiSIM for 100nm CMOS technology," *Proc. Int. Workshop on Compact Modeling*, pp. 18–23, Yokohama, Jan., 2004.

[18] H. Nyquist, "Thermal agitation of electric charge in conductors," *Phys. Rew.*, vol. 32, no. 7, pp. 110–113, 1928.

[19] F. M. Klaassen and J. Prins, "Thermal noise of MOS transistors," *Philips Res. Repts.*, vol. 22, pp. 505–514,1967.

[20] S. Hosokawa, D. Navarro, M. Miura-Mattausch, H. J. Mattausch, T. Ohguro, T. Iizuka, M. Taguchi, S. Kumashiro, and S. Miyamoto, "Gate-length and drain-voltage dependence of thermal drain noise in advanced metal-oxide-semiconductor field-effect transistors," Appl. Phys. Lett. 87, 092104, 2005.

[21] *HiSIM2.0.0 User's Manual*, Hiroshima University and STARC, 2005.

[22] S. Hosokawa, "Modeling of thermal noise and its verification with measurements for 100nm-MOSFETs," Dissertation for Master of Engineering (Japanese), Hiroshima University, February, 2004.

[23] A. A. Abidi, "High-frequency noise measurements on FET's with small dimensions," *IEEE Trans. Electron Devices*, vol. ED-33, no. 11, pp. 1801–

1805, 1986.

[24] G. Knoblinger, P. Klein, and M. Tiebout, "A new model for thermal channel noise of deep-submicron MOSFETs and its application in RF-CMOS design," *IEEE J. Solid-State Circuits*, vol. 36, no. 5, pp. 831–837, 2001.

[25] M. Jamal Deen and C.-H. Chen, "RF MOSFET noise parameter extraction and modeling," *Proc. Workshop on Compact Modeling*, pp. 694–697, Perto Rico, May, 2002.

[26] A. J. Scholten, L. F. Tiemeijer, R. v. Langevelde, R. J. Havens, A. T. A. Zegers-van Duijinhoven, and V. C. Venezia, "Noise modeling for RF CMS circuit simulation," *IEEE Trans. Electron Devices*, vol. 50, no. 3, pp. 618–632, 2003.

[27] M. Shoji, "Analysis of high-frequency thermal noise of enhancement mode MOS field-effect transistors," *IEEE Trans. Electron Devices*, vol. ED-13, no. 6, pp. 520–524, 1966.

[28] T. Warabino, M. Miyake, D. Navarro, Y. Takeda, G. Suzuki, T. Ezaki, M. Miura-Mattausch, H. J. Mattausch, T. Ohguro, T. Iizuka, M. Taguchi, S. Kumashiro, and S. Miyamoto, "Analysis and compact modeling of MOSFET high-frequency noise," *Proc. SISPAD*, pp. 158-161, Monterey, Sept., 2006.

[29] M. Miura-Mattausch, "MOSFET modeling beyond 100nm technology: Challenges and perspectives," *Proc. SISPAD*, pp. 1-6, Tokyo, Sept., 2005.

[30] Chih-Tang Sah, "Fundamentals of solid-state electronics," *World Scientific Publishing Co*, Singapore, 1991.

[31] *MEDICI User's Manual*, Synopsis Inc., Version 2002.2.

[32] *ATLAS User's Manual*, Silvaco, Nov., 1998.

[33] Y. P. Tsividis, "Operation and Modeling of the MOS Transistor (Second Edition)," *McGraw-Hill*, 1999.

[34] S.-Y. Oh, D. E. Ward, and R. W. Dutton, "Transient Analysis of MOS Transistors," *IEEE J. Solid-State Circ.*, vol. SC-15, no. 8, pp. 636–643, 1980.

[35] C. Turchetti, P. Mancini, and G. Masetti, "A CAD-oriented non-quasi-static approach for the transient analysis of MOS IC's," *IEEE J. Solid-State Circ.*, vol. SC-21, no. 5, pp. 827–836, 1986.

[36] H.-J. Park, P. K. Ko, C. Hu, "A charge conserving non-quasi-static (NQS) MOSFET model for SPICE transient analysis," *IEEE Trans. Computer-Aided Design*, vol. 10, no. 5, pp. 629–642, 1991.

[37] A. J. Scholten, L. F. Tiemeijer, P. W. H. de Vreede, and D. B. M. Klaassen, "A large signal non-quasi-static MOS model for RF circuit simulation," *IEDM Tech. Dig.*, pp. 163–166, Washington DC, Dec., 1999.

[38] T. Okagaki, M. Tanaka, H. Ueno, and M. Miura-Mattausch, "Importance of Ballistic Carriers for the Dynamic Response in Sub-100nm MOSFETs," *IEEE Electron Device Letters*, vol. 23, no. 3, pp. 154–156, 2002.

[39] N. Nakayama, D. Navarro, M. Tanaka, H. Ueno, M. Miura-Mattausch, H. J. Mattausch, T. Ohguro, S. Kumashiro, M. Taguchi, and S. Miyamoto, "Non-quasi-static model for MOSFET based on carrier-transit delay," *Electronics Letters*, vol. 40, no. 4, pp. 276–278, 2004.

[40] D. Navarró, Y. Takeda, M. Miyake, N. Nakayama, K. Machida, T. Ezaki, H. J. Mattausch, and M. Miura-Mattausch, "A carrier-tansit-delay-based nonquasi-static MOSFET model for circuit simulation and its application to hamonic distortion analysis," *IEEE Trans. Electron Devices*, vol. 53, no. 9, pp. 2025–2034, 2006.

[41] D. Navarro, N. Nakayama, K. Machida, Y. Takeda, H. Ueno, H. J. Mattausch, M. Miura-Mattausch, T. Ohguro, T. Iizuka, M. Taguchi, T. Kage, and S. Miyamoto, "Modeling of carrier transport dynamics at GHz-frequencies for RF circuit simulation," *Proc. SISPAD*, pp. 259-262, Munich, Sept., 2004.

[42] D. Navarro, N. Nakayama, K. Machida, Y. Takeda, M. Miura-Mattausch, H. J. Mattausch, T. Ohguro, T. Iizuka, M. Taguchi, and S. Miyamoto, "A carrier tansit time delay-based non-quasi-static MOSFET model for RF circuit simulation," *Proc. Int. Workshop on Compact Modeling* pp. 23–28, Shanghai, 2005.

[43] *BSIM4.0.0 MOSFET Model, User's Manual*, Department of Electrical Engineering and Computer Science, University of California, Berkeley CA, 2000.

[44] C. C. Enz and Y. Cheng, "MOS transistor modeling for RF IC design," *IEEE Trans. Solid-State Circ.*, vol. 35, no. 2, pp. 186–201, 2000.

[45] H. Kawano, M. Nishizawa, S. Matsumoto, S. Mitani, M. Tanaka, N. Nakayama, H. Ueno, M. Miura-Mattausch, and H. J. Mattausch, "A practical small-signal equivalent circuit model for RF-MOSFETs valid up to cut-off frequency," *Microwave Symposium Digest*, pp. 2121–2124, Washington, June, 2002.

[46] J. J. Paulos and D. A. Antoniadis, "Limitations of quasi-static capacitance models for the MOS transistor," *IEEE Electron Device Lett.*, vol. EDL-4, no. 7, pp. 221–224, 1983.

[47] Y. Niitsu, "Simple small-signal model for 3-port MOS transistors," *IEICE Trans. Electron.*, vol. E79-C, no. 12, pp. 1760–1765, 1996.

[48] S. Jinbou, H. Ueno, H. Kawano, K. Morikawa, N. Nakayama, M. Miura-Mattausch, and H. J. Mattausch, "Analysis of non-quasistatic contribution to small-signal response for deep sub-μm MOSFET technologies," *Ext. Abs. SSDM*, pp. 26–27, Tokyo, Sept., 2002.

[49] H. Ueno, S. Jinbou, H. Kawano, K. Morikawa, N. Nakayama, M. Miura-Mattausch, and H. J. Mattausch, "Drift-diffusion-based modeling of the non-quasistatic small-signal response for RF-MOSFET applications," *Proc. SISPAD*, pp. 71–74, Kobe, Sept., 2002.

[50] K. Machida, D. Navarro, M. Miyake, R. Inagaki, N. Sadachika, G. Suzuki, Y. Takeda, T. Ezaki, H. J. Mattausch, M. Miura-Mattausch, "Efficient NQS MOSFET Model for both Time-Domain and Frequency-Domain Analysis," Proc. SiRF, pp. 73–76, San Diego, Jan., 2006.

Chapter 7

Summary of HiSIM's Model Equations, Parameters, and Parameter-Extraction Method

In this chapter a short summary of the HiSIM model (valid up to version HiSIM2.4.0) is given. It is meant to serve as a quick and direct reference for those who want to skip most of the detailed modeling explanations, or those who know the HiSIM model already quite well and just want to check a certain issue with respect to the model constraction or the model usage.

7.1 Model Equations of HiSIM

7.1.1 *Physical Quantities*

[gate capacitance]

$$C_{\text{ox}} = \frac{\epsilon_{\text{ox}}}{T_{\text{ox,eff}}} = \frac{\mathbf{KAPPA}}{T_{\text{ox,eff}}}$$

[effective oxide thickness including Quantum Effects]

$$T_{\text{ox,eff}} = \mathbf{TOX} + \Delta T_{\text{ox}}$$

$$\Delta T_{\text{ox}} = \mathbf{QME1} \left\{ V_{\text{gs}} - V_{\text{th}}(T_{\text{ox,eff}} = \mathbf{TOX}) - \mathbf{QME2} \right\}^2 + \mathbf{QME3}$$

[effective gate voltage]

$$V_{\text{G}}' = V_{\text{gs}} - \mathbf{VFB} + \Delta V_{\text{th}}$$

[Deby length]

$$L_{\text{D}} = \sqrt{\frac{\epsilon_{\text{Si}}}{q\beta N_{\text{sub}}}}$$

319

[thermal voltage]

$$\beta = \frac{q}{kT}$$

[threshold voltage]

$$V_{th} = \mathbf{VFB} + 2\Phi_B + \frac{\sqrt{2\epsilon_{Si}qN_{sub}}}{C_{ox}}\sqrt{2\Phi_B}$$

[surface potential at threshold condition]

$$2\Phi_B = \frac{2}{\beta}\ln\left(\frac{N_{sub}}{n_i}\right)$$

[widly used HiSIM const0]

$$const0 = qN_{sub}L_D\sqrt{2}$$

7.1.2 *MOSFET Size*

[fabricated gate length]

$$L_{gate} = L_{drawn}$$

[fabricated gate width]

$$W_{gate} = W_{drawn}$$

[actual gate length]

$$L_{poly} = L_{gate} - 2 \cdot dL$$

$$dL = \frac{\mathbf{LL}}{(L_{gate} + \mathbf{LLD})^{\mathbf{LLN}}}$$

[actual gate width]

$$W_{poly} = W_{gate} - 2 \cdot dW$$

[effective gate length]

$$L_{eff} = L_{poly} - 2 \cdot \mathbf{XLD}$$

[effective gate width]

$$W_{eff} = W_{poly} - 2 \cdot \mathbf{XWD}$$

7.1.3 *Temperature Dependence*

[thermal voltage]

$$\beta = \frac{q}{kT}$$

[bandgap]

$$E_{\mathrm{g}} = \mathbf{EG0} - \mathbf{BGTMP1} \cdot T - \mathbf{BGTMP2} \cdot T^2$$

[intrinsic carrier density]

$$n_{\mathrm{i}} = n_{\mathrm{i0}} \cdot T^{\frac{3}{2}} \cdot \exp\left(-\frac{E_{\mathrm{g}}}{2q}\beta\right)$$

[mobility due to phonon scattering]

$$\mu_{\mathrm{PH}}(\mathrm{phonon}) = \frac{M_{\mathrm{uephonon}}}{(TT)^{\mathbf{MUETMP}} \cdot E_{\mathrm{eff}}^{\mathbf{MUEPH0}}}$$

[maximum carrier velocity]

$$V_{\max} = \frac{\mathbf{VMAX}}{1.8 + 0.4(TT) + 0.1(TT)^2 - \mathbf{VTMP} \cdot (1 - TT)}$$

[T: device temperature]

$$TT = T/\mathbf{TNOM}$$

7.1.4   *Substrate Impurity Concentration N_{sub}*

[Substrate impurity concentration]

$$N_{\text{sub}} = \frac{\textbf{NSUBC}(L_{\text{gate}} - \textbf{LP}) + N_{\text{subp}} \cdot \textbf{LP}}{L_{\text{gate}}} + \frac{\textbf{NPEXT} - \textbf{NSUBC}}{\left(\frac{1}{\textbf{xx}} + \frac{1}{\textbf{LPEXT}}\right) L_{\text{gate}}}$$

$$\textbf{xx} = 0.5 \cdot L_{\text{gate}} - \textbf{LP}$$

[increase of N_{sub} bottom due to pocket overlap]

$$N_{\text{subb}} = 2 \cdot N_{\text{subp}} - \frac{(N_{\text{subp}} - \textbf{NSUBC}) \cdot L_{\text{gate}}}{\textbf{LP}} - \textbf{NSUBC}$$

[pocket impurity concentration]

$$N_{\text{subp}}$$

[STI effect]

$$N_{\text{substi}} = \textbf{NSUBPSTI1} \left[1 + \frac{1}{1 + \textbf{NSUBPSTI2}} \cdot \left(\frac{\textbf{NSUBPSTI1}}{\textbf{LOD}}\right)^{\textbf{NSUBPSTI3}} \right]$$

7.1.5   *Threshold Voltage Shift ΔV_{th}*

[Threshold voltage shift]

$$\Delta V_{\text{th}} = \Delta V_{\text{th,SC}} + (\Delta V_{\text{th,R}}) + \Delta V_{\text{th,P}} + \Delta V_{\text{th,W}} - \phi_{\text{Spg}}$$

[short-channel effect]

$$\Delta V_{\text{th,SC}} = \frac{\epsilon_{\text{Si}}}{C_{\text{ox}}} \cdot W_{\text{d}} \frac{2(\textbf{VBI} - 2\Phi'_{\text{B}})}{(L_{\text{gate}} - \textbf{PARL2})^2}$$
$$\cdot \left(\textbf{SC1} + \textbf{SC2} \cdot V_{\text{ds}} + \textbf{SC3} \cdot \frac{2\Phi_{\text{B}} - V_{\text{bs}}}{L_{\text{gate}}} \right)$$

$$W_{\text{d}} = \sqrt{\frac{2\epsilon_{\text{Si}}(2\Phi_{\text{B}} - V_{\text{bs}})}{q N_{\text{sub}}}}$$

$$\Phi'_{\text{B}} = \Phi_{\text{B}} + \textbf{PTHROU} \cdot \left(\Phi''_{\text{B}}(V_{\text{gs}}) - \Phi_{\text{B}} \right)$$

$$\Phi''_{\text{B}}(V_{\text{gs}}) = V'_{\text{G}} + \left(\frac{const0}{C_{\text{ox}}} \frac{\beta}{2} \right)^2 \left[1 - \sqrt{1 - \frac{4\beta(V'_{\text{G}} - V_{\text{bs}} - 1}{\beta^2 \left(\frac{const0}{C_{\text{ox}}} \right)^2}} \right]$$

[pocket implant effect]

$$\Delta V_{\text{th,P}} = (V_{\text{th,P}} - V_{\text{th0}}) \frac{\epsilon_{\text{Si}}}{C_{\text{ox}}} W_{\text{d}} \frac{dE_{y,\text{P}}}{dy} - \frac{\textbf{SCP22}}{(\textbf{SCP21} + V_{\text{ds}})^2}$$

$$V_{\text{th,P}} = \textbf{VFB} + 2\Phi_{\text{B}} + \frac{Q_{\text{Bmod}}}{C_{\text{ox}}} + \log \left(\frac{N_{\text{subb}}}{\textbf{NSUBC}} \right)$$

$$V_{\text{th0}} = \textbf{VFB} + 2\Phi_{\text{BC}} + \frac{\sqrt{2q\textbf{NSUBC}\epsilon_{\text{Si}}(2\Phi_{\text{BC}} - V_{\text{bs}})}}{C_{\text{ox}}}$$

$$2\Phi_{\text{BC}} = \frac{2}{\beta} \ln \left(\frac{\textbf{NSUBC}}{n_{\text{i}}} \right)$$

$$\frac{dE_{y,\text{P}}}{dy} = \frac{2(\textbf{VBI} - 2\Phi_{\text{B}})}{\textbf{LP}^2} \left(\textbf{SCP1} + \textbf{SCP2} \cdot V_{\text{ds}} + \textbf{SCP3} \cdot \frac{2\Phi_{\text{B}} - V_{\text{bs}}}{\textbf{LP}} \right)$$

$$Q_{\text{Bmod}} = \sqrt{2q \cdot N_{\text{sub}} \cdot \epsilon_{Si} \cdot \left(2\Phi_{\text{B}} - V_{\text{bs}} - \frac{\textbf{BS1}}{\textbf{BS2} - V_{\text{bs}}} \right)}$$

[narrow width effect]

$$\Delta V_{\text{th,W}} = \left(\frac{1}{C_{\text{ox}}} - \frac{1}{C_{\text{ox}} + 2C_{\text{ef}}/(L_{\text{eff}} W_{\text{eff}})} \right) q N_{\text{sub}} W_{\text{d}} + \frac{\textbf{WVTH0}}{W_{\text{gate}} \cdot 10^4}$$

$$C_{\text{ef}} = \frac{2\textbf{KAPPA}}{\pi} L_{\text{eff}} \ln \left(\frac{2T_{\text{fox}}}{T_{\text{ox}}} \right) = \frac{\textbf{WFC}}{2} L_{\text{eff}}$$

[poly-depletion effect]

$$\phi_{\mathrm{Spg}} = \mathbf{PGD1} \left(\frac{N_{\mathrm{sub}}}{N_{\mathrm{subc}}} \right)^{\mathbf{PGD4}} \exp \left(\frac{V_{\mathrm{gs}} - \mathbf{PGD2} - \mathbf{PGD3} \cdot V_{\mathrm{ds}}}{V} \right)$$

7.1.6 *Mobility Model*

[high-field mobility]

$$\mu = \frac{\mu_0}{\left(1 + \left(\frac{\mu_0 E_{\mathrm{y}}}{V_{\mathrm{max}}} \right)^{\mathbf{BB}} \right)^{\frac{1}{\mathbf{BB}}}}$$

[maximum velocity]

$$V_{\mathrm{max}} = \mathbf{VMAX} \cdot \left(1 + \frac{\mathbf{VOVER}}{(L_{\mathrm{gate}} \cdot 10^4)^{\mathbf{VOVERP}}} \right)$$
$$\cdot \left(1 + \frac{\mathbf{VOVERWL}}{(L_{\mathrm{gate}} \cdot 10^4)^{\mathbf{VOVERWLP}}} \right)$$

[low-field mobility]

$$\frac{1}{\mu_0} = \frac{1}{\mu_{\mathrm{CB}}} + \frac{1}{\mu_{\mathrm{PH}}} + \frac{1}{\mu_{\mathrm{SR}}}$$

$$\mu_{\mathrm{CB}}(\text{Coulomb}) = \mathbf{MUECB0} + \mathbf{MUECB1} \frac{Q_{\mathrm{i}}}{q \cdot 10^{11}}$$

[phonon scattering]

$$\mu_{\mathrm{PH}}(\text{phonon}) = \frac{M_{\mathrm{uephonon}}}{E_{\mathrm{eff}}^{\mathbf{MUEPH0}}}$$

[surface-roughness scattering]

$$\mu_{\mathrm{SR}}(\text{surface roughness}) = \frac{\mathbf{MUESR1}}{E_{\mathrm{eff}}^{M_{\mathrm{uesurface}}}}$$

$$E_{\mathrm{eff}} = \frac{1}{\epsilon_{\mathrm{Si}}} \left(N_{\mathrm{dep}} \cdot Q_{\mathrm{B}} + N_{\mathrm{inv}} \cdot Q_{\mathrm{I}} \right)$$

$$N_{\mathrm{dep}} = \mathbf{NDEP} \left(1 - \frac{\mathbf{NDEPL}}{\mathbf{NDEPL} + (L_{\mathrm{gate}} \cdot 10^4)^{\mathbf{NDEPLP}}} \right)$$

$$N_{\text{inv}} = \textbf{NINV}\left(1 - \frac{\textbf{NINVL}}{\textbf{NINVL} + (L_{\text{gate}} \cdot 10^4)^{\textbf{NINVLP}}}\right)$$

$$M_{\text{uephonon}} = \textbf{MUEPH1} \cdot \left(1 + \frac{\textbf{MUEPHL}}{(L_{\text{gate}} \cdot 10^4)^{\textbf{MUEPLP}}}\right) \cdot$$

$$\cdot \left(1 + \frac{\textbf{MUEPHW}}{(W_{\text{gate}} \cdot 10^4)^{\textbf{MUEPWP}}}\right)$$

$$\cdot \textbf{MUESTI1}\left(1 + \frac{1}{1 + \textbf{MUESTI2}} \cdot \left(\frac{\textbf{MUESTI1}}{\text{LOD}}\right)^{\textbf{MUESTI3}} \cdot\right)$$

$$M_{\text{uesurface}} = \textbf{MUESR0} \cdot \left(1 + \frac{\textbf{MUESRL}}{(L_{\text{gate}} \cdot 10^4)^{\textbf{MUESLP}}}\right)$$

$$\cdot \left(1 + \frac{\textbf{MUESRW}}{(W_{\text{gate}} \cdot 10^4)^{\textbf{MUESWP}}}\right)$$

7.1.7 Drain Current I_{ds}

[drain current]

$$I_{\text{ds}} = \frac{W_{\text{eff}}}{L_{\text{eff}}} \cdot \mu \cdot \frac{I_{\text{dd}}}{\beta}$$

$$I_{\text{dd}} = C_{\text{ox}}(\beta V'_{\text{G}} + 1)(\phi_{\text{SL}} - \phi_{\text{S0}}) - \frac{\beta}{2} C_{\text{ox}}(\phi_{\text{SL}}^2 - \phi_{\text{S0}}^2)$$

$$- \frac{2}{3} const0 \Big[\{\beta(\phi_{\text{SL}} - V_{\text{bs}}) - 1\}^{\frac{3}{2}} - \{\beta(\phi_{\text{S0}} - V_{\text{bs}}) - 1\}^{\frac{3}{2}}\Big]$$

$$+ const0 \Big[\{\beta(\phi_{\text{SL}} - V_{\text{bs}}) - 1\}^{\frac{1}{2}} - \{\beta(\phi_{\text{S0}} - V_{\text{bs}}) - 1\}^{\frac{1}{2}}\Big]$$

[ΔV_{th}: threshold voltage shift]

$$V'_{\text{G}} = V_{\text{gs}} - \textbf{VFB} + \Delta V_{\text{th}}$$

7.1.8 Channel-Length Modulation

[ΔL: length of pinch-off region]

$$L_{\text{eff}} = L_{\text{eff}} - \Delta L$$

$$\Delta L = \frac{1}{2}\left[-\frac{1}{L_{eff}}\left(2\frac{I_{dd}}{\beta Q_i}z + 2\frac{N_{sub}}{\epsilon_{Si}}(\phi_S(\Delta L) - \phi_{SL})z^2 + E_0 z^2\right)\right.$$
$$+ \left\{\frac{1}{L_{eff}^2}\left(2\frac{I_{dd}}{\beta Q_i}z + 2\frac{N_{sub}}{\epsilon_{Si}}(\phi_S(\Delta L) - \phi_{SL})z^2 + E_0 z^2\right)^2\right.$$
$$\left.\left.-4\left(2\frac{N_{sub}}{\epsilon_{Si}}(\phi_S(\Delta L) - \phi_{SL})z^2 + E_0 z^2\right)\right\}^{\frac{1}{2}}\right]$$

[surface potential at p/n junction]

$$\phi_S(\Delta L) = (1 - \mathbf{CLM1}) \cdot \phi_{SL} + \mathbf{CLM1} \cdot (\phi_{S0} + V_{ds})$$
$$z = \frac{\epsilon_{Si}}{\mathbf{CLM2} \cdot Q_b + \mathbf{CLM3} \cdot Q_i}$$
$$E_0 = 10^5$$

7.1.9   Shallow-Trench-Isolation (STI) Effect

[shallow-trench-isolation (STI) effect]

$$I_{ds,STI} = 2\frac{W_{STI}}{L_{eff} - \Delta L}\mu\frac{Q_{i,STI}}{\beta}\left[1 - \exp(-\beta V_{ds})\right]$$

[$\Delta V_{th,SCSTI}$: threshold voltage due to STI]

$$V'_{gs,STI} = V_{gs} - V_{fb} + \mathbf{VTHSTI} + \Delta V_{th,SCSTI}$$

[short-channel effect]

$$\Delta V_{th,SCSTI} = \frac{\epsilon_{Si}}{C_{ox}}W_{d,STI}\frac{dE_y}{dy}$$
$$\frac{dE_y}{dy} = \frac{2(\mathbf{VBI} - 2\Phi_B)}{(L_{gate} - \mathbf{PARL2})^2}$$
$$\cdot\left(\mathbf{SCSTI1} + \mathbf{SCSTI2} \cdot V_{ds} + \mathbf{SCSTI3} \cdot \frac{2\Phi_B - V_{bs}}{L_{gate}}\right)$$
$$W_{STI} = \mathbf{WSTI}\left(1 + \frac{\mathbf{WSTIL}}{(L_{gate} \cdot 10^4)^{\mathbf{WSTILP}}}\right)$$

7.1.10   *Capacitances*

[Overlap Capacitances]

Bias Dependence Option 1

i) $y_n \leq L_{\text{over}}$

$$\frac{Q_{\text{gxo}}}{WC_{\text{ox}}}$$

ii) $y_n > L_{\text{over}}$

$$\frac{Q_{\text{gxo}}}{WC_{\text{ox}}} = V'_{\text{gs}} \cdot \textbf{XLDOV} - \frac{a}{3}\{(\textbf{XLDOV} - y_n)^3 + y_n^3\}$$

$$y_n = \left(-\frac{\phi_{S0} + V_{\text{ds}} - \phi_S(\Delta L)}{a(\simeq -1 \times 10^{11})}\right)^{\frac{1}{2}} \qquad (V_{\text{ds}}=0 \text{ for x=S})$$

Bais Dependence Option 2

$$\frac{Q_{\text{gxo}}}{WC_{\text{ox}}} = \frac{\textbf{XLDOV} - (\textbf{OVMAG} + V'_{\text{gs}})\textbf{OVSLP}(\textbf{VBI} - \phi_{S0})}{V'_{\text{gs}}}$$

Bias independence Option

$$\frac{Q_{\text{gxo}}}{WC_{ox}} = V'_{\text{gs}} \cdot \textbf{XLDOV}$$

i) x=S

$$V'_{\text{gs}} = V_{\text{gs}}$$

ii) x=D

$$V'_{\text{gs}} = V_{\text{gs}} - V_{\text{ds}}$$

$$C_{\text{gxo}} = \frac{dQ_{\text{go}}}{dV_{\text{x}}}; \quad Q_{\text{go}} = Q_{\text{gso}} + Q_{\text{gdo}}.$$

[Longitudinal-Field-Induced Capacitance]

$$C_{\text{Qy}} = \epsilon_{\text{Si}} W_{\text{eff}} W_{\text{d}} \left(\frac{\dfrac{d\phi_{S0}}{dV_{\text{ds}}} + 1 - \dfrac{d\{\phi_S(\Delta L)\}}{dV_{\text{ds}}}}{\textbf{XQY}}\right)$$

$$W_{\text{d}} = \sqrt{\frac{2\epsilon_{\text{Si}}}{qN_{\text{sub}}}(\phi_S(\Delta L) - V_{\text{bs}})}$$

[Fringing Capacitance]

$$C_{\text{f}} = \frac{\epsilon_{\text{ox}}}{\pi/2} W_{\text{gate}} \ln\left(1 + \frac{\textbf{TPOLY}}{T_{\text{ox,eff}}}\right)$$

7.1.11 *Leakage Currents*

[Substrate Current]

$$I_{\text{sub}} = SUB1\big(\phi(\Delta L') - \phi(0)\big) I_{\text{ds}} \exp\left(-\frac{\textbf{SUB2}}{\phi(\Delta L') - \phi(0)}\right)$$

$$SUB1 = \textbf{SUB1} + \frac{\textbf{SUB4}}{L_{\text{eff}}}$$

[Gate Current]

$$I_{\text{gate}} = q \cdot \textbf{GLEAK1}\frac{E^2}{E_{\text{gp}}^{\frac{1}{2}}} \exp\left(-\frac{E_{\text{gp}}^{\frac{3}{2}}\textbf{GLEAK2}}{E}\right)\sqrt{\frac{Q_{\text{i}}}{const0}}W_{\text{eff}}L_{\text{eff}}$$

$$E = \frac{V_{\text{G}} - \textbf{GLEAK3} \cdot \phi_{\text{S}}(\Delta L)}{T_{\text{ox}}}$$

$$V_{\text{G}} = V_{\text{gs}} - V_{\text{fb}} + \textbf{GLEAK4} \cdot \Delta V_{\text{th}} \cdot (L_{\text{gate}} \cdot 10^4)$$

$$I_{\text{gb}} = \textbf{GLKB1} \cdot E_{\text{gb}}^2 \cdot \exp\left(-\frac{\textbf{GLKB}}{E_{\text{gb}}}\right)W_{\text{eff}}L_{\text{eff}}$$

$$I_{\text{gs}} = sign \cdot \textbf{GLKSD1} \cdot E_{\text{gs}}^2$$
$$\cdot \exp\left(T_{\text{ox,eff}}(-\textbf{GLKDS2} \cdot V_{\text{gs}} + \textbf{GLKSD3})\right)W_{\text{eff}}$$

$$I_{\text{gd}} = sign \cdot \textbf{GLKSD1} \cdot E_{\text{gd}}^2$$
$$\cdot \exp\left(T_{\text{ox,eff}}(\textbf{GLKSD} \cdot (-V_{\text{gs}} + V_{\text{ds}}) + \textbf{GLKSD3})\right)W_{\text{eff}}$$

[GIDL Current]

$$I_{\text{GIDL}} = q \cdot \textbf{GIDL1} \cdot \frac{E^2}{E_{\text{g}}^{\frac{1}{2}}} \cdot \exp\left(-\textbf{GIDL2} \cdot \frac{E_{\text{g}}^{\frac{3}{2}}}{E}\right) \cdot W_{\text{eff}}$$

$$E = \frac{\textbf{GIDL3} \cdot (V_{\text{ds}} + \textbf{GIDL4}) - V_{\text{G}}'}{T_{\text{ox,eff}}}$$

7.1.12 *Noise*

[1/f Noise Model]

$$S_{I_{\text{d}}} = \frac{I_{\text{ds}}^2 \textbf{NFTRP}}{\beta f(L_{\text{eff}} - \Delta L)W_{\text{eff}}}\left[\frac{1}{(N_0 + N^*)(N_L + N^*)}\right.$$
$$\left. + \frac{2v\textbf{NFALP}}{N_L - N_0}\ln\left(\frac{N_L + N^*}{N_0 + N^*}\right) + (v\textbf{NFALP})^2\right]$$

[Thermal Noise Model]

$$S_{I_d} = 4kT\frac{W_{\text{eff}}C_{\text{ox}}V_gV_t}{(L_{\text{eff}} - \Delta L)}\frac{\mu}{15(1+\eta)\mu_{\text{av}}^2}\left[(1 + 3\eta + 6\eta^2)\mu_d^2\right.$$
$$\left. + (3 + 4\eta + 3\eta^2)\mu_d\mu_s + (6 + 3\eta + \eta^2)\mu_s\right]$$

$$\eta = 1 - \frac{(\phi_{\text{SL}} - \phi_{\text{S0}}) + \chi(\phi_{\text{SL}} - \phi_{\text{S0}})}{V_gV_t}$$

$$\chi = 2\frac{const0}{C_{\text{ox}}}\left[\left(\frac{2}{3}\frac{1}{\beta}\frac{\{\beta(\phi_{\text{SL}} - V_{\text{bs}}) - 1\}^{\frac{3}{2}} - \{\beta(\phi_{\text{S0}} - V_{\text{bs}}) - 1\}^{\frac{3}{2}}}{\phi_{\text{SL}} - \phi_{\text{S0}}}\right)\right.$$
$$\left. - \sqrt{\beta(\phi_{\text{S0}} - V_{\text{bs}}) - 1}\right]$$

[Induced Gate Noise]

$$S_{I_g} = \frac{16}{135}kT\frac{\omega^2 C^2}{g_{\text{ds0}}}\frac{4\xi_0 + 20\xi_0^{\frac{3}{2}}\xi_L^{\frac{1}{2}} + 42\xi_0\xi_L + 20\xi_0^{\frac{1}{2}}\xi_L^{\frac{3}{2}} + 4\xi_L}{(\xi_0^{\frac{1}{2}} + \xi_L^{\frac{1}{2}})} \cdot (2\pi f)^2$$

7.1.13   *Non-Quasi-Static (NQS) Effects*

$$\frac{1}{\tau} = \frac{1}{\tau_{\text{diff}}} + \frac{1}{\tau_{\text{cond}}}$$

$$\tau_{\text{diff}} = \mathbf{DLY1}$$

$$\tau_{\text{cond}} = \mathbf{DLY2} \cdot \frac{Q_i}{I_{\text{ds}}}$$

$$\tau_B = \mathbf{DLY3} \cdot C_{\text{ox}}$$

7.1.14   *Parasitic Resistances*

[Source/Drain Resistances]

$$V_{\text{gs,eff}} = V_{\text{gs}} - I_{\text{ds}} \cdot R_{\text{s}}$$
$$V_{\text{ds,eff}} = V_{\text{ds}} - I_{\text{ds}} \cdot (R_{\text{s}} + R_{\text{d}})$$
$$V_{\text{bs,eff}} = V_{\text{bs}} - I_{\text{ds}} \cdot R_{\text{s}}$$
$$R_{\text{s}} = \mathbf{RS} + \mathbf{NRS} \cdot \mathbf{RSH}$$
$$R_{\text{d}} = \mathbf{RD} + \mathbf{NRD} \cdot \mathbf{RSH}$$

[Gate Resistance]

$$R_g = \frac{\text{RSHG} \cdot \left(\text{XGW} + \frac{W_{\text{eff}}}{3 \cdot \text{NGCON}}\right)}{\text{NGCON} \cdot \left(L_{\text{drawn}} - \text{XGL}\right) \cdot \text{NF}}$$

7.2   Model Flags and Exclusion of Modeled Effects

7.2.1   *Parameter setting for Exclusion of Certain Model Parts*

Table 7.1   HiSIM model parameter setting to exclude specific modeled effects.

Short-Channel Effect	**SC1 = SC2 = SC3** = 0
Reverse-Short-Channel Effect	**LP** = 0
Quantum-Mechanical Effect	**QME1 = QME2 = QME3** = 0
Poly-Depletion Effect	**PGD1 = PGD2 = PGD3** = 0
Channel-Length Modulation	**CLM1 = CLM2 = CLM3** = 0
Narrow-Channel Effect	**WFC = MUEPHW** = 0
Small-Size Effect	**WL2** = 0

7.2.2   *Flags for Setting Model Options*

1. Inclusion of Contact Resistances R_s, R_d
   CORSRD = 0: no (default)
   CORSRD = 1: yes, as internal resistances
   CORSRD = −1: yes, as external resistances

2. Selection of Overlap Capacitance Model
   COOVLP = 0: constant overlap capacitance (default)
   COOVLP = 1: yes

3. Selection of Substrate current I_{sub} Calculation
   COISUB = 0: no (default)
   COISUB = 1: yes

4. Selection of Gate current I_{gate} Calculation
   COIIGS = 0: no (default)
   COIIGS = 1: yes

5. Selection of GIDL current I_{GIDL} Calculation
   COGIDL = 0: no (default)
   COGIDL = 1: yes

6. Selection of STI Leakage Current $I_{ds,STI}$ Calculation
   COISTI = 0: no (default)
   COISTI = 1: yes

7. Addition of field Induced and Overlap Charges/Capacitances:

> **COADOV** = 0: no
> **COADOV** = 1: yes (default)

8. Selection of Effective Bias for Smoothing Conductances

> **COSMBI** = 0: no (default)
> **COSMBI** = 1: yes

9. Selection of Non-Quasi-Static Mode

> **CONQS** = 0: no (default)
> **CONQS** = 1: yes

10. Inclusion of Contact Resistance

> **CORG** = 0: no (default)
> **CORG** = 1: yes

11. Invoking of Substrate resistance network is invoked:

> **CORBNET** = 0: no (default)
> **CORBNET** = 1: yes

12. Calcation of $1/f$ Noise

> **COFLICK** = 0: no (default)
> **COFLICK** = 1: yes

13. Calculation of Thermal Noise

> **COTHRML** = 0: no (default)
> **COTHRML** = 1: yes

14. Calculation of Induced Gate Noise

> **COIGN** = 0 ‖ **COTHRML** = 0: no (default)
> **COIGN** = 1 && **COTHRML** = 1: yes

7.3 Model Parameters and their Meaning

The HiSIM model parameters are summerized in the fllowing tables according to the different modeling purposes of these parameters. The parameters with the symbol as their leading symbol areminor parameters, which should normarry not be changed from their default values.

Table 7.2 Basic device parameters.

TOX	physical oxide thickness
XLD	gate-overlap length
XWD	gate-overlap width
XLDOV	gate-overlap length for overlap capacitance
TPOLY	height of the gate poly-Si
LL	coefficient of gate length modification
LLD	coefficient of gate length modification
LLN	coefficient of gate length modification
WL	coefficient of gate width modification
WLD	coefficient of gate width modification
WLN	coefficient of gate width modification
RS	source-contact resistance in LDD region
RD	drain-contact resistance in LDD region
RSH	source/drain sheet resistance
NRS	number of source squares
NRD	number of drain squares
NSUBC	substrate-impurity concentration
NSUBP	maximum pocket concentration
*NSUBP0	modification of pocket concentration for narrow width
*NSUBWP	modification of pocket concentration for narrow width
*NPEXT	maximum concentration of pocket tail
*LPEXT	extension length of pocket tail
VFB	flat-band voltage
VBI	built-in potential
LP	pocket penetration length
XQY	distance from drain junction to maximum electric field point
KAPPA	dielectric constant for gate dielectric
EG0	bandgap
BGTMP1	temperature dependence of bandgap
BGTMP2	temperature dependence of bandgap
TNOM	temperature selected as a nominal temperature value

Table 7.3 Velocity.

VMAX	saturation velocity
VOVER	velocity overshoot effect
VOVERP	L_{eff} dependence of velocity overshoot
*VTMP	temperature dependence of the saturation velocity

Table 7.4   Quantum effect.

QME1	V_{gs} dependence
QME2	V_{gs} dependence
QME3	minimum T_{ox} modification

Table 7.5   Poly depletion.

PGD1	strength of poly depletion
PGD2	threshold voltage of poly depletion
PGD3	V_{ds} dependence of poly depletion
∗PGD4	L_{gate} dependence of poly depletion

Table 7.6   Short channel.

PARL2	depletion width of channel/contact junction
SC1	magnitude of short-channel effect
SC2	V_{ds} dependence of short-channel effect
∗SC3	V_{bs} dependence of short-channel effect
SCP1	magnitude of short-channel effect due to pocket
SCP2	V_{ds} dependence of short-channel due to pocket
∗SCP3	V_{bs} dependence of short-channel effect due to pocket
∗SCP21	short-channel-effect modification for small V_{ds}
∗SCP22	short-channel-effect modification for small V_{ds}
∗BS1	body-coefficient modification by impurity profile
∗BS2	body-coefficient modification by impurity profile

Table 7.7   Mobility.

MUECB0	Coulomb scattering
MUECB1	Coulomb scattering
MUEPH0	phonon scattering
MUEPH1	phonon scattering
MUETMP	temperature dependence of phonon scattering
∗MUEPHL	L_{gate} dependence of phonon mobility reduction
∗MUEPLP	L_{gate} dependence of phonon mobility reduction
MUESR0	surface-roughness scattering
MUESR1	surface-roughness scattering
∗MUESRL	L_{gate} dependence of surface roughness scattering
∗MUESLP	L_{gate} dependence of surface roughness scattering
NDEP	depletion-charge contribution on effective-electric field
NDEPL	L_{gate} dependence of depletion charge on effective-electric field
NDEPLP	L_{gate} dependence of depletion charge on effective-electric field
NINV	inversion charge contribution on effective-electric field
BB	high-field-mobility degradation

Table 7.8   Narrow channel.

WFC	threshold voltage change due to capacitance change
∗WVTH0	threshold voltage shift
∗NSUBP0	modification of pocket concentration
∗NSUBWP	modification of pocket concentration
∗MUEPHW	W_{gate} dependence of phonon scattering
∗MUEPWP	W_{gate} dependence of phonon scattering
∗MUESRW	W_{gate} dependence of surface roughness scattering
∗MUESWP	W_{gate} dependence of surface roughness scattering
∗VTHSTI	threshold voltage shift due to STI
SCSTI1	the same effect as **SC1** but at STI edge
SCSTI2	the same effect as **SC2** but at STI edge
SCSTI3	the same effect as **SC3** but at STI edge
NSTI	substrate-impurity concentration at the STI edge
WSTI	width of the high-field region at STI edge
WSTIL	channel-length dependence of **WSTI**
WSTILP	channel-length dependence of **WSTI**
LOD	diffusion-region length between gate and STI
NSUBPSTI1	pocket concentration change due to **LOD**
NSUBPSTI2	pocket concentration change due to **LOD**
NSUBPSTI3	pocket concentration change due to **LOD**
MUESTI1	mobility change due to **LOD**
MUESTI2	mobility change due to **LOD**
MUESTI3	mobility change due to **LOD**

Table 7.9   Small size.

WL2	magnitude of small-size effect
WL2P	magnitude of small-size effect
∗MUEPHS	mobility modification
∗MUEPSP	mobility modification
∗VOVERS	modification of maximum velocity
∗VOVERSP	modification of maximum velocity

Table 7.10   Channel-length modulation.

CLM1	hardness coefficient of channel/contact junction
CLM2	coefficient for depletion-charge contribution
CLM3	coefficient for inversion-charge contribution

Table 7.11   Substrate current.

SUB1	magnitude of substrate current
SUB2	coefficient of electric field
SUB3	modification of electric field
SUB4	L_{gate} dependence of substrate current

Table 7.12   Gate Leakage Current.

GLEAK1	magnitude of gate to channel current
GLEAK2	coefficient of electric field for gate to channel current
GLEAK3	modification of electric field for gate to channel current
GLEAK4	modification of electric field for gate to channel current
GLKSD1	magnitude of gate to source/drain current
GLKSD2	coefficient of electric field for gate to source/drain current
GLKSD3	modification of electric field for gate to source/drain current
GLKB1	magnitude of gate to bulk current
GLKB2	coefficient of electric field for gate to bulk current

Table 7.13   GIDL current.

GIDL1	magnitude of GIDL
GIDL2	field dependence of GIDL
GIDL3	V_{ds} dependence of GIDL

Table 7.14   Conservation of the symmetry at $V_{\text{ds}} = 0$ for short-channel MOSFETs.

VZADD0	symmetry conservation coefficient
PZADD0	symmetry conservation coefficient

Table 7.15   Source/bulk and drain/bulk diodes.

JS0	saturation current density
JS0SW	sidewall saturation current density
NJ	emission coefficient
NJSW	sidewall emission coefficient
XTI	temperature coefficient for forward current densities
XTI2	temperature coefficient for reverse current densities
CISB	reverse biased saturation current
CVB	bias dependence coefficient of **CISB**
CTEMP	temperature coefficient of reverse currents
CISBK	reverse biased saturation current (at low temperature)
CVBK	bias dependence coefficient of **CISB** (at low temperature)
DIVX	reverse current coefficient
VDIFFJ	diode threshold voltage between source/drain and substrate

Table 7.16 $1/f$ noise.

NFALP	contribution of the mobility fluctuation
NFTRP	ratio of trap density to attenuation coefficient
*CIT	capacitance caused by the interface trapped carriers

Table 7.17   Subthreshold swing.

*PTHROU	correction for subthreshold swing

Table 7.18   Non-quasi-static model.

DLY1	coefficient for delay due to diffusion of carriers
DLY2	coefficient for delay due to conduction of carriers
DLY3	coefficient for RC delay of bulk carriers

Table 7.19   Overlap capacitance.

OVSLP	model parameter for overlap capacitance
OVMAG	model parameter for overlap capacitance
CGSO	gate-to-source overlap capacitance if COOVLP=0
CGDO	gate-to-drain overlap capacitance if COOVLP=0
CGBO	gate-to-bulk overlap capacitance if COOVLP=0

7.4   Default Values of the Model Parameter

The maximum and minimum limits of the parameters are recommended values. These values may be violated in individual cases.

Table 7.20   Default parameters and limits of the parameter values.

parameter	unit	min	max	default	remarks
TOX	[m]			3n	
XLD	[m]	0	50n	0	
XWD	[m]	-10n	100n	0	
LL				0	
LLD	[m]			0	
LLN				0	
WL				0	
WLD	[m]			0	
WLN				0	
RS	$[\mathrm{V\,A^{-1}m}]$	0	100μ	0	
RD	$[\mathrm{V\,A^{-1}m}]$	0	100μ	0	
RSH	$[\mathrm{V\,A^{-1}square}]$	0	100μ	0	
RSHG	$[\mathrm{V\,A^{-1}square}]$	0	100μ	0	
NSUBC	$[\mathrm{cm^{-3}}]$	1×10^{16}	1×10^{19}	5×10^{17}	
VFB	[V]	-1.2	-0.8	-1.0	
KAPPA	[—]			3.9	
EG0	[eV]	1.0	1.3	1.1785	
BGTMP1	$[\mathrm{eV\,K^{-1}}]$	50μ	100μ	90.25μ	fixed
BGTMP2	$[\mathrm{eV\,K^{-2}}]$	-1μ	1μ	0.1μ	
TNOM	[°C]			300.15	
QME1	$[\mathrm{V\,m^{-2}}]$	0	300n	0	
QME2	[V]	0	3.0	1.0	
QME3	[m]	0	800p	0	

Table 7.21 Default parameters and limits of the parameter values.

parameter	unit	min	max	default	remarks
PGD1	[V]	0	100m	10m	
PGD2	[V]	0	1.5	1.0	
PGD3	[—]	0	1.0	0.8	
*PGD4	[—]	0	3.0	1.0	
PARL2	[m]	0	50n	10n	
VBI	[V]	1.0	1.1	1.0	
SC1	[—]	0	200	1.0	
SC2	[V^{-1}]	0	50	1.0	
*SC3	[$V^{-1}m$]	0	1m	0	
LP	[m]	0	300n	15n	
NSUBP	[cm^{-3}]	1×10^{16}	1×10^{19}	1×10^{18}	
*NSUBP0	[cm^{-3}]			0	
*NSUBWP				1.0	
*NPEXT	[cm^{-3}]	1×10^{16}	1×10^{18}	6×10^{17}	
*LPEXT	[m]	1×10^{-50}	10×10^{-6}	1×10^{-50}	
SCP1	[—]	0	50	1.0	
SCP2	[V^{-1}]	0	50	0.1	
*SCP3	[$V^{-1}m$]	0	1m	0	
*SCP21	[V]	0	5.0	0	
*SCP22	[V^4]	0	50m	0	
*BS1	[V^2]	0	100m	0	
*BS2	[V]	0.5	1.0	0.9	
*PTHROU	[—]	0	50m	0	
MUECB0	[$cm^2V^{-1}s^{-1}$]	100	100K	1K	
MUECB1	[$cm^2V^{-1}s^{-1}$]	15	10K	100	
MUEPH0	[—]	0.25	0.35	0.3	fixed
MUEPH1	[$cm^2V^{-1}s^{-1}(V\,cm^{-1})^{MUEPH0}$]	2K	30K	25K	
MUETMP	[—]	0.5	2.0	1.5	
*MUEPHL	[—]			0	
*MUEPLP	[—]			1.0	
MUESR0	[—]	1.8	2.2	2.0	
MUESR1	[$cm^2V^{-1}s^{-1}(V\,cm^{-1})^{MUESR0}$]	1×10^{14}	1×10^{16}	1×10^{15}	
*MUESRL				0	
*MUESLP				1.0	
NDEP	[—]	0	1.0	1.0	
*NDEPL	[—]			0	
*NDEPLP	[—]			1.0	
NIN	[—]			0.5	
BB	[—]			2.0(nMOS), 1.0(pMOS)	fixed

Table 7.22   Default parameters and limits of the parameter values.

parameter	unit	min	max	default	remarks
VMAX	$[\mathrm{cm\,s^{-1}}]$	1MEG	20MEG	10MEG	
VOVER	$[\mathrm{cm}^{VOVERP}]$	0	1.0	0.3	
VOVERP	$[-]$	0	2	0.3	
∗VOVERS	$[-]$			0	
∗VOVERSP	$[-]$			0	
∗VTMP	$[\mathrm{cm\,s^{-1}}]$	-2.0	1.0	0	
WFC	$[\mathrm{F\,cm^{-2}m^{-1}}]$	-5.0×10^{-15}	1×10^{-6}	0	
∗WVTH0				0	
∗MUEPHW				0	
∗MUEPWP				1.0	
∗MUESRW				0	
∗MUESWP				1.0	
∗VTHSTI				0	
SCSTI1				0	
SCSTI2				0	
SCSTI3				0	
NSTI	$[\mathrm{cm^{-3}}]$			1×10^{17}	
WSTI	$[\mathrm{m}]$			0	
WSTIL				0	
WSTILP				1.0	
NSUBPSTI1	$[\mathrm{m}]$			0	
NSUBPSTI2	$[\mathrm{m}]$			0	
NSUBPSTI3	$[\mathrm{m}]$			1.0	
MUESTI1				0	
MUESTI2				0	
MUESTI3				1.0	
WL1				0	
WLP				1.0	
∗MUEPHS				0	
∗MUEPSP				1.0	
CLM1	$[-]$	0.5	1.0	0.7	
CLM2	$[-]$	1.0	2.0	2.0	
CLM3	$[-]$	1.0	5.0	1.0	

Table 7.23   Default parameters and limits of the parameter values.

parameter	unit	min	max	default	remarks
TPOLY	[m]			0	
XQY	[m]	0	50n	0	
XLDOV	[—]			1.0	
OVSLP	[—]			2.1×10^{-7}	
OVMAG	[V]			0.6	
CGSO	[F m^{-1}]	0		0	
CGDO	[F m^{-1}]	0		0	
CGBO	[F m^{-1}]	0		0	
GBMIN	[—]			1×10^{-12}	
RBPB	[Ω]			50	
RBPD	[Ω]			50	
RBPS	[Ω]			50	
RBDB	[Ω]			50	
RBSB	[Ω]			50	

Table 7.24   Default parameters and limits of the parameter values.

parameter	unit	min	max	default	remarks
SUB1	[V^{-1}]			3×10^{-9}	
SUB2	[V]			25.0	
SUB1L	[—]			0	
SUB1LP	[—]			1.0	
SUB2L	[—]	0	1.0	2×10^{-6}	
SUB2LP	[—]	0	1.0	0.8	
SVDS	[—]			0.7	
SLG	[—]			3×10^{-8}	
SLGL	[—]			0	
SLGLP	[—]			1.0	
SVBS	[—]			0.5	
SVBSL	[—]			0	
SVBSLP	[—]			1.0	
SVGS	[—]			0.8	
SVGSL	[—]			0	
SVGSLP	[—]			1.0	
SVGSW	[—]			0	
SVGSWP	[—]			1.0	
FN1	[—]			5×10^{-3}	
FN2	[—]			170×10^{-6}	
FN3	[—]			0.8	
FVBS	[—]			12×10^{-3}	

Table 7.25   Default parameters and limits of the parameter values.

parameter	unit	min	max	default	remarks
GLEAK1	$[\mathrm{A\,V^{-3/2}C^{-1}}]$			50K	
GLEAK2	$[\mathrm{V^{-1/2}m^{-1}}]$			20MEG	
GLEAK3	$[-]$			60×10^{-3}	
GLEAK4	$[\mathrm{m^{-1}}]$			300×10^{-3}	
*****GLEAK5**	$[\mathrm{V\,m^{-1}}]$			3×10^{6}	
*****GLEAK6**	$[\mathrm{V}]$			0.7	
*****GLEAK7**	$[\mathrm{m^{2}}]$			1×10^{-6}	
*****IGTEMP1**	$[\mathrm{V}]$			0	
*****IGTEMP2**	$[\mathrm{V\,K}]$			0	
*****IGTEMP3**	$[\mathrm{V\,K^{2}}]$			0	
GLEAKUNIT	$[-]$			1.0	
GLKSD1	$[\mathrm{A\,m\,V^{-2}}]$			50f	
GLKSD2	$[\mathrm{V^{-1}m^{-1}}]$			5MEG	
GLKSD3	$[\mathrm{m^{-1}}]$			0.55	
GLKB1	$[\mathrm{A\,V^{-2}}]$			0	
GLKB2	$[\mathrm{m\,V^{-1}}]$			0	
GIDL1	$[\mathrm{A\,V^{-3/2}C^{-1}m}]$			2.0	
GIDL2	$[\mathrm{V^{-2}m^{-1}F^{-3/2}}]$			3×10^{7}	
GIDL3	$[-]$			0.9	
GIDL4	$[\mathrm{V}]$			0.9	
GIDL5	$[-]$			0.2	
VZADD0	$[\mathrm{V}]$			10m	fixed
PZADD0	$[\mathrm{V}]$			5m	fixed
JS0	$[\mathrm{A\,m^{-2}}]$			0.5×10^{-6}	
JS0SW	$[\mathrm{A\,m^{-1}}]$			0	
NJ	$[-]$			1.0	
NJSW	$[-]$			1.0	
XTI	$[-]$			2.0	
XTI2	$[-]$			0	
DIVX	$[\mathrm{V^{-1}}]$			0	
CISB	$[-]$			0	
CVB	$[-]$			0	
CTEMP	$[-]$			0	
CISBK	$[\mathrm{A}]$			0	
CVBK	$[-]$			0	
CJ	$[\mathrm{F\,m^{-2}}]$			5×10^{-4}	
CJSW	$[\mathrm{F\,m^{-1}}]$			5×10^{-10}	
CJSWG	$[\mathrm{F\,m^{-1}}]$			5×10^{-10}	
MJ	$[-]$			0.5	
MJSW	$[-]$			0.33	
MJSWG	$[-]$			0.33	
PB	$[\mathrm{V}]$			1.0	
PBSW	$[\mathrm{V}]$			1.0	
PBSWG	$[\mathrm{V}]$			1.0	
VDIFFJ	$[\mathrm{V}]$			0.6×10^{-3}	
NFALP	$[\mathrm{cm\,s}]$			1×10^{-16}	
NFTRP	$[\mathrm{V^{-1}cm^{-2}}]$			10G	
*****CIT**	$[\mathrm{F\,cm^{-2}}]$			0	
DLY1	$[\mathrm{s}]$			100×10^{-12}	
DLY2	$[-]$			0.7	
DLY3	$[\Omega]$			0.8×10^{-6}	

Table 7.26   HiSIM instance parameters.

parameter	unit	min	max	default	remarks
L	[m]			5μ	
W	[m]			5μ	
AD	[m^2]			0	
AS	[m^2]			0	
PD	[m]			0	
PS	[m]			0	
NRS	[m]			1	
NRD	[m]			1	
XGW	[m]			0	
XGL	[m]			0	
NF	[m]			0	
NGCON	[m]			0	
LOD	[m]			10μ	

7.5   Parameter Extraction Method

HiSIM is a complete surface-potential-based MOSFET model and does not include the threshold voltage V_{th} as a model parameter. This means at the same time that basic device parameters such as the oxide thickness T_{ox} and the substrate impurity concentration N_{sub} mostly determine the device features. These basic device parameters in particular determine the carrier mobility and the short-channel effects through the surface potentials. Therefore, accurate extraction of these basic device parameters is indispensable for the subsequent extraction of other model parameters.

In HiSIM, where MOS transistor characteristics are strongly dependent on the basic device parameter values, the parameter-value extraction has to be repeated with measured characteristics of different device size in a specific sequence until the extracted parameter values reproduce all device characteristics consistently and reliably. To achieve such reliable results, it is recommended to start with initial parameter values according to the recommendations listed in the table 7.27. Since some of the model parameters such as T_{ox} are difficult to extract, they are expected to be determined directly by dedicated measurements. Threshold voltage measurements allow to derive a rough extraction for the model parameters referred to as "basic device parameters". The parameters identified with the symbol "*" in the model parameter tables (see section 7.3) are initially fixed to zero. Table 7.28 summarizes the parameter settings at the beginning of an ex-

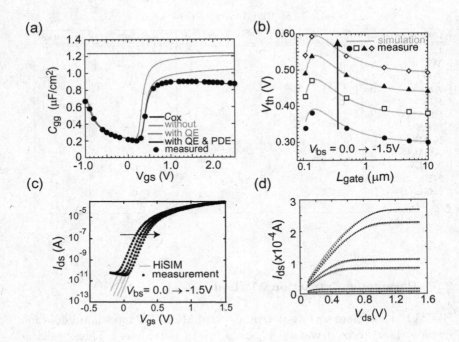

Fig. 7.1 Measurements used for parameter extraction. (a) $C - V$ measurements used for extracting poly-depletion effect (PDE) and quantum-mechanical effect (QE), (b) $V_{th} - L_{gate}$ measurements used for rough extraction of the basic device parameters, (c) $I_{ds} - V_{gs}$ measurements used for precise extraction of device parameters and universal-mobility parameters, and (d) $I_{ds} - V_{ds}$ measurements used for extracting the mobility parameters.

traction procedure.

The extraction sequence of HiSIM's model parameters consists of 7 basic steps and is summarized in Table 7.28. In the 1st step all parameters are initialized according to the recommendation in the Table 7.27. A rough extraction of the basic device and short-channel parameters is then carried out with the data of the V_{th}-dependence on L_{gate} (see Fig. 7.1b) and the quantum and poly-depletion-effect parameters are determined from the gate-capacitance measurements as a function of gate voltage (see Fig. 7.1a).

The 2nd parameter extraction step uses data from long and wide MOS transistors (see Fig. 7.1c) to fit the subthreshold characteristics and to

Table 7.27   Recommended parameter setting at the beginning of an extraction procedure.

Parameter recommended to be determined by dedicated measurements (fixed during extraction procedure)	Default values (see section 7.3 of this chapter) are used initially for the groups of parameters listed below
TOX	basic device parameters:
XPOLYD	out of those listed on left side
XDIFFD	velocity
LL	quantum effect
LLD	poly-depletion effects
LLN	short channel
WL	mobility
WLD	narrow channel
WLN	small size
KAPPA	channel-length modulation
	substrate current
	gate leakage
	GIDL
	source/bulk and drain/bulk diodes
	noise
	subthreshold swing
	non-quasi-static model
	overlap capacitances

obtain a rough extraction of the mobility parameters for low and high drain voltages V_{ds}.

In the 3rd extraction step the measured data of medium/short length transistors with large width is used to determine the pocket-implantation parameters (see Fig. 7.1c), to obtain improved short-channel parameters (see Fig. 7.1b), to refine the mobility parameters for low V_{ds} (see Fig. 7.1c), to extract the carrier-velocity parameters for high V_{ds} (see Fig. 7.1c) and to determine the channel-length-modulation parameters along with the source/drain resistance parameters (see Fig. 7.1d).

The 4th extraction step is deveted to the fitting of the width dependencies of long-channel transistors by focussing on the subthreshold data and determining also the width-dependence parameters of the mobility.

The width dependencies of short-channel transistors are determined in the 5th parameter extraction step, again from subthreshold measurement data.

In the 6th parameter extraction step the parameters for small-geometry (small width and small length) effects are determined from conventional $I_{\mathrm{ds}} - V_{\mathrm{ds}}$ data (see Fig. 7.1d). These small geometry effects mainly require

Table 7.28   Summary of the 7 steps of HiSIM's Parameter Extraction Procedure.

Step 1: Initial preparation and rough extraction

1-1.	Initialize all parameters to their default values	
1-2.	Use the measured gate-oxide thickness for TOX	TOX
1-3.	Rough extraction with V_{th}-dependence on L_{gate} [Fig. 7.1b]	NSUBC, VFB, SC1, SC2 SC3, NSUBP, LP, SCP1 SCP2, SCP3 NPEXT, LPEXT
1-4.	Quantum and poly-depletion effects [Fig. 7.1a]	QME1, QME2, QME3 PGD1, PGD2

Step 2: Extraction with long and wide transistors

2-1.	Fitting of sub-threshold characteristics [Fig. 7.1c]	NSUBC, VFB, MUECB0 MUECB1
2-2.	Determination of mobility parameters for low V_{ds} [Fig. 7.1c]	MUEPH0, MUEPH1 MUESR0, MUESR1
2-3.	Determination of mobility parameters for high V_{ds} [Fig. 7.1c]	NINVPH, NINVSR NDEP

Step 3: Extraction with medium/short length and large width transistors

3-1.	Pocket-parameter extraction with medium length transistors [Fig. 7.1c]	NSUBP, LP SCP1, SCP2, SCP3 NPEXT, LPEXT
3-2.	Short-channel-parameter extraction with short-length transistors [Fig. 7.1b]	SC1, SC2, SC3 PARL2, XLD
3-3.	Mobility-parameter refinement for low V_d [Fig. 7.1c]	MUEPHL, MUEPLP MUESRL, MUESLP
3-4.	Velocity parameter extraction for high V_d [Fig. 7.1c]	VMAX, VOVER, VOVERP
3-5.	Parameters for channel-length modulation [Fig. 7.1d]	CLM1, CLM2, CLM3
3-6.	Source/drain resistances [Fig. 7.1d]	RS, RD, RSH, NRS, NRD

Step 4: Extraction of the width dependencies for long transistors

4-1.	Fitting of sub-threshold width dependencies [Fig. 7.1c]	WFC, XWD, WVTH0
4-2.	Fitting of mobility width dependencies [Fig. 7.1c]	MUEPHW, MUEPWP MUESRW, MUESWP

Step 5: Extraction of the width dependencies for short transistors

5-1.	Fitting of sub-threshold dependencies [Fig. 7.1c]	NSUBP0, NSUBWP

Step 6: Extraction of small-geometry effects

6-1.	Effective channel-length corrections	WL2, WL2P
6-2.	Mobility and velocity [Fig. 7.1d]	MUEPHS, MUEPSP VOVERS VOVERSP

Step 7: Extraction of temperature dependence with long-channel transistors

7-1.	Sub-threshold dependencies [Fig. 7.1c]	BGTMP1, BGTMP2 EG0
7-2.	Mobility and maximum carrier-velocity dependencies [Fig. 7.1c]	MUETMP, VTMP

---

effective channel-length, mobility and carrier velocity corrections.

The final 7th parameter extraction step extracts the temperature-dependence parameters and for this purpose the subthreshold data of long-channel transistors (see Fig. 7.1c) are normally sufficient.

After finishing step 7, it is recommended to recheck the extraction results, starting from step1, but without initialization, and carry out additional parameter adjustments, if necessary.

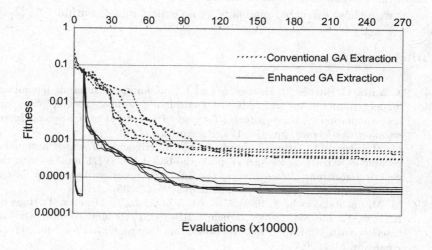

Fig. 7.2 Fitness (error) vs. evaluations (iterations) for five independent genetic algorithm (GA) runs. The conventional-GA-extraction algorithm extracts all 32 model parameters at the same time. On the contrary the enhanced-GA-extraction performs the GA extraction in two steps. At the first extraction step the most sensitive model parameters are extracted, and all other model parameters are then extracted together with the parameters extracted in the first step.

Accurate extraction of model parameters is a key for accurate prediction of circuit performances. Not only the $I - V$ characteristics has to be correctly reproduced, but also accurate prediction of non-linear device characteristics such as the harmonic distortion are necessary, especially in RF-circuit application. For fullfilling these necessities, the model core has to be physical, namely model parameters should be connected to device parameters and the number of model parameters cannot be huge. The surface-potential-based models, such as HiSIM, have been developed as an answer to these necessifies. However, it is known that the parameter extraction requires a lot efforts or experiences to meet real MOSFET characteristies.

To realize accurate and reliable parameter extraction with minimized effort and less experience, a genetic algorithm (GA)-based parameter extraction method has also been investigated [1–3]. Experimental results demonstrate that extraction of 32 model parameters, which are core model parameter in HiSIM, can be completed within 23 hours with a conventional PC (Athlon XP 2500). Fig. 7.2 shows extraction speeds as a function of iterations for two extraction algorithms. The extracted model parameters agree with those extracted manually by an expert mostly within 10% [2,4].

Bibliography

[1] J. Watts, C. Bittner, D. Heaberlin, and J. Hoffmann, "Extraction of compact model parameters for ULSI MOSFETs using a genetic algorithm," *Modeling and Simulation of Microsystems, Technical Proceedings of the Second International Conference*, pp. 176–179, Puerto Rico, 1999.

[2] M. Murakawa, M. Miura-Mattausch, and T. Higuchi, "Towards automatic parameter extraction for surface-potential-based MOSFET models with the genetic algorithm," *Proceedings of the 2005 conference on Asia South Pacific design automation*, pp. 204–207, Shanghai, Jan, 2005.

[3] M. Miura-Mattausch, N. Sadachika, M. Murakawa, S. Mimura, T. Higuchi, K. Itoh, R. Inagaki, and Y. Iguchi, "RF-MOSFET Model-Parameter Extraction with HiSIM," *Proc. Workshop on Compact Modeling*, pp. 69–74, Anaheim, May, 2005.

[4] M. Miura-Mattausch, D. Navarro, N. Sadachika, G. Suzuki, Y. Takeda, M. Miyake, T. Warabino, K. Machida, T. Ezaki, H.J. Mattausch, T. Ohguro, T. Iizuka, M. Taguchi, S. Kumashiro, R. Inagaki and S. Miyamoto, "Advanced Compact MOSFET Model HiSIM2 Based on Surface Potentials with a Minimum Number of Approximation," *Technical Proceedings of the 2006 NSTI Nanotechnology Conference and Trade Show*, vol. 3, pp. 638 - 643, Boston, May, 2006.

Index

Mitiko Miura-Mattausch received the Dr. Sc. Degree from Hiroshima Univerisity. She joined the Max-Planck-Institute for solid-state physics in Stuttgart, Germany as a researcher from 1981 to 1984. From 1984 to 1996, she was with Corporate Research and Development, Siemens AG, Munich, Germany, working on hot-electron problems in MOSFETs, the development of bipolar transistors, and analytical modeling of deep submicron MOSFETs for circuit simulation. Since 1996, she has been a professor in Department of Electrical Engineering, Graduate School of Advanced Sciences of Matter at Hiroshima University, leading the ultra-scaled devices laboratory.

Hans Jürgen Mattausch received the Dr. rer. nat. degree from the University of Stuttgart, Germany in 1981.

In 1982 he joined the Research Laboratories of Siemens AG in Munich, Germany, where he was involved in the development of MOS technology as well as memory and telecommunication circuit design. From 1990 he led a research group working on power semiconductor devices, including device design, compact modeling and packaging. In 1995 he joined the Siemens Semiconductor Group as Department Head for Product Analysis and Improvement in the Chip Card IC Division.

Since 1996 he is with Hiroshima University, Higashi-Hiroshima, Japan, where he is a Professor at the Research Center for Nanodevices and Systems and at the Graduate School for Advanced Sciences of Matter. His main interests are circuit design and device modeling issues related to effective utilization of nanodevices and nanotechnology.

Dr. Mattausch is a senior member of IEEE and a member of IEICE.

Tatsuya Ezaki received the Ph.D. degree in electronic engineering from Osaka University, Japan, in 1997. In 1998, he joined the Silicon Systems Research Laboratories, NEC Corporation, Sagamihara, Japan. He has been engaged in research and development of ULSI device modeling since 1998. In 2005, he joined the Graduate School of Advanced Sciences of Matter, Hiroshima University, Japan. His research interests includes compact modeling of MOSFETs and quantum transports in nano-scale devices.

Printed in the United States
By Bookmasters